VOLUME FOUR HUNDRED AND FORTY-FOUR

Methods in

ENZYMOLOGY

Angiogenesis: *In Vivo*
Systems, Part A

METHODS IN ENZYMOLOGY

Editors-in-Chief

JOHN N. ABELSON AND MELVIN I. SIMON

Division of Biology
California Institute of Technology
Pasadena, California

Founding Editors

SIDNEY P. COLOWICK AND NATHAN O. KAPLAN

VOLUME FOUR HUNDRED AND FORTY-FOUR

METHODS IN
ENZYMOLOGY

Angiogenesis: *In Vivo* Systems, Part A

EDITED BY

DAVID A. CHERESH
University of California, San Diego
Moore's Cancer Center
La Jolla, CA, USA

ELSEVIER

AMSTERDAM • BOSTON • HEIDELBERG • LONDON
NEW YORK • OXFORD • PARIS • SAN DIEGO
SAN FRANCISCO • SINGAPORE • SYDNEY • TOKYO
Academic Press is an imprint of Elsevier

Academic Press is an imprint of Elsevier
525 B Street, Suite 1900, San Diego, CA 92101-4495, USA
30 Corporate Drive, Suite 400, Burlington, MA 01803, USA
32 Jamestown Road, London NW1 7BY, UK

For information on all Elsevier Academic Press publications
visit our Web site at elsevierdirect.com

ISBN-13: 978-0-12-374313-8

PRINTED IN THE UNITED STATES OF AMERICA
08 09 10 11   9 8 7 6 5 4 3 2 1

Contents

Contributors

Edith Aguilar
Department of Cell Biology, The Scripps Research Institute, La Jolla, California

Niren Angle
Section of Vascular and Endovascular Surgery, Department of Surgery, School of Medicine, University of California, San Diego, San Diego, California

Samuel Barillas
Section of Vascular and Endovascular Surgery, Department of Surgery, School of Medicine, University of California, San Diego, San Diego, California

Ann F. Chambers
Department of Oncology, University of Western Ontario, London, Ontario, Canada

Vesselina G. Cooke
Division of Matrix Biology, Department of Medicine, Beth Israel Deaconess Medical Center and Harvard Medical School, Boston, Massachusetts

Elena I. Deryugina
The Scripps Research Institute, La Jolla, California

Michael I. Dorrell
Department of Cell Biology, The Scripps Research Institute, La Jolla, California

Ann M. Dvorak
Department of Pathology, Beth Israel Deaconess Medical Center, and Harvard Medical School, Boston, Massachusetts

Harold F. Dvorak
Department of Pathology, Beth Israel Deaconess Medical Center, and Harvard Medical School, Boston, Massachusetts

David Friedlander
Department of Cell Biology, The Scripps Research Institute, La Jolla, California

Martin Friedlander
Department of Cell Biology, The Scripps Research Institute, La Jolla, California

Frank J. Giordano
Cardiovascular Medicine, Department of Medicine, and Vascular Biology and Translation Program, Yale University School of Medicine, New Haven, Connecticut

Joshua I. Greenberg
Section of Vascular and Endovascular Surgery, Department of Surgery, School of Medicine, University of California, San Diego, San Diego, California

Fritjof Helmchen
Department of Neurophysiology, Brain Research Institute, University of Zurich, Zurich, Switzerland

Yan Huang
Cardiovascular Medicine, Department of Medicine, and Vascular Biology and Translation Program, Yale University School of Medicine, New Haven, Connecticut

Ruth A. Jacobson
Department of Cell Biology, The Scripps Research Institute, La Jolla, California

Audra Johnson
Department of Cell Biology, The Scripps Research Institute, La Jolla, California

Raghu Kalluri
Department of Biological Chemistry and Molecular Pharmacology, Harvard Medical School, Boston, Massachusetts, and Division of Matrix Biology, Department of Medicine, Beth Israel Deaconess Medical Center and Harvard Medical School, Boston, Massachusetts, and Harvard–MIT Division of Health Sciences and Technology, Boston, Massachusetts

David Kleinfeld
Department of Physics, University of California, San Diego, La Jolla, California

David Lyden
Department of Pediatrics and Department of Cell and Developmental Biology, Weill Cornell Medical College, New York, and Memorial Sloan-Kettering Cancer Center, New York

Ian C. MacDonald
Department of Medical Biophysics, University of Western Ontario, London, Ontario, Canada

Milan Makale
Moores Cancer Center, University of California, San Diego, La Jolla, California

Valentina Marchetti
Department of Cell Biology, The Scripps Research Institute, La Jolla, California

Mary C. McKinney
Laboratory of Molecular Genetics, National Institute of Child Health and Human Development, Bethesda, Maryland

Stacey K. Moreno
Department of Cell Biology, The Scripps Research Institute, La Jolla, California

Walter L. Murfee
Assistant Professor Department of Biomedical Engineering, Tulane University, New Orleans, Louisiana

Janice A. Nagy
Department of Pathology, Beth Israel Deaconess Medical Center, and Harvard Medical School, Boston, Massachusetts

Marianna Papaspyridonos
Department of Pediatrics and Department of Cell and Developmental Biology, Weill Cornell Medical College, New York

James P. Quigley
The Scripps Research Institute, La Jolla, California

Matthew R. Ritter
Department of Cell Biology, The Scripps Research Institute, La Jolla, California

Geert W. Schmid-Schönbein
Professor Department of Bioengineering, University of California, San Diego, La Jolla, California

Shou-Ching Shih
Department of Pathology, Beth Israel Deaconess Medical Center, and Harvard Medical School, Boston, Massachusetts

Ahmed Suliman
Section of Vascular and Endovascular Surgery, Department of Surgery, School of Medicine, University of California, San Diego, San Diego, California

Noel Weidner
Department of Pathology, University of California, San Diego, San Diego, California

Brant M. Weinstein
Laboratory of Molecular Genetics, National Institute of Child Health and Human Development, Bethesda, Maryland

Sara M. Weis
Moores Cancer Center, University of California, San Diego, La Jolla, California

Wendy H. Wong
Department of Pathology, Beth Israel Deaconess Medical Center, and Harvard Medical School, Boston, Massachusetts

Preface

A Tribute to Dr. Judah Folkman

The field of angiogenesis has recently lost its pioneer and leader, Dr. Judah Folkman. This was a tremendous loss to many of us who knew him and to the field in general. Dr. Folkman inspired a generation of scientists in efforts to translate basic discoveries toward new therapeutics for a wide range of diseases including cancer, blinding eye disease, and inflammatory disease. Due in large part to Dr. Folkman's efforts and direction, we now have the first generation of therapeutics that disrupt angiogenesis in patients suffering from cancer and macular degeneration. While Dr. Folkman clearly passed away before his time, he did live long enough to observe that many thousands of patients are now better off due to antiangiogenic therapy.

I had a rather interesting initiation to the field of antiangiogenesis that was wholly inspired by Dr. Folkman. In the mid–1980s as a junior faculty at the Scripps Research Institute, I was studying what many of us in the field were beginning to appreciate were a family of cell adhesion receptors, later termed "integrins." I had developed a monoclonal antibody (LM609) to the vitronectin receptor later referred to as integrin $\alpha v \beta 3$. During the course of my work, LM609 was used to stain a variety of diseased and normal tissues. To my surprise, LM609 reacted strongly with blood vessels in tumors and inflammatory sites, but failed to react with blood vessels in normal tissues. After seeing this result, I began to read up on the emerging field of angiogenesis research. It was clear that most of the literature in the field came from Dr. Folkman or one of his disciples. I immediately contacted Dr. Folkman. By the time I finished describing our results, I realized that he was excited as I was about our studies. In fact, before I could ask him any questions, he suggested that I visit his lab to learn the chick chorioallantoic membrane (CAM) assay to determine whether LM609 might have an impact on angiogenesis in a quantitative animal model.

Naturally I arranged a trip to the Folkman lab within the next couple of weeks. I had never been to Harvard, and was a bit intimidated by the place. I introduced myself to his administrative assistant, who welcomed me and indicated that Dr. Folkman was expecting me. Within minutes, Dr. Folkman, clad in a lab coat greeted me and suggested that we get started. At this point, I assumed he was going to introduce me to one of his students or technicians who would then proceed to show me the CAM assay step by step. To my surprise, Dr. Folkman led me to a hood, sat down, and immediately started to instruct me in how to induce angiogenesis on the CAM. In fact, the next thing I knew, I was sitting at the hood next to Dr. Folkman going through the procedure in detail. Therefore, I can say I learned the technique from the master. Ultimately, Dr. Folkman introduced me to several members of the Folkman lab, including Drs. Donald Inber, Pat D'Amore, and Mike Klagsburn. I remember how enthusiastic and communicative all of these folks were. In fact, I am happy to say that I still maintain close contact with them and have had many opportunities over the years to discuss science and reminisce about the past. In fact, Don, Pat, and Mike have all kindly contributed chapters to *Methods in Enzymology* volumes on angiogenesis.

While on the airline flight home from the Folkman lab, I began to realize that my career was about to take a change in course. From that point forward, I began to focus on the role of adhesion receptors in angiogenesis and began to realize that blocking angiogenesis with integrin antagonists could have a very impressive impact on the growth of tumors in mice. Importantly, two of the agents we developed, including humanized LM609, have shown clinical activity in patients with late-stage cancer.

Since my initiation to the field, I have since followed Dr. Folkman's work and have attended dozens of his lectures. Listening to a Folkman lecture is like watching one of your favorite movies—you can watch it over and over again and still find something interesting to think about. It was difficult for anyone to attend his lecture and not come away excited about science in general and angiogenesis in particular. The field of angiogenesis has matured over the past 25 years due in large part to Dr. Folkman's drive, enthusiasm, perseverance, and kindness. Dr. Folkman's leadership has helped to recruit many scientists and physicians from the academic and private sectors to focus on new approaches to develop angiogenesis inhibitors.

In the early days, there were a limited number of technological approaches to measure and study angiogenesis. The CAM assay was among the first quantitative approaches to measure the growth of newly forming blood vessels. From this humble beginning, the field has exploded and as a result we now have a wide range of techniques, approaches, and animal models designed to monitor and study the growth of new blood vessels in development, tissue remodeling, and disease. These methods are described in detail in this volume by many of the current leaders of the field.

METHODS IN ENZYMOLOGY

VOLUME 435. Oxygen Biology and Hypoxia
Edited by HELMUT SIES AND BERNHARD BRÜNE

VOLUME 436. Globins and Other Nitric Oxide-Reactive Protiens (Part A)
Edited by ROBERT K. POOLE

VOLUME 437. Globins and Other Nitric Oxide-Reactive Protiens (Part B)
Edited by ROBERT K. POOLE

VOLUME 438. Small GTPases in Disease (Part A)
Edited by WILLIAM E. BALCH, CHANNING J. DER, AND ALAN HALL

VOLUME 439. Small GTPases in Disease (Part B)
Edited by WILLIAM E. BALCH, CHANNING J. DER, AND ALAN HALL

VOLUME 440. Nitric Oxide, Part F Oxidative and Nitrosative Stress in Redox
Regulation of Cell Signaling
Edited by ENRIQUE CADENAS AND LESTER PACKER

VOLUME 441. Nitric Oxide, Part G Oxidative and Nitrosative Stress in Redox
Regulation of Cell Signaling
Edited by ENRIQUE CADENAS AND LESTER PACKER

VOLUME 442. Programmed Cell Death, General Principles for Studying Cell
Death (Part A)
Edited by ROYA KHOSRAVI-FAR, ZAHRA ZAKERI, RICHARD A. LOCKSHIN,
AND MAURO PIACENTINI

VOLUME 443. Angiogenesis: *In Vitro* Systems
Edited by DAVID A. CHERESH

VOLUME 444. Angiogenesis: *In Vivo* Systems, (Part A)
Edited by DAVID A. CHERESH

Molecular Mechanism of Type IV Collagen–Derived Endogenous Inhibitors of Angiogenesis

Vesselina G. Cooke* *and* Raghu Kalluri*,†,‡

Contents

Abstract

Angiogenesis, the process of new blood vessel formation, is regulated on both genetic and molecular levels. Pro- and anti-angiogenic stimuli maintain the angiogenic balance, and the tipping of that balance toward pro-angiogenic activity is critical for tumor growth and survival. Endogenous inhibitors of

* Division of Matrix Biology, Department of Medicine, Beth Israel Deaconess Medical Center and Harvard Medical School, Boston, Massachusetts
† Department of Biological Chemistry and Molecular Pharmacology, Harvard Medical School, Boston, Massachusetts
‡ Harvard–MIT Division of Health Sciences and Technology, Boston, Massachusetts

Methods in Enzymology, Volume 444
ISSN 0076-6879, DOI: 10.1016/S0076-6879(08)02801-2

angiogenesis, many of which are fragments from large extracellular matrix proteins, counter the effect of growth factors and keep angiogenesis in check. This chapter will discuss the molecular mechanisms of endogenous inhibitors derived from type IV collagen and review the *in vitro* and *in vivo* assays available to study their role in angiogenesis.

1. INTRODUCTION

A key event in the growth and progression of tumors is shifting the balance between pro- and anti-angiogenic stimuli. Angiogenesis is the process of new blood vessel formation from either pre-existing capillaries or blood vessel precursor cells, in which case it is termed *vasculogenesis*. The idea that tumor growth is dependent on angiogenic signals was first intro-duced in the 1970s (Folkman, 1971), and by the late 1980s and 1990s angiogenic inhibitors such as TNP-470 and interferon alpha were used to inhibit a wide variety of tumors (Sidky and Borden, 1987). Subsequently, the discovery of endogenous inhibitors paved the way for the development of the "angiogenic switch" theory (Augustin, 2003; Carmeliet, 2003).

In healthy adults, angiogenesis is normally suppressed and is temporarily activated during processes such as the female reproductive cycle (Folkman and Shing, 1992). During the angiogenic switch, pro- and anti–angiogenic signals can regulate the angiogenic process by stimulating it, suppressing it, or keeping it in balance. Switching to an angiogenic phenotype is a very complex process and still is not completely understood. It is thought to require both upregulation of angiogenic activators and downregulation of angiogenic inhibitors (Folkman, 1995; Nyberg *et al.*, 2005).

Vascular endothelial growth factor (VEGF) and basic fibroblast growth factor (bFGF) are the most widely studied angiogenic inducers (Carmeliet, 2003). Endogenous inhibitors include variety of molecules such as small peptides, hormone metabolites, and apoptosis stimulators (Cao, 2001; Folkman, 1995; Grant and Kalluri, 2005; Nyberg *et al.*, 2005). Their modes of action can vary from antagonism of angiogenic growth factors to inhibition of angiogenic proteinases, as well as suppression of endothelial cell proliferation, migration and tube formation (Grant and Kalluri, 2005). Endogenous angiogenic inhibitors can be found either in the circulation or sequestered in the extracellular matrix (ECM). Many such inhibitors are, in fact, fragments of larger ECM molecules that become released upon prote-olysis of the ECM and the vascular basement membrane (VBM) by tumor enzymes (Sund *et al.*, 2005).

The ECM is a complex structure composed of various glycoproteins, proteoglycans, and hyluronic acid (Mundel and Kalluri, 2007). A specialized form of ECM, the basement membrane, is stabilized and assembled by a structural network formed by type IV collagen, laminins, entactins, elastin, fibrilin, fibrin, and fibronectin (Kuhn *et al.*, 1981; Timpl *et al.*, 1981). Six different types of IV collagen α chains exist (α 1 to 6), with α1 and α2 chains being predominantly found in most basement membranes, while α3 to α6 chains have a more restricted pattern of expression (Sudhakar *et al.*, 2003). All type IV–collagen α chains share the same structure with an N–terminal 7S domain, a middle triple helical domain, and a C-terminal, globular noncollagenous domain (Sudhakar *et al.*, 2003). VBM degradation by tumor-associated enzymes leads to the release of type IV collagen fragments that function as endogenous angiogenic inhibitors, named *arresten, canstatin,* and *tumstatin* (Sudhakar *et al.*, 2003; Sund *et al.*, 2005). Arresten, a 26-kDA, C-terminal, globular noncollagenous domain of the α1 chain of type IV collagen was initially isolated from human placental basement membrane (Colorado *et al.*, 2000). Canstatin, a 24-kDa NC domain of the α2 chain of type IV collagen was recombinantly produced in *Escherichia coli* and 293 human embryonic kidney cells (Kamphaus *et al.*, 2000). Finally, tumstatin, which was purified from basement membranes of kidney, placenta, and testis, is a 28-kDa NC domain of the α3 chain of type IV collagen (Maeshima *et al.*, 2000b). In this chapter, we will provide details on the assays used to test the role of these molecules in angiogenesis, and will show that all three of them exhibit distinct antiangiogenic activity mediated by integrins.

2. MOLECULAR MECHANISM OF TYPE IV COLLAGEN–DERIVED ENDOGENOUS INHIBITORS

2.1. Antiangiogenic activity of arresten

Arresten was first isolated from human placenta and produced as a recombinant fusion protein with a COOH-terminal, 6-histidine tag in *E. coli* using a bacterial expression plasmid (Colorado *et al.*, 2000). Human arresten was also produced as a secreted soluble protein in 293 embryonic kidney cells, using the pcDNA 3.1 eukaryotic vector (Colorado *et al.*, 2000). Arresten functions as an antiangiogenic molecule by inhibiting basic fibroblast growth factor (bFGF)–induced endothelial cell proliferation, migration, tube formation, and Matrigel neovascularization. Arresten isolated from *E. coli* to dose-dependently inhibits bFGF-induced, endothelial cell proliferation (Colorado *et al.*, 2000), and arresten from both *E. coli* and 29 kidney cells inhibited endothelial tube formation and blood vessel formation (Colorado *et al.*, 2000). Arresten reduces the size of established human xenograft tumors in nude mice, although tumor growth returns 12 days after arresten treatment had been stopped (Colorado *et al.*, 2000). Additionally, MV38/MUC1 metastatic cancer cell produce fewer pulmonary nodules in arresten-treated mice, suggesting that arresten inhibits the development of tumor metastases (Colorado *et al.*, 2000). Its mechanism of action is proposed to be mediated by integrin $\alpha_1\beta_1$ via inhibition of MAPK signaling in endothelial cells (Colorado *et al.*, 2000; Sudhakar *et al.*, 2005). These findings suggest that arresten is a potent inhibitor of angiogenesis with a potential therapeutic use.

2.2. Molecular mechanism of arresten

Since integrins $\alpha_1\beta_1$ and $\alpha_2\beta_1$ bind type IV collagen isolated from Engelbreth-Holm-Swarm mouse sarcoma tumors (Senger *et al.*, 1997), a cell adhesion assay using endothelial cells was performed to test if they can also bind to arresten (Colorado *et al.*, 2000). In this assay, function-blocking antibodies to α_1 and β_1 integrin subunits were added to endothelial cells grown in arresten-coated plates. Results revealed that α_1 antibody inhibited endothelial cell attachment to arresten-coated plates by 60%, and β_1 antibody caused 70% reduction in endothelial cells binding to arresten, suggesting that arresten is capable of binding to both α_1 and β_1 integrins (Colorado *et al.*, 2000). Since the expression of these integrins on endothelial cells is induced by VEGF, it was speculated that arresten may downregulate VEGF-induced proliferation and migration of endothelial cells when it binds to α_1 and β_1 integrins on these cells (Bloch *et al.*, 1997). Migration, proliferation, and tube formation assays using VEGF as a growth factor revealed that arresten significantly inhibited the ability of endothelial cells to migrate, proliferate, and form tubes, which are all necessary processes for angiogenesis (Colorado *et al.*, 2000; Sudhakar *et al.*, 2005). Additionally, depleting arresten of its potential binding sites to $\alpha_1\beta_1$ integrin by preincubating it with soluble $\alpha_1\beta_1$ integrin protein led to reversal of its antiangiogenic action (Sudhakar *et al.*, 2005), providing further evidence that arresten's antiangiogenic mechanism of action involves integrins. These studies also indicated that $\alpha_1\beta_1$ integrin may be the functional receptor for arresten; which was confirmed when arresten was immunoprecipitated with an antibody to $\alpha_1\beta_1$ integrin from endothelial cell lysates treated with different concentrations of arresten (Sudhakar *et al.*, 2005).

Besides their role in cell attachment, integrins play a unique role in signal transduction pathways by sending signals from the ECM to the cell (outside-in signaling) and from the cell to the ECM (inside-out signaling) (Clark and Brugge, 1995). During outside-in signaling, biochemical signals across the cell membrane activate intracellular signaling pathways, such as ERK1/2, Ras, MEK1/2, Raf, and p38 MAPK family members (Clark and Brugge, 1995). To determine whether the binding of arresten to $\alpha_1\beta_1$ integrin has an effect on the integrin's effector kinases, endothelial cells were preincubated with arresten and plated on type IV collagen matrix; cell extracts were analyzed on SDS-PAGE and immunoblotted for phosphorylated and unphosphorylated kinases (Sudhakar *et al.*, 2005). Similar to results with other collagen NC1 domains, tumstatin and endostatin (Sudhakar *et al.*, 2003), arresten was found to inhibit phosphorylation of FAK, suggesting that arresten's mechanism of action involves integrin–mediated signaling. Next, the ability of arresten to block FAK phosphorylation was tested in mouse lung endothelial cells (MLECs) isolated from α_1 integrin knockout mice. Interestingly, in these arresten-treated MLECs, inhibition of FAK phosphorylation did not occur, further confirming the significance of integrin–mediated signaling in arresten function (Sudhakar *et al.*, 2005). Downstream of FAK, the mitogen-activated protein kinase (MAPK) signaling pathway plays an important role in endothelial cell survival and migration (Clark and Brugge, 1995). Pretreatment of MLECs with arresten significantly blocked phosphorylation of Raf, MEK, ERK1, and p38 in these cells (Sudhakar *et al.*, 2005), suggesting a role for the MAPK and p38 pathways in $\alpha_1\beta_1$ integrin-mediated inhibition of EC migration on type IV collagen. In contrast, treatment of MLECs prepared from α_1 integrin knockout mice had no effect on phosphorylation of c-Raf, MEK1/2, ERK1, and p38 on type IV collagen, strongly suggesting that arresten-induced signaling is mediated by $\alpha_1\beta_1$ integrin (Sudhakar *et al.*, 2005).

To further delineate the mechanism of action of arresten, the next set of experiments aimed at understanding how binding of arresten to integrin $\alpha_1\beta_1$ and subsequent activation of MAPK pathway might be affecting VEGF expression. In a tumor environment, endothelial cell migration, proliferation, and tube formation leading to angiogenesis can be stimulated by hypoxic conditions, where HIF-1α transcriptionally induces VEGF expression (Semenza, 2000). To understand whether $\alpha_1\beta_1$ integrin and arresten can modulate HIF-1α and VEGF expression, arresten-treated wild-type (WT) and α_1 integrin knockout endothelial cells were placed under hypoxic conditions for 12 to 24 h, and the effect on signaling molecules was investigated by immunoblot. The results demonstrated that when WT endothelial cells were treated with arresten, the expression of ERK1/2, p38, HIF-1α, and VEGF was strongly inhibited (Sudhakar *et al.*, 2005). In comparison, arresten had no effect on hypoxia-induced HIF-1α and VEGF expression in hypoxic α_1 integrin knockout endothelial cells, suggesting that when arresten binds to $\alpha_1\beta_1$ integrin, it inhibits MAPK signaling, resulting in inhibition of HIF-1α expression (Sudhakar *et al.*, 2005). Altogether, these results demonstrate that arresten binds to integrin $\alpha_1\beta_1$, resulting in inhibition of MAPK signaling pathway and downstream inhibition of HIF-1α and VEGF (Fig. 1.1). This mechanism of action of arresten is further validated by the ability of arresten to inhibit tumor vasculature which ultimately leads to the regression of tumor growth (Sudhakar *et al.*, 2005). Xenographs of SCC-PSA-1 (teratocarcinoma) in mice that were given arresten regressed and had a decreased number of CD31-positive blood vessels compared to tumors in α_1 integrin knockout mice given arresten or WT mice given PBS

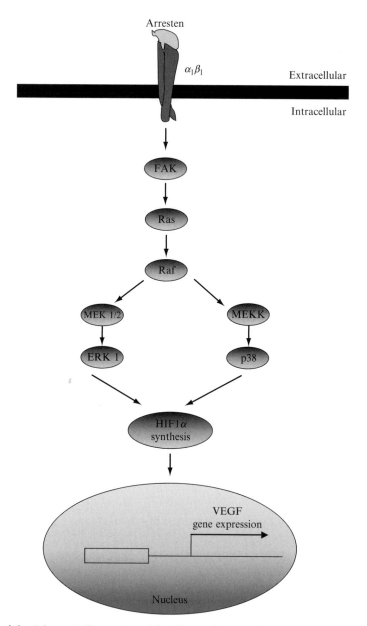

Figure 1.1   Schematic illustration of signaling pathway mediated by arresten. Arresten binds to $\alpha_1\beta_1$ integrin and inhibits phosphorylation of FAK. This leads to inhibition of Raf/MEK/ERK1/2/p38 MAPK pathways with subsequent downregualtion of HIF-1α and VEGF expression. As a result, endothelial cell migration, proliferation, and tube formation in proliferating endothelial cells are inhibited. (Modified from Sudhakar, A., Nyberg, P., Keshamouni, V. G., Mannam, A. P., Li, J., Sugimoto, H., Cosgrove, D., and Kalluri, R. (2005). Human alpha1 type IV collagen NC1 domain exhibits distinct antiangiogenic activity mediated by alpha1beta1 integrin. *J. Clin. Invest.* **115**, 2801–2810.)

(Sudhakar *et al.*, 2005). These findings confirm the significance of integrin-mediated signaling in arresten function and support the potential use of arresten in antiangiogenic therapy.

2.3. Antiangiogenic activity of canstatin

Canstatin is the 24-kDa NC1 domain of the α_2 chain of type IV collagen, and human canstatin was produced in *E. coli* as a fusion protein and in 293 embryonic kidney cells as a secreted soluble protein (Kamphaus *et al.*, 2000). While canstatin has no effect on nonendothelial cells, it induces apoptosis of endothelial cells and significantly inhibits endothelial cell proliferation, migration, and tube formation (He *et al.*, 2003; Hou *et al.*, 2004; Kamphaus *et al.*, 2000). The molecular mechanism of canstatin involves binding to integrins $\alpha_v\beta_3$ and $\alpha_v\beta_5$, which initiates apoptotic pathways via activation of procaspase 8 and 9, leading to activation of caspase 3 (Kamphaus *et al.*, 2000; Magnon *et al.*, 2005; Panka and Mier, 2003).

2.4. Molecular mechanism of canstatin

Normal capillaries consist of a single layer of quiescent flat endothelial cells. During angiogenesis, endothelial cells acquire a spindle-shaped structure, begin to dissolve the basement membrane and extracellular matrix, migrate, proliferate, and form new tubes (Carmeliet, 2003). To understand the antiangiogenic properties of canstatin, canstatin purified from either 293 kidney cells or *E. coli* was administered to endothelial cells, and its effect on their migration, proliferation, and tube formation was investigated by using a Boyden chamber migration assay, thymidine incorporation, and methylene blue colorimetric methods, and endothelial tube formation assay on Matrigel, respectively (Kamphaus *et al.*, 2000). In each assay, endothelial cells showed reduced ability to form tubes, and exhibited decreased migration and proliferation when incubated with canstatin (Kamphaus *et al.*, 2000). In contrast, neither a control protein, bovine serum albumin (BSA), nor the NC1 domain of type IV collagen α_5 chains, had an effect on these endothelial cell properties, suggesting that canstatin is an inhibitor of multiple stages of the angiogenic process.

As mentioned earlier, downstream signal transduction pathways that lead to endothelial cell migration and proliferation involve members of the MAPK family of kinases such as ERK, c-jun terminal kinase (JNK), and p38 (Clark and Brugge, 1995). Canstatin, however, did not cause an immediate decrease in ERK phosphorylation induced by 20% fetal bovine serum or endothelial mitogens. ERK phosphorylation was only decreased at later time points, suggesting that the mechanism of action of canstatin may not involve the early steps of growth factor–mediated signaling leading to ERK activation (Kamphaus *et al.*, 2000). It was speculated that the mode of action of canstatin is to induce cell death rather than block pro-survival signals.

To understand whether canstatin can induce apoptosis in endothelial cells, annexin V-FITC labeling of externalized phosphatidylserine was employed. Disruption of the cell membrane with subsequent externalization of phosphatidylserine (PS) is a hallmark of apoptosis (Fadok et al., 2000). Because externalized PS can bind with high affinity to annexin V (van Engeland et al., 1998), the number of apoptotic cells can be found by counting the number of annexin V-labeled cells by flow cytometry (Kamphaus et al., 2000). Exposure to canstatin induced a significant amount of apoptosis in endothelial cells, while no effect was seen on human embryonic kidney cells, suggesting that canstatin specifically induces apoptosis of endothelial cells. This apoptotic effect of canstatin on endothelial cells was additionally confirmed by measuring the amounts of FLIP and FAS ligand expression. FAS ligand activates apoptotic signaling through a cytoplasmic death domain that interacts with signaling adaptors, including FAS-associated protein with death domain (FADD) to activate caspase proteolytic cascade (Nagata, 1996). FLIP protein, on the other hand, binds to FADD and caspase-8, inhibiting death receptor–mediated apoptosis (Irmler et al., 1997). Canstatin increased the amount of Fas ligand and decreased the amount of FLIP protein, as measured by western blot analysis, not only confirming the apoptotic properties of canstatin, but also suggesting that canstatin-induced apoptosis is mediated through a death receptor–dependent pathway (Kamphaus et al., 2000; Panka and Mier, 2003). This pathway was further amplified in the mitochondria where exposure of endothelial cells to canstatin caused reduction of the membrane potential, leading to increased caspase 3, 8, and 9 activity (Magnon et al., 2005; Panka and Mier, 2003). The ability of canstatin to induce apoptosis is not only restricted to endothelial cells. Using MDA-MB-231 xenografts on nude mice, Magnon and colleagues (2005) showed that canstatin is able to localize at the extracellular surface on the tumor cells, increase mitochondrial caspase-9 activity, and induce apoptosis of these tumor cells. Altogether, these findings indicate that canstatin can inhibit endothelial cell migration, proliferation, and tube formation via induction of apoptosis in endothelial cells.

Antagonists against $\alpha_v\beta_3$ integrins have been shown to inhibit endothelial cell proliferation, migration, and apoptosis (Magnon et al., 2005). To understand whether integrins play a role in mediating the apoptotic activity of canstatin, siRNA against α_v, β_3, and β_5 was used to decrease the amount of integrin expression in endothelial cells (Magnon et al., 2005). The specific inhibition of these proteins significantly reduced the amount of caspase 9 activity, while no effect on caspase 8 activity was observed, suggesting that the interaction of canstatin with these proteins selectively triggers the caspase 9 cascade. Further fluorescence-activated cell sorting (FACS) analysis showed that both $\alpha_v\beta_3$ and $\alpha_v\beta_5$ integrins were expressed at high levels on HUVEC cells with high levels of $\alpha_v\beta_5$ detected on tumor cells as well (Magnon *et al.*, 2005). Immunoprecipitation studies using mAbs against $\alpha_v\beta_3$ or $\alpha_v\beta_5$ further confirmed that canstatin binds to both integrins on the surface of both human endothelial and tumor cells (Magnon *et al.*, 2005).

Most integrins recruit FAK and participate in signal transduction pathways such as the MAPK, JUN, and Akt pathways (Guo and Giancotti, 2004). Additionally, phosphorylation of FAK/PI3K/Akt is known to inhibit caspase 9 activity, thereby inactivating the mitochondrial apoptotic pathway (Frisch *et al.*, 1996). When canstatin was added to endothelial cells, it significantly inhibited the phosphorylation of FAK, Akt, and downstream targets such as mTOR, 4E-BP1, and p70s6k, suggesting that canstatin activates caspase 9 activity through inhibition of the FAK/Akt pathway (Panka and Mier, 2003). Collectively, these results demonstrate that the molecular mechanism of canstatin involves two distinct pathways of apoptosis in endothelial cells: the α_v integrin-FAK/PI3K/caspase9 pathway and the Fas/Fas ligand/caspase 8/caspase 9 cascade. Both of these pathways increase mitochondria-associated events that lead to caspase 3 activation (Magnon *et al.*, 2005) (Fig. 1.2).

2.5. Antiangiogenic properties of tumstatin

Tumstatin is a 28-kDa NC1 domain of the $\alpha3$ chain of type IV collagen, purified from MMP-degraded basement membrane preparations from kidney, placenta, and testis (Maeshima *et al.*, 2000a). Recently, it was shown that p53-mediated upregulation of α_2 collagen prolyl-4-hydroxylase is capable of secreting full-length collagens, which are then proteolytically processed and produce antiangiogenic peptides, one of which is tumstatin (Teodoro *et al.*, 2006). Tumstatin inhibits bFGF-induced HUVEC proliferation (Maeshima *et al.*, 2000a), proliferation of melanoma cells (Han *et al.*, 1997), neovascularization on Matrigel plug assays, and tumor growth in many different mouse cancer models (Maeshima *et al.*, 2000a,b, 2001, 2002). The molecular mechanism of tumstatin involves binding to integrin $\alpha_v\beta_3$ utilizing an RGD-independent mechanism. Binding of tumstatin to $\alpha_v\beta_3$ inhibits CAP-dependent protein translation via downregulation of mTOR in proliferating endothelial cells (Maeshima *et al.*, 2002). This activity is localized to amino acids 69–98 of tumstatin referred to as T3. Binding of tumstatin to integrin $\alpha_v\beta_3$ is also dependent on the PTEN/Akt pathway (Kawaguchi *et al.*, 2006).

2.6. Molecular mechanism of tumstatin

Standard methods of endothelial cell proliferation and tube formation have shown that tumstatin is an inhibitor of angiogenesis (Han *et al.*, 1997; Maeshima *et al.*, 2000b). Similar to arresten and canstatin, tumstatin was

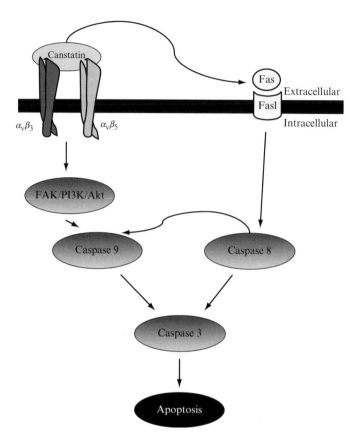

Figure 1.2 Schematic illustration of signaling pathways induced by canstatin in endothelial cells. Canstatin binds to $\alpha_v\beta_3$ and $\alpha_v\beta_5$ on the endothelial cell surface and initiates two apoptotic pathways, leading to activation of caspase 3. Canstatin activates caspase 9, not only directly through inhibition of the FAK/PI3K/Akt pathway, but also by indirectly amplifying the mitochondrial pathway through Fas-dependent, caspase 8 activation. (Modified from Magnon, C., Galaup, A., Mullan, B., Rouffiac, V., Bouquet, C., Bidart, J. M., Griscelli, F., Opolon, P., and Perricaudet, M. (2005). Canstatin acts on endothelial and tumor cells via mitochondrial damage initiated through interaction with alphavbeta3 and alphavbeta5 integrins. *Cancer Res.* **65**, 4353–4361.)

originally proposed to exert its mode of action through integrins because synthetic peptides corresponding to the C-terminal of tumstatin were shown to bind to $\alpha_v\beta_3$ and inhibit tumor cell proliferation (Shahan *et al.*, 1999). Maeshima and colleagues (2000a) demonstrated that $\alpha_v\beta_3$ antibody inhibited HUVECs attachment to tumstatin coated dishes by 80% and cell binding was blocked by 91% when $\alpha_v\beta_3$ and β_1 antibodies were used together, suggesting that these integrins may play a role in the antiangiogenic properties of tumstatin. A cell-adhesion assay and various deletion mutants of tumstatin were used to identify its binding site for integrin $\alpha_v\beta_3$.

Within its N-terminus, tumstatin has a RGD sequence in amino acid 7–9, which is generally considered an important binding site for integrin $\alpha_v\beta_3$. A deletion mutant of tumstatin that lacks its RGD sequence, however, is still capable of binding to integrin $\alpha_v\beta_3$ (Maeshima et al., 2000a). Additionally, a synthetic peptide of tumstatin that contains the RGD sequence does not have an effect on cell attachment (Maeshima et al., 2000a), suggesting that the binding property of tumstatin to endothelial cells via integrins is RGD-independent (Maeshima et al., 2000a). Integrin $\alpha_v\beta_3$ was further shown to be the binding partner of tumstatin since preincubation of tumstatin with either $\alpha_v\beta_3$ or $\alpha_v\beta_1$ to block their binding sites led to a decrease in cell proliferation only when $\alpha_v\beta_3$ was used, and had no effect when $\alpha_v\beta_1$ was used.

To further investigate the mechanism of action of tumstatin, its effect on apoptosis was investigated by measuring levels of PS and caspase 3. Endothelial cells showed both increased levels of annexin-5 staining and caspase 3 when incubated with tumstatin, suggesting that tumstatin stimulates apoptosis in endothelial cells (Maeshima et al., 2000b). Apoptosis is generally associated with inhibition of cap-dependent synthesis; therefore, the effect of tumstatin on protein synthesis in endothelial cells was investigated (Maeshima et al., 2002). Both full-length tumstatin and three active subfragments of tumstatin (Tum5, T3, and T7) were found to inhibit protein synthesis only on endothelial cells, and did not have any effect on PC-3 prostate carcinoma cells, 786-O renal carcinoma cells, NIH 3T3 fibroblasts, primary human renal epithelial cells, or human melanoma cells (Maeshima et al., 2002). Whether tumstatin causes cap-dependent or cap-independent protein synthesis was investigated by transfecting WT or integrin-$\beta3$ null endothelial cells with a plasmid-encoding dicistronic reporter mRNA consisting of luciferase (LUC) and chloramphenicol acetyltransferase (CAT), separated by internal ribozomal entry site (IRES), and under the control of the cytomegalovirus promoter (Maeshima et al., 2002). In eukaryotes, protein synthesis can only be initiated at the $5'$ end of the mRNA (cap-dependent synthesis); however, IRES can initiate translation in the middle of the mRNA avoiding the need for $5'$ capping. The presence of IRES in the dicistronic mRNA therefore allows for cap-independent protein synthesis to occur. Tumstatin peptides decreased the cap-dependent translation of the LUC mRNA in WT cells by 38%, but had no effect on the cap-independent translation of the CAT mRNA, clearly demonstrating that tumstatin-mediated inhibition of protein synthesis is cap dependent. Tumstatin had no effect on cap-dependent protein synthesis in β_3-null endothelial cells and mouse embryonic fibroblasts expressing $\alpha_v\beta_3$, suggesting that integrin expression is essential but not sufficient for tumstatin activity (Maeshima et al., 2002).

Inhibition of cap-dependent translation by tumstatin involves downstream regulators of $\alpha_v\beta_3$ such as FAK, PI3K, and Akt whose activation was decreased upon treatment of endothelial cells with tumstatin.

The mammalian target of rapamycin (mTOR), which is activated by Akt, directly phosphorylates eukaryotic initiation factor 4E (eIF4E)-binding protein (4E-BP1). When 4E-BP1 is dephosphorylated, it interacts with eIF4E and inhibits cap-dependent protein synthesis. Tumstatin peptides are able to inhibit mTOR activity and phosphorylation of 4E-BP1, thereby enhancing 4E-BP1 binding to eIF-4E and inhibiting cap-dependent translation (Maeshima *et al.*, 2002). The inhibition of cap-dependent translation of protein synthesis by tumstatin was reversed by overexpression of constitutively active Akt, consistent with the hypothesis that tumstatin peptides inhibit endothelial protein synthesis by negative regulation of mTOR signaling (Maeshima *et al.*, 2002) (Fig. 1.3).

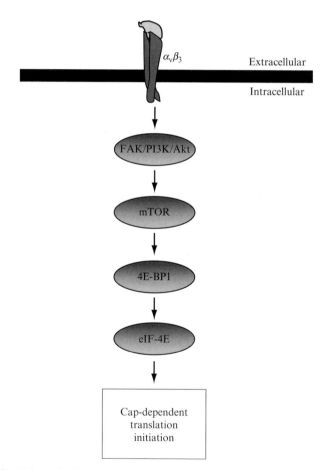

Figure 1.3   Schematic illustration of signaling pathway mediated by tumstatin (T3) in endothelial cells. Tumstatin binds to $\alpha_v\beta_3$ and inhibits the FAK/PI3K/Akt pathway, which leads to inhibition of mTOR, 4E-BP1, and eIF-4E. Cap-dependent translation is subsequently halted.

Based on these findings, Kawaguchi and colleagues (2006) speculated that perhaps tumors in which tumstatin has no effect, such as WM-164 melanoma and PC-3 prostate adenocarcinoma, have a constitutively active Akt, thereby making them resistant to tumstatin. In fact, they showed that several integrin $\alpha_v\beta_3$-expressing glioma cells, which have a mutated PTEN and subsequently high levels of Akt, did not respond to tumstatin (T3 fragment) treatment. However, in glioma cells with functional PTEN/low levels of Akt, tumstatin was able to suppress tumor growth (Kawaguchi et al., 2006). These studies therefore suggest that tumstatin could be used as a therapeutic agent when combined with a pre-screen of patients for PTEN/Akt levels.

3. ASSESSING THE ROLE OF TYPE IV COLLAGEN–DERIVED ENDOGENOUS INHIBITOR DURING ANGIOGENESIS

Studying angiogenesis *in vivo* has been accomplished by using a number of well-established methods. Here we will review some of the major methods used to reveal the roles of type IV–derived endogenous inhibitors in angiogenesis.

3.1. Migration assay

Migration of cells was investigated by the standard Boyden chamber assay (Chen, 2005). In this assay, cells are placed in the upper compartment, and are allowed to migrate through the pores of the membrane into the lower compartment in which chemotactic agents are present. After an appropriate incubation time, the membrane between the two compartments is fixed and stained, and the number of cells that have migrated to the lower side of the membrane is determined (Chen, 2005). This method provided information about the inhibitory roles of arresten and canstatin on endothelial cells (Kamphaus et al., 2000; Sudhakar et al., 2005). The basic technique of this assay performed with arresten and canstatin as inhibitors of endothelial cell migration is described below.

1. Dilute 10,000 to 60,000 endothelial cells in media and plate them on the upper chamber well supplemented with or without arresten or canstatin (0.01 or 1 μg/ml).
2. Cover the bottom wells of the Boyden chamber with medium containing 10% FCS and 4 to 10 ng/ml VEGF as chemoattractant.
3. Incubate the chamber for 4.5 to 6 h at 37 °C with 5% CO_2 and 95% humidity. Polycarbonate filters of 8-μm pore size separated the cell-containing compartments.

4. Wash wells with PBS and discard nonmigrated cells. Scrape the remaining cells with a plastic blade, fix them in 4% formaldehyde in PBS, and place on a glass slide.
5. Analyze migrating cells by taking pictures of several independent homogenous populations. Count cells either manually or with the help of imaging software.

3.2. Proliferation assay

A proliferation assay measures the ability of cells to grow in the presence of proliferating or inhibitory agents. The most common assay used to measure proliferation rates of cells is the [³H] thymidine assay. In this assay, metabolic incorporation of tritiated thymidine into cellular DNA is used to monitor rates of DNA synthesis and cell proliferation. When arresten was used as an inhibitor of endothelial cell proliferation, the media was supplemented with both endothelial mitogen to induce proliferation and arresten to study its effect on inhibiting proliferation (Sudhakar et al., 2005). After cells are plated (usually on a 96-well plate), they are incubated for 24 h at 37 °C. Next, cells are incubated with [³H]thymidine for another 24 h, at which point the cells are harvested, lysed, washed, and passed through a filter membrane to collect intact DNA. Next, the filter membrane is dried and the [³H]thymidine incorporation is measured by a scintillation counter. Since [³H]thymidine is incorporated in the cells' DNA during each cell division; the higher the proliferation rate of the cells, the more radioactive thymidine will be incorporated into the DNA. The presence of an inhibitor will therefore decrease the amount of radio-activity incorporated into the cells. Accumulating evidence, however, suggests that [³H]thymidine can induce cell cycle arrest and apoptosis, suggesting that it may not truly represent the proliferation rates of cells (Hu et al., 2002). Alternatively, safe detection of proliferation and cyto-toxicity of cells can be determined by determination of their ATP levels. Bioluminescent measurement of ATP, present in all metabolically active cells is achieved by a luminometer; thus, the use of tritiated thymidine uptake is avoided (Lonza).

3.3. Endothelial tube assay

The ability of endothelial cells to form tube-like structures has been studied in both two-dimensional (2D) and three-dimensional (3D) culture systems. In 2D systems, endothelial cells are usually cultured on a matrix consisting of fibrin, collagen, or Matrigel, which stimulates the attachment, migration, and differentiation of endothelial cells into tubules (Lawley and Kubota, 1989). Endothelial cells can be plated on either regular Matrigel (originally derived from the mouse Engelbreth-Holm-Swarm sarcoma) or on growth factor-reduced Matrigel in which the levels of cytokines and growth factors have been reduced. The use of growth factor–reduced Matrigel therefore allows for comparing the ability of endothelial cells to form tubes in the presence or absence of different growth factors. Typically, growth factors such as bFGF are used at a concentration of 10 ng/ml, although this amount may vary based on the type of endothelial cells used. In the 3D system, the endothelial cells are placed between layers of matrix and allowed to form tubes over a period of time. Initially, endothelial cells form tubules in the horizontal plane, but over a period of 12 or more days, the endothelial tubes begin to branch upward and penetrate the gel to form a 3D network of tubules (Gagnon *et al.*, 2002). The advantage of the 3D assay is that it is more *in vivo*-like. Quantification of this assay, however, is more challenging since its analysis requires taking pictures at different depths of the matrix gel. Another challenge for both the 2D and 3D systems is the fact that whether the tubes formed by the endothelial cells in these matrices have a lumen or not is still questionable. Alternatively, endothelial cells could be induced to form tubes by co-culturing them with stromal cells such as fibroblasts, smooth muscle cells, or aortic ring explants (Bishop *et al.*, 1999; Montesano *et al.*, 1993). This type of assay has been shown to produce tubules that contain lumen and relies on the ability of the stromal cells to secrete the necessary matrix to provide a scaffold for the tubes to form. The disadvantage of this assay is that it is not very well characterized, and unknown interactions between endothelial and stromal cells could interfere with endothelial-specific reagents such as inhibitors.

3.4. Matrigel plug assay

The Matrigel plug assay is a very useful *in vivo* model of angiogenesis, and it has been described and used extensively (Akhtar *et al.*, 2002). In this model, Matrigel is subcutaneously injected into the flank of 4- to 6-week-old mice. The host endothelial cells then migrate and form a capillary network in the Matrigel plugs, which are normally avascular. Each Matrigel plug can contain a combination of growth factors, angiogenic inhibitors, or other reagents. When the antiangiogenic properties of tumstatin were tested, Matrigel was mixed with 20 units/ml of heparin, 150 ng/ml bFGF, and 1 μg/ml tumstatin (*E. coli*---produced, soluble) (Maeshima *et al.*, 2000b). Generally, after 7 to 14 days post-Matrigel injection, mice can be sacrificed and Matrigel plugs evaluated for angiogenesis. The optimal time for angiogenic response should be determined based on the mouse strain and growth factors used. In the tumstatin studies, Matrigel plugs were kept for 14 days. The angiogenic response can be evaluated by embedding the plugs in paraffin with subsequent sectioning and hematoxylin and eosin (H&E) staining where sections are examined by light microscopy. Alternatively, plugs could be embedded in freezing medium (such as OCT), sectioned, and immunostained for endothelial-specific markers such as CD31, VE-cadherin, vWF, VEGFR2, or CD34. The angiogenic response could also be evaluated by measuring the hemoglobin content using the Drabkin method (Sigma).

3.5. *In vivo* tumor studies

The effects of many different antiangiogenic and anticancer treatments have been studied by utilizing *in vivo* tumor models. In these models, tumors are grown subcutaneously, orthotopically (in the tissue of origin), or as xenografts in immunodeficient rodents, and the effect of the test treatment on tumor size and animal survival evaluated (Staton *et al.*, 2004). Tumor volumes are normally calculated using a standard formula length × width squared × 0.52. Depending on the purpose of the study, animals could be sacrificed and tumors examined by various histologic techniques such as H&E staining, which assesses general morphology. Alternatively, noninvasive imaging of tumors could also be a very useful, tool and new ways to achieve that are being developed. Noninvasive imaging could help in assessing a particular treatment efficacy by visualizing tumor metabolism or apoptosis. Direcks and colleagues (2008), for example, have developed a rat model of human breast cancer in which a positron emission tomography (PET) scan is used to detect potential functional changes that occur before any visible anatomic changes. Noninvasive imaging therefore can greatly impact medicine, since it aims to look at changes associated with tumor progression rather than the resulting endpoint change.

4. Concluding Remarks

Type IV collagen–derived endogenous inhibitors of angiogenesis are molecules that are naturally produced in the body. It is now well established that endogenous inhibitors play a significant role in maintaining the angiogenic balance, which is crucial for tumor growth and vascularization. Similar to other endogenous inhibitors, arresten, canstatin, and tumstatin are derived from ECM, and depend on the tumor-associated proteolytic enzymes to initiate ECM degradation. Thus, the activity of these enzymes could regulate the availability of the endogenous inhibitors, and is another factor in the maintenance of the angiogenic balance. Understanding the mechanism of action of all three inhibitors required dissection of the molecules involved in cell-signaling pathways, such as MAPK, Akt, and apoptosis pathways. Knowing how these inhibitors work is important for determining their potential use as therapeutic agents.

REFERENCES

Akhtar, N., Dickerson, E. B., and Auerbach, R. (2002). The sponge/Matrigel angiogenesis assay. *Angiogenesis* **5,** 75–80.

Augustin, H. G. (2003). [Angiogenesis research—quo vadis?]. *Ophthalmologe* **100,** 104–110.

Bishop, E. T., Bell, G. T., Bloor, S., Broom, I. J., Hendry, N. F., and Wheatley, D. N. (1999). An *in vitro* model of angiogenesis: Basic features. *Angiogenesis* **3,** 335–344.

Bloch, W., Forsberg, E., Lentini, S., Brakebusch, C., Martin, K., Krell, H. W., Weidle, U. H., Addicks, K., and Fassler, R. (1997). Beta 1 integrin is essential for teratoma growth and angiogenesis. *J. Cell Biol.* **139,** 265–278.

Cao, Y. (2001). Endogenous angiogenesis inhibitors and their therapeutic implications. *Int. J. Biochem. Cell Biol.* **33,** 357–369.

Carmeliet, P. (2003). Angiogenesis in health and disease. *Nat. Med.* **9,** 653–660.

Chen, H. C. (2005). Boyden chamber assay. *Methods Mol. Biol.* **294,** 15–22.

Clark, E. A., and Brugge, J. S. (1995). Integrins and signal transduction pathways: The road taken. *Science* **268,** 233–239.

Colorado, P. C., Torre, A., Kamphaus, G., Maeshima, Y., Hopfer, H., Takahashi, K., Volk, R., Zamborsky, E. D., Herman, S., Sarkar, P. K., Ericksen, M. B., Dhanabal, M., *et al.* (2000). Anti-angiogenic cues from vascular basement membrane collagen. *Cancer Res.* **60,** 2520–2526.

Direcks, W. G., van Gelder, M. B., Lammertsma, A. A., and Molthoff, C. F. (2008). A new rat model of human breast cancer for evaluating efficacy of new anti-cancer agents *in vivo*. *Cancer Biol. Ther.* **3**(4), 7.

Fadok, V. A., Bratton, D. L., Rose, D. M., Pearson, A., Ezekewitz, R. A., and Henson, P. M. (2000). A receptor for phosphatidylserine-specific clearance of apoptotic cells. *Nature* **405,** 85–90.

Folkman, J. (1971). Tumor angiogenesis: Therapeutic implications. *N. Engl. J. Med.* **285,** 1182–1186.

Folkman, J. (1995). Angiogenesis in cancer, vascular, rheumatoid and other disease. *Nat. Med.* **1,** 27–31.

Folkman, J., and Shing, Y. (1992). Angiogenesis. *J. Biol. Chem.* **267,** 10931–10934.

Frisch, S. M., Vuori, K., Ruoslahti, E., and Chan-Hui, P. Y. (1996). Control of adhesion-dependent cell survival by focal adhesion kinase. *J. Cell Biol.* **134,** 793–799.

Gagnon, E., Cattaruzzi, P., Griffith, M., Muzakare, L., LeFlao, K., Faure, R., Beliveau, R., Hussain, S. N., Koutsilieris, M., and Doillon, C. J. (2002). Human vascular endothelial cells with extended life spans: *In vitro* cell response, protein expression, and angiogenesis. *Angiogenesis* **5,** 21–33.

Grant, M. A., and Kalluri, R. (2005). Structural basis for the functions of endogenous angiogenesis inhibitors. *Cold Spring Harb. Symp. Quant. Biol.* **70,** 399–410.

Guo, W., and Giancotti, F. G. (2004). Integrin signalling during tumour progression. *Nat. Rev. Mol. Cell Biol.* **5,** 816–826.

Han, J., Ohno, N., Pasco, S., Monboisse, J. C., Borel, J. P., and Kefalides, N. A. (1997). A cell binding domain from the alpha3 chain of type IV collagen inhibits proliferation of melanoma cells. *J. Biol. Chem.* **272,** 20395–20401.

He, G. A., Luo, J. X., Zhang, T. Y., Wang, F. Y., and Li, R. F. (2003). Canstatin-N fragment inhibits *in vitro* endothelial cell proliferation and suppresses *in vivo* tumor growth. *Biochem. Biophys. Res. Commun.* **312,** 801–805.

Hou, W. H., Wang, T. Y., Yuan, B. M., Chai, Y. R., Jia, Y. L., Tian, F., Wang, J. M., and Xue, L. X. (2004). Recombinant mouse canstatin inhibits chicken embryo chorioallantoic membrane angiogenesis and endothelial cell proliferation. *Acta Biochim. Biophys. Sin. (Shanghai)* **36,** 845–850.

Hu, V. W., Black, G. E., Torres-Duarte, A., and Abramson, F. P. (2002). ^{3}H-thymidine is a defective tool with which to measure rates of DNA synthesis. *FASEB J.* **16,** 1456–1457.

Irmler, M., Thome, M., Hahne, M., Schneider, P., Hofmann, K., Steiner, V., Bodmer, J. L., Schroter, M., Burns, K., Mattmann, C., Rimoldi, D., French, L. E., *et al.* (1997). Inhibition of death receptor signals by cellular FLIP. *Nature* **388,** 190–195.

Kamphaus, G. D., Colorado, P. C., Panka, D. J., Hopfer, H., Ramchandran, R., Torre, A., Maeshima, Y., Mier, J. W., Sukhatme, V. P., and Kalluri, R. (2000). Canstatin, a novel matrix-derived inhibitor of angiogenesis and tumor growth. *J. Biol. Chem.* **275,** 1209–1215.

Kawaguchi, T., Yamashita, Y., Kanamori, M., Endersby, R., Bankiewicz, K. S., Baker, S. J., Bergers, G., and Pieper, R. O. (2006). The PTEN/Akt pathway dictates the direct alphaVbeta3-dependent growth-inhibitory action of an active fragment of tumstatin in glioma cells *in vitro* and *in vivo. Cancer Res.* **66,** 11331–11340.

Kuhn, K., Wiedemann, H., Timpl, R., Risteli, J., Dieringer, H., Voss, T., and Glanville, R. W. (1981). Macromolecular structure of basement membrane collagens. *FEBS Lett.* **125,** 123–128.

Lawley, T. J., and Kubota, Y. (1989). Induction of morphologic differentiation of endothelial cells in culture. *J. Invest. Dermatol.* **93,** 59S–61S.

Maeshima, Y., Colorado, P. C., and Kalluri, R. (2000a). Two RGD-independent alpha vbeta 3 integrin binding sites on tumstatin regulate distinct anti-tumor properties. *J. Biol. Chem.* **275,** 23745–23750.

Maeshima, Y., Colorado, P. C., Torre, A., Holthaus, K. A., Grunkemeyer, J. A., Ericksen, M. B., Hopfer, H., Xiao, Y., Stillman, I. E., and Kalluri, R. (2000b). Distinct antitumor properties of a type IV collagen domain derived from basement membrane. *J. Biol. Chem.* **275,** 21340–21348.

Maeshima, Y., Manfredi, M., Reimer, C., Holthaus, K. A., Hopfer, H., Chandamuri, B. R., Kharbanda, S., and Kalluri, R. (2001). Identification of the anti-angiogenic site within vascular basement membrane–derived tumstatin. *J. Biol. Chem.* **276,** 15240–15248.

Maeshima, Y., Sudhakar, A., Lively, J. C., Ueki, K., Kharbanda, S., Kahn, C. R., Sonenberg, N., Hynes, R. O., and Kalluri, R. (2002). Tumstatin, an endothelial cell-specific inhibitor of protein synthesis. *Science* **295,** 140–143.

Magnon, C., Galaup, A., Mullan, B., Rouffiac, V., Bouquet, C., Bidart, J. M., Griscelli, F., Opolon, P., and Perricaudet, M. (2005). Canstatin acts on endothelial and tumor cells via mitochondrial damage initiated through interaction with alphavbeta3 and alphavbeta5 integrins. *Cancer Res.* **65,** 4353–4361.

Montesano, R., Pepper, M. S., and Orci, L. (1993). Paracrine induction of angiogenesis *in vitro* by Swiss 3T3 fibroblasts. *J. Cell Sci.* **105**(Pt 4), 1013–1024.

Mundel, T. M., and Kalluri, R. (2007). Type IV collagen-derived angiogenesis inhibitors. *Microvasc. Res.* **74,** 85–89.

Nagata, S. (1996). Fas-mediated apoptosis. *Adv. Exp. Med. Biol.* **406,** 119–124.

Nyberg, P., Xie, L., and Kalluri, R. (2005). Endogenous inhibitors of angiogenesis. *Cancer Res.* **65,** 3967–3979.

Panka, D. J., and Mier, J. W. (2003). Canstatin inhibits Akt activation and induces Fas-dependent apoptosis in endothelial cells. *J. Biol. Chem.* **278,** 37632–37636.

Semenza, G. L. (2000). HIF-1 and human disease: One highly involved factor. *Genes Dev.* **14,** 1983–1991.

Senger, D. R., Claffey, K. P., Benes, J. E., Perruzzi, C. A., Sergiou, A. P., and Detmar, M. (1997). Angiogenesis promoted by vascular endothelial growth factor: Regulation through alpha1beta1 and alpha2beta1 integrins. *Proc. Natl. Acad. Sci. USA* **94,** 13612–13617.

Shahan, T. A., Ziaie, Z., Pasco, S., Fawzi, A., Bellon, G., Monboisse, J. C., and Kefalides, N. A. (1999). Identification of CD47/integrin-associated protein and alpha (v)beta3 as two receptors for the alpha3(IV) chain of type IV collagen on tumor cells. *Cancer Res.* **59,** 4584–4590.

Sidky, Y. A., and Borden, E. C. (1987). Inhibition of angiogenesis by interferons: Effects on tumor- and lymphocyte-induced vascular responses. *Cancer Res.* **47,** 5155–5161.

Staton, C. A., Stribbling, S. M., Tazzyman, S., Hughes, R., Brown, N. J., and Lewis, C. E. (2004). Current methods for assaying angiogenesis *in vitro* and *in vivo*. *Int. J. Exp. Pathol.* **85,** 233–248.

Sudhakar, A., Nyberg, P., Keshamouni, V. G., Mannam, A. P., Li, J., Sugimoto, H., Cosgrove, D., and Kalluri, R. (2005). Human alpha1 type IV collagen NC1 domain exhibits distinct antiangiogenic activity mediated by alpha1beta1 integrin. *J. Clin. Invest.* **115,** 2801–2810.

Sudhakar, A., Sugimoto, H., Yang, C., Lively, J., Zeisberg, M., and Kalluri, R. (2003). Human tumstatin and human endostatin exhibit distinct antiangiogenic activities mediated by alpha v beta 3 and alpha 5 beta 1 integrins. *Proc. Natl. Acad. Sci. USA* **100,** 4766–4771.

Sund, M., Hamano, Y., Sugimoto, H., Sudhakar, A., Soubasakos, M., Yerramalla, U., Benjamin, L. E., Lawler, J., Kieran, M., Shah, A., and Kalluri, R. (2005). Function of endogenous inhibitors of angiogenesis as endothelium-specific tumor suppressors. *Proc. Natl. Acad. Sci. USA* **102,** 2934–2939.

Teodoro, J. G., Parker, A. E., Zhu, X., and Green, M. R. (2006). p53-mediated inhibition of angiogenesis through up-regulation of a collagen prolyl hydroxylase. *Science* **313,** 968–971.

Timpl, R., Wiedemann, H., van Delden, V., Furthmayr, H., and Kuhn, K. (1981). A network model for the organization of type IV collagen molecules in basement membranes. *Eur. J. Biochem.* **120,** 203–211.

van Engeland, M., Nieland, L. J., Ramaekers, F. C., Schutte, B., and Reutelingsperger, C. P. (1998). Annexin V-affinity assay: A review on an apoptosis detection system based on phosphatidylserine exposure. *Cytometry* **31,** 1–9.

CHAPTER TWO

Chick Embryo Chorioallantoic Membrane Models to Quantify Angiogenesis Induced by Inflammatory and Tumor Cells or Purified Effector Molecules

Elena I. Deryugina *and* James P. Quigley

Contents

The Scripps Research Institute, La Jolla, California

Methods in Enzymology, Volume 444
ISSN 0076-6879, DOI: 10.1016/S0076-6879(08)02802-4

Abstract

Angiogenesis plays a critical role in many normal physiological processes as well as in tumor neovascularization associated with cancer progression. Among various animal model systems designed to study the mechanisms underlying angiogenesis, chick embryo models have been useful tools in analyzing the angiogenic potential of purified factors and intact cells. The chorioallantoic membrane (CAM), a specialized, highly vascularized tissue of the avian embryo, serves as an ideal indicator of the anti- or pro-angiogenic properties of test compounds. In this chapter, we describe a number basic chick embryo CAM models of angiogenesis. A special emphasis is on the model system employing three-dimensional (3D) collagen grafts planted on the CAM, referred herein as onplants. This collagen onplant model allows for unambiguous quantification of angiogenesis and also for in-depth analysis of the cellular and biochemical mechanisms by which specific cells of different origin or purified effector molecules induce or inhibit the angiogenic process.

1. INTRODUCTION

Angiogenesis is a progressive, multistep physiological process by which new blood vessels are generated from pre-existing vasculature. Adult vasculature is maintained mostly in an angiostatic state that must be switched off to allow for new blood vessel formation. This angiogenic switch is a part of normal physiologic responses, for example to tissue injury, as well as a critical step in the pathology of tumor progression. It is commonly accepted that specific mechanisms underlining the angiogenic switch involve a selective remodeling of the extracellular matrix (ECM) by proteolytic enzymes and the induction, generation or release of angiogenic growth factors, which induce endothelium sprouting, followed by reorganization and formation of new blood vessels.

During cancer progression, the newly formed tumor-associated blood vessels serve first as feeding/nurturing tubes for a growing tumor and next, as conduits for dissemination of tumor cells that escaped from an established primary tumor. Therefore, control of tumor angiogenesis has became a central issue in the fight against cancer progression since anticancer therapy could be ineffective once tumor cells reach favored secondary organs and generate metastatic foci.

To analyze the mechanisms underlying normal and pathological angiogenesis, numerous *in vivo* angiogenic assays have been established employing different species of laboratory animals, including mammals (mouse, rat, hamster, and rabbit), birds (chicken and quail), and fish (mainly zebra fish). In this chapter, we will focus on major models of angiogenesis in the chick embryo. The use of chick embryo models for angiogenic studies is facilitated by the existence in avian species of a specialized respiratory tissue, named the chorioallantoic membrane (CAM) that allows for gas exchange between the embryo and the atmosphere surrounding the egg and in effect performs the function of a lung during embryonic life (Romanoff, 1960).

In the chick embryo, the chorioallantois is formed between days 4 and 5 of development, when the outer mesodermal layer of the allantois fuses with the mesodermal lining of the chorion, and a network of blood vessels is gradually formed between the two layers. The central portion of the CAM is fully developed by day 8 to 10 at which time it becomes capable of sustaining tissue grafts, while the outskirts of the CAM are still developing and expanding until the CAM fully envelopes the embryo at day 12 of incubation. Histologically, the CAM consists of three germ layers, that is, ectoderm, mesoderm, and endoderm. The ectoderm faces the shell membrane and is underlined by the respiratory capillary plexus, which starts to form between days 5 and 6 of embryonic development by both angiogenesis and vasculogenesis (Melkonian et al., 2002). This capillary plexus is very dense and appears as a honeycomb network of tiny capillaries originating from terminal capillaries (Fig. 2.1). The mesoderm of the chorioallantois is a collagen-rich embryonic connective tissue transversed by blood vessels belonging to the arteriolar and venous systems. The mesoderm is underlined with a thin endoderm layer, which separates the CAM from the allantoic cavity (Romanoff, 1960; Tufan and Satiroglu-Tufan, 2005).

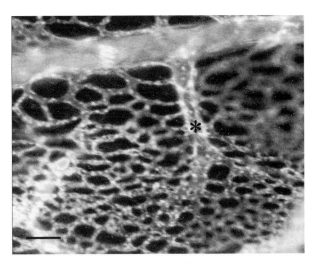

Figure 2.1 Ectoderm capillary plexus. Chick embryos on day 12 of development were injected intravenously with rhodamin-conjugated Lens *culinaris* agglutinin (0.1 ml of 1 mg/ml solution in PBS per embryo), which specifically binds to the chick endothelium. After 30 min of incubation at 37 °C, the embryos were sacrificed and the CAMs were detached from the shell membrane and visualized with a fluorescent microscope equipped with a digital camera. The image shows a terminal capillary (*), which radiates into the honeycomb-like network of capillaries constituting the ectoderm plexus. Bar, 50 μm.

Until day 11 or 12 of chick embryo development, the blood vessel system of the CAM is highly angiogenic, that is, undergoing maturation through a constant generation of new blood vessels as well as establishment of new blood vessel anastomoses. Therefore, between day 8 and day 10, the developing CAM vasculature is ready to sprout in response to additional proangiogenic stimuli and, in turn, is very responsive to antiangiogenic factors. This feature renders the chick embryo CAM models well suited for experimental validation of pro- and anti-angiogenic compounds. The response of the CAM to angiogenic stimuli is relatively rapid and most assays require only 3 to 5 days. In addition, recent modifications of the originally described assays readily allow for quantitation of the angiogenic process, revolutionizing the use of CAM assays in angiogenesis studies. In addition, the chick embryo is naturally immunoincompetent until embryonic day 17, thus allowing for grafting of cells of different species origin, such as human tumor cells, and therefore providing a useful tool for analysis of the proangiogenic potential of test cells.

2. Overview of CAM Angiogenesis Models

Several CAM angiogenic assays have been introduced since almost a century ago when rat Jensen sarcoma cells, implanted into the CAM on the day 6 of incubation, were demonstrated to develop large tumors showing signs of tumor-induced angiogenesis (Murphy, 1913). All modifications of the original angiogenic assay in the chick embryo involve grafting of test material onto developing CAM. The grafting is often performed through a window cut in the egg shell over the CAM. The angiogenic material is usually introduced in the form of small disks soaked in angiogenic factors or small pieces of polymerized materials such as gelatin sponges or biologically inert synthetic polymers, containing either purified angiogenic factors or impregnated with tumor cells. Another less traumatic way of introducing angiogenic material onto the CAM involves the use of shell-less embryos grown *ex ovo*, which makes the CAM more accessible for repetitive manipulations, for quantitation of angiogenesis, and for direct visualization of the angiogenic process under a stereoscope. Below, we will briefly describe major CAM angiogenic assays employing both in-shell and shell-less chick embryos and different sources of angiogenic factors such as purified molecules and tumor cells.

2.1. CAM filter disk assays

The factor-containing disks placed directly on the CAM are widely used in validation of the pro- and anti-angiogenic properties of test compounds (Beckers *et al.*, 1997; Brooks *et al.*, 1998; Eliceiri *et al.*, 1999; Han *et al.*, 2001; Hood *et al.*, 2003; Miller *et al.*, 2004; Murugesan *et al.*, 2007; Sahni *et al.*, 2006). The filter disks can be cut from nitrocellulose membrane or

Whatman paper. The CAM is prepared for grafting by making an air sac, usually on day 3 of egg incubation, by sucking out approximately 3 ml of albumen with a syringe. This procedure allows for the development of the intact, noninjured CAM, which becomes experimentally accessible between day 8 and 9 through a window cut in the egg shell. The fact that the CAM develops intact is an important issue since any injury to the CAM provides potent hypertrophic or atrophic reactions that may alter the angiogenic potential of test material. However, the egg shell dust generated while making the window to access the CAM might serve as inflammatory stimuli and cause a change in the angiogenesis readout. An alternative method of separating the CAM from the shell membrane involves opening a small window in the egg shell at days 7 to 10 of embryonal development above the CAM after removing 2 to 3 ml of albumen or after creating a false air sac (Brooks *et al.*, 1994; Hood *et al.*, 2003).

Filter disks are used to confine the test material to a defined area of the CAM. The disks are pre-soaked in test compounds and usually are dried before grafting on the CAM. Following disk application, the window in the egg shell is closed with a piece of tape or glass and the eggs are placed back into the incubator. The available CAM area under the window is only enough for the application of one disk per embryo, thus dictating the use of large groups of animals to get reliable, quantitative readouts. In this assay, quantitation is performed usually 3 days after implantation and involves counting the number of CAM vessels in the area of filter disk. In response to proangiogenic stimuli, the newly formed blood vessels appear converging toward the disk in a wheel-spoke pattern. Inhibition of angiogenesis by antiangiogenic compounds results in the lack of new blood vessel formation and sometimes in disappearance of pre-existing vessel networks. Angiogenesis levels can be also determined by counting branch points in the vessels adjacent to the disks.

Time course of new blood vessel formation in the CAM can be readily performed by analyzing the images taken at different time points after grafting of the disk using a stereoscope equipped with a digital camera. More sophisticated techniques have been designed recently to perform reliable quantitative evaluation of vascular density, endothelial proliferation, and protein expression in response to angiogenic agents released from the filter disks to the underlying CAM (Miller *et al.*, 2004). These techniques include in ovo cell proliferation, layered expression scanning to visualize the protein of interest, and fluorescent confocal microscopy of new blood vessel formation in the CAM at the site of filter application. The major disadvantage of this angiogenic assay is that introducing a disk alone, even without growth factors, can induce a high angiogenic response, therefore obscuring any proangiogenic properties of the test compounds. Thus, more commonly filter disk assays are employed for evaluation of antiangiogenic potential of test materials since the diminishment or lack of vascular convergence towards the disk is easier to appreciate and quantify (Ribatti *et al.*, 2000).

2.2. CAM assays employing various gelated materials

A number of CAM angiogenesis assays are based on testing purified factors and intact cells incorporated into gelated materials such as methylcellulose, Matrigel, or sodium alginate. Although disks containing gelated material impose the same mechanical injury as filter disks directly soaked in test compounds, the use of gels allows a slow, but efficient release of anti- or pro-angiogenic factors into the underlying CAM tissue.

Preparation of methylcellulose disks involves spreading and drying of the factor-containing mixtures on Teflon surfaces (Yang and Moses, 1990), glass surfaces (Ribatti et al., 1995), or parafilm (Hagedorn et al., 2004) before their application on the CAMs of in-shell or shell-less individual embryos (Struman et al., 1999). Otherwise, methylcellulose mixture can be dried on nylon meshes, which provide a support for the disks (Cao et al., 1998). Three to 5 days after disk implantation, the CAMs are examined by stereomicroscope for new blood vessel formation or inhibition of angiogenesis within the field of the implanted disks.

Matrigel mixtures can be distributed in small volumes directly onto the CAM where rapid polymerization occurs. Alternatively, defined aliquots of Matrigel supplemented with test compounds can be pre-gelated at 37 °C on nylon meshes and then placed onto the CAM (Vazquez et al., 1999; Watanabe et al., 2004). Qualitative and quantitative variations in the growth factors intrinsically present in different preparations of Matrigel poses a serious concern. Therefore, the use of Matrigel with reduced amounts of growth factors is recommended in order to negate lot-to-lot variations.

Slow release of purified growth factors or angiogenic factors produced by test cells can be achieved by incorporating the test components into alginate pellet (Riboldi et al., 2005). The pellets are prepared by mixing a solution of sodium alginate with cells or purified molecules to achieve desirable final concentrations, followed by a dropwise releasing of the mixture into a $CaCl_2$-contating solution. The calcium ions cause immediate gelling of the alginate droplets, which, after washing, can be implanted onto the CAM.

New models are constantly introduced aiming to improve various aspects of in-shell CAM angiogenesis models. One of such models is a cylinder model designed to assess the vascularization potential of engineered tissues (Borges et al., 2003). In this model, cell-containing matrices are applied within specially constructed plastic cylinders, allowing for continual observation of graft vascularization using a light microscope. This model requires a variety of specialized technical devices and the experimental setup is complex. In addition, it lacks a straightforward approach to quantify the implant-induced angiogenic response.

2.3. Gelatin sponge CAM assay

One of the major limitations of the filter disk angiogenesis assays is that the support material does not allow for a proper maintenance of tumor cell inoculums. This problem is overcome in a modification of the CAM angiogenic assay employing gelatin sponges impregnated with tumor cells. The sponge assay usually involves the use of small pieces of polymerized gelatin pre-soaked with test ingredients or filled with tumor cells (Ribatti et al., 2000, 2001, 2006). Gelatin sponges can be implanted on the growing CAM on day 8 of embryonic development. Since the sponges firmly adhere to the CAM surface, the test substances or cell suspensions are confined to the site of administration. The in-shell chick embryos are prepared for the sponge CAM assay in a similar way as described for the filter disk assay (Section 2.1). Similarly, the number of converging blood vessels or branch points is determined at the end of the experiment. Therefore, the same readout limitation of the filter disk assay is applicable to the CAM sponge assay, that is, the ambiguity in the level of new versus pre-existing vessel convergence. However, despite this limitation, the sponge assay is regarded as more advanced and reliable compared to the filter disk CAM assay. Furthermore, the implantation of gelatin sponges is regarded to be better tolerated because it causes less nonspecific inflammatory reaction than filer disk grafting. In addition, new truly angiogenic blood vessels growing vertically into the sponge can be quantified by morphometric evaluation of histologic CAM sections (Ribatti et al., 2000). Recently, the gelatin sponge CAM assay was used to demonstrate the role of aquaporin-1 in tumor-induced angiogenesis (Camerino et al., 2006).

2.4. CAM collagen onplant model

Many of the disadvantages of the above-described methods have been negated with the introduction of a shell-less modification of the CAM angiogenesis assay, employing viable embryos grown ex ovo and grafted with 3D gridded collagen onplants. These modifications led to a major breakthrough in quantitation of the CAM angiogenesis methods. This method will be described in detail because it offers many practical advantages over filter disk, Matrigel, and sponge CAM assays (see Sections 2.1 to 2.3) and allows for unambiguous scoring of new blood vessels, and also direct imaging, in situ analysis, dissection, and ex vivo biochemical analysis. The method was first introduced by Judah Folkman and colleagues in 1994 (Nguyen et al., 1994) and is based on the scoring of new capillary CAM blood vessels grown vertically from the pre-existing vasculature into a pepsin-solubilized bovine dermal collagen (vitrogen) through two parallel nylon meshes. Angiogenesis was induced either by purified basic fibroblast growth factor, bFGF or FGF-2, or by tumor cells incorporated into gels.

New blood vessels are induced to grow upright into collagen gels from the underlying CAM, and therefore can be clearly discriminated from the background vascular network, providing a clear advantage for angiogenesis scoring. This represents a modification of the CAM angiogenesis method to offer a straightforward approach for quantifying angiogenesis (Nguyen et al., 1994).

Our modification of this assay (Seandel et al., 2001) involves the use of 3D grafts, each of which consists of two nylon grid meshes embedded into native, nonpepsinized, type I fibrillar collagen. As a support for angiogenic blood vessels, collagen is superior to filter disks, sponges and even Matrigel because collagen is the natural major protein component of stromal tissue where angiogenesis occurs (Tufan and Satiroglu-Tufan, 2005). Neutralized collagen in solution is premixed with defined amounts of factors and/or cells to be tested for their pro- or anti-angiogenic properties. The collagen mixture is distributed over the grid meshes in a small, defined volume (usually 30 μl per onplant) and allowed to polymerize in a 37 °C thermostat. Polymerization occurs within minutes, which ensures that the incorporated cells are distributed evenly throughout the 3D collagen and not settled at the bottom of the onplants. Additional 30 to 45 min of incubation allow the collagen to completely solidify, after which time the collagen onplants are placed with forceps on the CAMs of 10-day-old shell-less embryos incubated ex ovo in a stationary incubator. This method of embryo preparation excludes the proinflammatory effects of shell dust induced by making a shell window. Moreover, since the entire CAM in the shell-less embryos develops and expands facing the top of a culture dish, large areas (about 30 to 40 cm^2) are available for the onplant grafting. Thus, several grafts (onplants) can be placed on the CAM of each embryo (routinely from 4 to 7), which greatly improves statistical power of the assay, allowing 20 to 42 individual angiogenic determinations when only 5 to 6 embryos are used for a given variable.

After receiving onplants, the embryos are returned immediately to the incubator for an additional 3 to 4 days. During this period of incubation, the embryos are readily available for repeated viewing under the light stereoscope or for intravital microscopy since the CAM is relatively thin, varying between 50 and 110 μm in thickness (Reizis et al., 2005). In addition, pharmacological intervention is simple: The grafts can be treated directly with chemicals of choice, or the embryos can be systemically treated by intravenous inoculation through the allantoic vein, which is clearly visible and accessible on the expanded CAM, or more locally treated through delivery of chemicals under the CAM into the allantoic cavity. This ability to readily modulate the angiogenic response in shell-less embryos provides an invaluable tool to analyze biochemically the mechanisms involved in angiogenesis induced by factors or cells originally incorporated into the collagen mixture.

Within 2 to 4 days of incubation, collagen onplants are infiltrated with newly formed blood vessels making an anastomosized tubular network filled with circulating blood. The blood vessels, which are visualized with a stereomicroscope above the lower nylon mesh, are regarded as angiogenic since only newly formed vessels infiltrate the collagen graft from the pre-existing CAM vasculature, which is located below the nylon mesh of the graft. Importantly, during scoring of angiogenesis, the ectoderm capillary plexus and the mesodermal vascular network are off the focus plane of the viewer, and therefore do not interfere with angiogenic scoring. Since the meshes incorporated into collagen are gridded, angiogenesis can be quantified as a ratio of the grid areas filled with distinct blood vessels over the total number of grid areas observed in an individual onplant. This approach makes the CAM onplant assay unambiguous in determination of the angiogenic response induced in each individual onplant, each individual embryo and, finally, in each set of embryos designated for a certain variable.

Intrinsically, this method allows for a wide variety of technical modifications, including different support material (any gelating material vs. originally introduced native collagen, such as fibrinogen, noncleavable mutant collagen, or collagen impregnated with various ECM proteins), and use of combinations of very small and importantly, defined amounts of incorporated effector molecules, that is, nanogram quantities of purified growth factors, cytokines, inhibitors, or matrix-modifying molecules such as proteinases. In addition, various types of cells can be tested in the assay, including tumor cells or inflammatory cells, alone or in combination with each other or with defined chemicals.

After angiogenesis scoring, the onplants are available for a variety of analyses, including histological, immunohistochemical, and biochemical examinations. Histological analysis has previously demonstrated that collagen onplants are rapidly integrated into the CAM tissue. Twenty-four hours following grafting, the onplants tightly attach to the CAM, and by 48 hours they are already covered with the ectoderm and have become an integral part of the mesoderm. As soon as 2 h after grafting, inflammatory cells, first heterophils (avian analog of mammalian neutrophils) and then monocyte/macrophages, infiltrate the onplants. These cells deliver pro- and anti-inflammatory factors and cytokines, as well as important modifiers of the ECM, that is, matrix metalloproteinases (MMPs). Avian neutrophils (heterophils) were demonstrated to import MMP-9, a major gelatinase (Zijlstra et al., 2006), while monocyte/macrophages were associated with delivery of MMP-13, a potent collagenase (Zijlstra et al., 2004). The original observation of defined inflammatory cell influx into onplants allowed us to begin a systematic investigation of the role of inflammatory MMPs in angiogenesis.

3. Assessing the Role of Purified Effector Molecules and Cells in Angiogenesis Using CAM Collagen Onplant Model

The role of human MMPs, including that of the neutrophil MMP-9, was thoroughly analyzed using the CAM collagen onplant assay. Recently, we demonstrated that neutrophils appear to provide an important proangiogenic factor, that is, MMP-9 to sites of physiological and tumor-induced angiogenesis (Ardi et al., 2007). The proangiogenic activity of neutrophil MMP-9 was mechanistically linked to its TIMP-free status, activation, and catalytic activity.

Neutrophil MMP-9 is stored in secretory granules and released upon stimulation in a zymogen form (proMMP-9). Pro-MMP-9 purified from neutrophil granule contents was shown to constitute a distinctly potent proangiogenic moiety, inducing angiogenesis in the CAM collagen onplant assay at subnanogram levels. Uniquely, neutrophils produce MMP-9 as a TIMP-free proenzyme that is readily available for proteolytic activation. Both MMP-9 proenzyme activation and the catalytic activity of the activated enzyme were required to induce an angiogenic response. That the high angiogenic potency of neutrophil proMMP-9 is associated with its unique TIMP-free status, was confirmed when a purified stoichiometric complex of neutrophil proMMP-9 with TIMP-1 failed to induce angiogenesis. Recombinant human pro-MMP-9, operationally free of TIMP-1, also induced angiogenesis at subnanomolar levels, but lost its proangiogenic potential when stoichiometrically complexed with TIMP-1. Similar pro-MMP-9/TIMP-1 complexes that are naturally produced by human monocytic U937 cells and HT-1080 fibrosarcoma cells did not stimulate angiogenesis. These findings have provided the first biochemical evidence that infiltrating neutrophils, in contrast to other cell types, deliver a potent proangiogenic moiety, that is, the unencumbered TIMP-free MMP-9 (Ardi et al., 2007). The cellular and biochemical evidence supporting these conclusions was generated solely because the CAM collagen onplant method allowed for the direct addition of intact neutrophils, their crude released contents (releasate) and purified releasate components such as MMP-9.

When tumor cells are incorporated into collagen onplants as low as 1×10^4 cells per onplant, they can provide proangiogenic stimuli (e.g., HT-1080 fibrosarcoma, HEp-3 epidermoid carcinoma, PC-3 prostate carcinoma) or be essentially inert, that is, not inducing additional levels of angiogenesis beyond basal control levels (e.g., HeLa cells). If tumor-containing onplants are left for 6 to 7 days on the CAM, they produce small, confined, and well-vascularized tumors implicating angiogenesis as an essential part of primary tumor establishment. Combining tumor cells with an exogenous source of inflammatory cells or their products allows for a detailed study of the role of inflammation on tumor-induced angiogenesis. In this modification of the CAM-collagen onplant method, the influx into onplants of endogenous, host inflammatory cells could be suppressed by the treatment of embryos with anti-inflammatory drugs such as ibuprofen. The shell-less embryos bearing the onplants are easily manipulated, allowing direct addition of the anti-inflammatory drugs. Treatment of the embryos with anti-inflammatory drugs significantly inhibits the high levels of angiogenesis induced by various tumor cell lines. In addition, incorporation of various function-blocking antibodies into the onplants containing tumor cells also decreases angiogenesis levels, thus indicating the involvement of the antigens targeted by the antibodies in tumor-induced angiogenesis. All of the above-discussed examples illustrate the versatile nature of the collagen onplant assay in analyzing the mechanisms of angiogenesis under physiological and pathological conditions.

4. CAM COLLAGEN ONPLANT ASSAY PROTOCOL

Provided in the following section is a detailed angiogenesis protocol employing 3D collagen grafts planted on the CAM (onplants). This assay is performed using shell-free chick embryos, essentially as described in Seandel *et al.* (2001), Zijlstra *et al.* (2004), Zijlstra *et al.* (2006), and Ardi *et al.* (2007), with some modifications.

4.1. Preparation of chick embryos: Shell-less *ex ovo* cultures

Fertilized COFAL-negative White Leghorn chicken eggs can be obtained from Charles River Labs (North Franklin, CT) or any other source providing specific pathogen-free animals. Upon arrival, the eggs should be placed either into a humidified refrigerator set at 7 to 10 °C for storage or into a rotary thermostat at 37.5 °C and 70 to 75% relative humidity for incubation. At day 3, the eggs are cleaned with 70% alcohol solution, and the egg shell is carefully cut using a wide wheel of a portable drill (e.g., Dremel) (Fig. 2.2A and B). The entire contents of the egg are transferred to plastic weigh boats (Fig. 2.2C). The boats should be pre-soaked in 70% ethanol and then dried under UV in the laminar hood to prevent microbial and particularly, fungal, contamination. The weigh boats with the egg contents are covered with square Petri dishes (Fig. 2.2D) and placed into the stationary incubator at 37.5 °C and 70 to 75% relative humidity. The embryos are allowed to develop for the next 7 days, at which time, (i.e., on embryonic day 10) the CAMs are developed enough to be able to sustain angiogenesis in collagen onplants (Fig. 2.3).

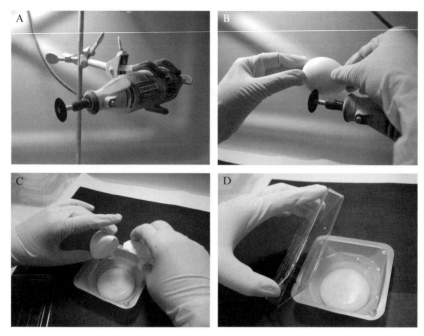

Figure 2.2 Preparation of shell-less chick embryos. At day 3 of embryonal development, the shell of the eggs is cut using a portable drill (A). Three incisions are made in the shell with a wide wheel (B), and the contents of the egg are transferred into sterilized weigh boats (C). The *ex ovo* cultures are covered with a square Petri dish and placed into a stationary incubator (D).

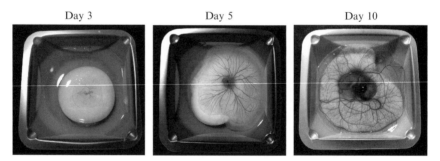

Figure 2.3 Development of chick embryos in shell-less culture *ex ovo*. Sequential images depict a day-3 chick embryo immediately after cracking the egg and placing the contents into a dish and after 2 and 7 additional days of incubation *ex ovo*.

4.2. Preparation of grid meshes

Nitex nylon mesh with 180-μm grid size can be purchased from Sefar America, Inc. (Kansas City, MO). The layers are cut into 4 × 4-mm (lower mesh) and 3 × 3-mm (upper mesh) pieces with an Ingento paper

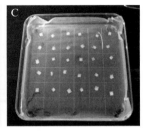

Figure 2.4 Preparation of meshes for collagen onplants. Nytex nylon meshes are cut into 4 × 4 mm and 3 × 3-mm pieces with a paper cutter (A) and sterilized by autoclaving. Large pieces are distributed into a Petri dish layered with parafilm and then covered with smaller pieces, making sandwiches (B). Meshes designated for collagen onplants with one test variable are assembled in an individual dish (C).

cutter (Fig. 2.4A). Cut pieces are placed in glass Petri dishes and sterilized in an autoclave. The larger lower meshes are placed into a square plastic Petri dish layered with parafilm (sprayed with alcohol and air-dried), which facilitates adhesion of meshes. Lower meshes are then covered by the smaller upper meshes, making double-gridded sandwiches (Fig. 2.4B) ready to be embedded into a drop of collagen mixture (Fig. 2.4C).

4.3. Preparation of test effector molecules and cell suspensions

Effector molecules should be prepared and kept on ice (unless indicated otherwise) to be readily available for incorporation into the neutralized collagen solution. Cells such as inflammatory cells are isolated from proper sources, such as neutrophils from peripheral blood, and resuspended in PBS or serum-free culture medium. Tumor cells are harvested by the desirable method, such as enzymatic or nonenzymatic detachment of adherent cell layers, washed and resuspended in serum-free medium or PBS. Cell suspensions should be prepared at desirable concentrations and kept on ice. A 5- to 15-× concentrated stock solutions of ingredients will allow for incorporating defined quantities of purified molecules or cells in a small volume into the neutralized collagen, such as 0.1 to 0.2 ml per 1.0 to 1.2 ml of collagen.

4.4. Preparation of collagen onplants and angiogenesis scoring

Native, nonpepsinized, type I rat tail collagen (BD Biosciences, Bedford, MA) is used in most of the CAM collagen onplant assays. Collagen should be neutralized (see below) before mixing with the effector molecules or cells. All components should be kept on ice to prevent premature polymerization of the collagen mixture. Importantly, even on ice, collagen will polymerize within 1 to 2 h after neutralizing. Therefore, all components of the collagen mixture should be prepared in advance and kept handy on ice. It is also recommended that all plastic material, including tubes, serological pipettes, and plastic tips are precooled on ice or in the freezer.

The original commercial collagen is provided in acid solution and therefore should be first neutralized to achieve a suitable pH of 7.4 to 7.6. To neutralize collagen, 8 parts (by volume) of original collagen preparation are mixed with 1 part of $10\times$ PBS or $10\times$ Eagle's minimal essential medium (EMEM) and 1 part of NaOH solution. If not indicated otherwise, we would recommend the use of $10\times$ EMEM supplemented with phenol red, since the color of the final mixture would give a good indication of whether pH 7.4 to 7.6 is reached during neutralization. Therefore, collagen is first carefully, without vortexing or agitation, but rather swirling by hand, mixed with $10\times$ EMEM. The mixture will become intensely yellow, indicating the acidic nature of the original collagen preparation. Then, 1 M NaOH solution is added dropwise to collagen/EMEM mixture until it reaches pH 7.4 to 7.6. If not all volume of 1 part is used, the rest is supplemented with sterile water. The concentration of collagen after neutralization is 80% of original concentration. Importantly, this concentration of neutralized collagen should be higher than the final concentration of collagen used in onplants (usually 2.0 to 2.1 mg/ml), allowing for addition of supplements and test molecules and/or cell suspensions. There is a considerable lot-to-lot variability in the original concentration of collagen. Therefore, the amount of collagen to be neutralized to prepare a desirable volume of mixture with the appropriate final concentration will depend on the original concentration of collagen. Taking into consideration a final volume of collagen solution, neutralized collagen can be supplemented with 10 to 25 mM HEPES buffer (pH 7.2 to 7.4) (Invitrogen, Carlsbad, CA) and, if desired, with 0.1 to 0.25 mg/ml BSA (Fraction V, Sigma). The collagen mixture should be further diluted with DMEM or PBS to 2.2 to 2.4 mg/ml, allowing for a final collagen concentration of 2.0 to 2.1 mg/ml after addition of test chemicals or cell suspensions.

Before mixing with test material, equal volumes of collagen is distributed into small tubes kept on ice and then various test molecules or cell suspensions are added to the tubes and gently but efficiently mixed with the tube contents. The use of equal volumes of collagen and test solutions or cell suspensions throughout collagen mixture preparation is recommended since it minimizes inconsistency in final collagen concentrations between experimental variables. Test molecules and cell suspensions should be prepared at 5 to $15\times$ concentrations to allow for desirable final concentrations in the onplants. For example, a convenient volume to mix is 1.0 ml of collagen solution at 2.4 mg/ml and 0.2 ml of $6\times$ test component, providing a final 1.2-ml collagen mixture, which is enough for 40 onplants of 30 μl volume at a final concentration of collagen of 2.0 mg/ml and 1× of effector molecule. Similar considerations are applicable to incorporation of cells. For example, 1.1 ml of 2.18 mg/ml collagen solution could be mixed with 0.1 ml cell suspension prepared at 12.0×10^6 cells/ml to make 40 onplants, each containing 3×10^4 cells in 30 μl (1×10^6/ml) and collagen at a final concentration of 2.0 mg/ml.

To assemble onplants, 30 μl of the final collagen mixture are placed atop two gridded nylon meshes (Fig. 2.5A). The collagen onplants containing the same test molecule or cells are allowed to polymerize at 37 °C in a Petri dish (Fig. 2.5B). After solidifying for 30 to 45 min, the onplants are lifted individually with fine-end forceps and placed on the CAM in areas containing fine vessel networks, between large blood vessels (Fig. 2.5C). The CAM should not be mechanically irritated during placement of onplants. Four to

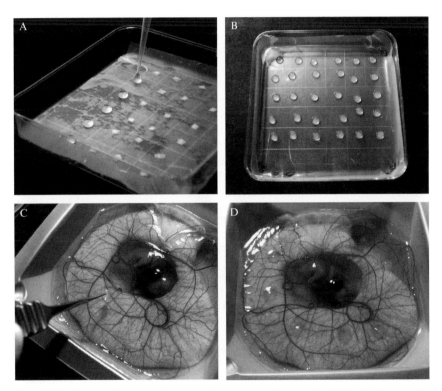

Figure 2.5 Preparation of collagen onplants and grafting on the CAM. Collagen mixture with test cells or purified effector molecules is distributed at 30-μl volumes over the meshes (A). Collagen onplants are allowed to polymerize in a Petri dish (B) placed into 37 °C incubator for 30 to 45 min. Using fine forceps, individual onplants are placed onto the CAM of day-10 embryos in areas located between large blood vessels (C). Individual embryos usually receive six collagen onplants containing the same test compounds (D).

eight onplants per embryo are grafted onto the CAM of each 10-day-old shell-less embryo (Fig. 2.5D). After onplants are grafted on the CAM, the chick embryos are immediately returned to the incubator for 3 to 4 days. Groups of four to six embryos should be used for a given variable to provide enough data points for a reliable statistical analysis of angiogenesis differentials.

4.5. Treatment of shell-less embryos carrying onplants with test compounds

During incubation, onplants are readily available for pharmacological intervention. Test chemicals can be applied topically onto the onplants, or injected intravenously into one of large allantoic veins with a glass capillary. The latter procedure is quite complicated technically and can be substituted with the injection of test components under the CAM into the allantois cavity. However, the permeability of CAM endoderm for an individual effector molecule should be considered. Incorporation of 5% DMSO into the 0.1- to 0.2-ml inoculums might increase CAM permeability for the test compounds.

4.6. Angiogenesis scoring

Angiogenesis is usually scored at 70 to 90 h after onplant grafting using a stereomicroscope (Fig. 2.6A). The plane of focus is chosen above the ectodermal capillary plexus and the network of pre-existing blood vessels. This assures that only newly formed, angiogenic vessels will be scored. Distinct blood-carrying vessels visualized in the grids of the upper mesh are regarded as angiogenic (Fig. 2.6B). Angiogenesis level is determined

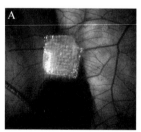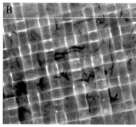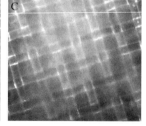

Figure 2.6 Scoring of angiogenesis in collagen onplants. Angiogenesis is scored with a stereomicroscope 70 to 90 h after onplant grafting. At that time point, the onplants appear more opaque as they become integrated into the CAM tissue (A). Newly formed blood vessels are scored in the focus plane of the upper mesh (B), that is, above vascular networks of the underlying CAM, including the ectoderm capillary plexus. In the absence of proangiogenic stimuli, control onplants have few if any angiogenic blood vessels (C).

either as an angiogenic index, that is, ratio of angiogenic grids (number of grids containing blood vessels over the total number of grids scored), or as a fold difference between a variable (intact cells or cell components added into collagen) over control (collagen alone) (Fig. 2.6C).

4.7. Analysis of cellular and protein composition of onplants

At different time points after grafting, collagen onplants can be subjected to histological, immunohistochemical or biochemical analyses. For histological purposes, the onplants are excised from the CAM along with the underlying tissues with fine scissors and washed in PBS. For paraffin sectioning, the onplants are fixed in a proper fixative, such as Zn-formalin, and processed further following standard protocols (Fig. 2.7). For cryosectioning, the onplants should be carefully aligned in the cryomolds filled with the O.C.T. embedding compound (Tissue-Tek, Miles Laboratories, Naperille, IL), frozen on dry ice and kept at −70 to −80 °C until use. Tissue-staining protocols vary and depend on the antigen or molecule of interest and might require antigen retrieval. For biochemical purposes, the onplants are harvested with or without underlying CAM tissue, and immediately placed into the corresponding buffer, such as 5 to 10× SDS sample buffer for zymography or Western blot analyses, or extraction buffers, such as

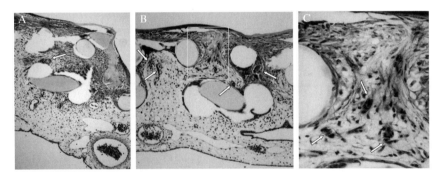

Figure 2.7 Histological analysis of collagen onplants for angiogenesis. After angiogenesis scoring, collagen onplants were excised from the CAM, fixed in Zn-formalin and processed for paraffin sectioning. Tissue sections were deparaffinized and stained by hematoxylin-eosin by standard procedure. Two rows of circular structures devoid of tissue represent the cross areas of lower and upper gridded meshes. Only a few blood vessels (arrows) grew into the control collagen onplant that was not supplemented with proangiogenic stimuli (A). In contrast, in the collagen onplant supplemented with a proangiogenic moiety, that is, recombinant MMP-9 (B), numerous angiogenic blood vessels could be identified between the meshes (*arrows*). Larger magnification of the boxed area (C) allows one to appreciate that angiogenic vessels are filled with blood (nucleated erythrocyes are stained dark pink with eosin) and lined with a continuous layer of endothelium.

modified RIPA for ELISA analyses. Along with angiogenesis scoring, the immunohistological and biochemical analyses add valuable information about the cell origin or time of appearance or activation status of a particular molecule of interest.

5. Validation of CAM Angiogenesis Findings in Collagen Implant Mouse Model

Findings generated with 3D collagen grafts in the avian angiogenesis model system can be validated in a mammalian model employing grafting of collagen-filled silicon tubes (angiotubes) under the skin of immunodeficient mice (Fig. 2.8A). Similar to the CAM assay, in this model system, collagen can be left nonsupplemented (Fig. 2.8B) or supplemented with different pro- and anti-angiogenic molecules as well as with tumor or inflammatory cells (Fig. 2.8C). Within 10 to 15 days after implantation, silicon tubes are infiltrated with newly formed angiogenic blood vessels originating from the pre-existing blood vessels converging towards the tube openings (Fig. 2.8C). Angiogenesis in individual tubes can be measured by hemoglobin content or by the level of endothelial cells (through indirect measurements of labeled ligand binding to endothelial cell–specific receptors).

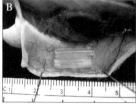
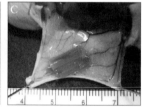

Figure 2.8   Mouse angiogenesis model system to study angiogenic potential of purified proteins and tumor cell variants selected *in vivo* (A). Inert silicon tubes were filled with native type I collagen and surgically inserted under the skin of immunodeficient mice (two per each side of the back). Twelve days later, the mice were sacrificed and tubes under the skin exposed and excised to determine levels of angiogenesis, such as by determining hemoglobin content in the lysed contents of the tubes. In addition, the levels of blood vessel convergence toward tube openings correlate well with the levels of angiogenesis in the implants. Thus, little convergence is observed toward control implants, which were not supplemented with any additional angiogenic factors (B). In contrast, when collagen mixture was supplemented with a highly disseminating variant of human HT-1080 fibrosarcoma, pre-existing blood vessels appear to converge toward tube openings and sprout into a fine network of angiogenic capillaries (C). Note that the tubes are filled with blood, which apparently leaks from the angiogenic capillaries that had grown into the tubes.

ACKNOWLEDGMENT

The authors would like to dedicate this chapter to the memory of Judah Folkman, a pioneer in the development of angiogenesis assays.

REFERENCES

Ardi, V. C., Kupriyanova, T. A., Deryugina, T. A., and Quigley, J. P. (2007). Human neutrophils uniquely release TIMP-free MMP-9 to provide a potent catalytic stimulator of angiogenesis. *Proc. Natl. Acad. Sci. USA* **104**, 20262–20267.

Beckers, M., Gladis-Villanueva, M., Hamann, W., Schmutzler, W., and Zwadlo-Klarwasser, G. (1997). The use of the chorio-allantoic membrane of the chick embryo as test for anti-inflammatory activity. *Inflamm Res.* **46**, S29–S30.

Borges, J., Tegtmeier, F. T., Padron, N. T., Mueller, M. C., Lang, E. M., and Stark, G. B. (2003). Chorioallantoic membrane angiogenesis model for tissue engineering: A new twist on a classic model. *Tissue Eng.* **9**, 441–450.

Brooks, P. C., Montgomery, A. M., Rosenfeld, M., Reisfeld, R. A., Hu, T., Klier, G., and Cheresh, D. A. (1994). Integrin alpha v beta 3 antagonists promote tumor regression by inducing apoptosis of angiogenic blood vessels. *Cell* **79**, 1157–1164.

Brooks, P. C., Silletti, S., von Schalscha, T. L., Friedlander, M., and Cheresh, D. A. (1998). Disruption of angiogenesis by PEX, a noncatalytic metalloproteinase fragment with integrin binding activity. *Cell* **92**, 391–400.

Camerino, G. M., Nicchia, G. P., Dinardo, M. M., Ribatti, D., Svelto, M., and Frigeri, A. (2006). *In vivo* silencing of aquaporin-1 by RNA interference inhibits angiogenesis in the chick embryo chorioallantoic membrane assay. *Cell Mol. Biol. (Noisy-le-grand)* **52**, 51–56.

Cao, Y., Linden, P., Farnebo, J., Cao, R., Eriksson, A., Kumar, V., Qi, J. H., Claesson-Welsh, L., and Alitalo, K. (1998). Vascular endothelial growth factor C induces angiogenesis *in vivo. Proc. Natl. Acad. Sci. USA* **95**, 14389–14394.

Eliceiri, B. P., Paul, R., Schwartzberg, P. L., Hood, J. D., Leng, J., and Cheresh, D. A. (1999). Selective requirement for Src kinases during VEGF-induced angiogenesis and vascular permeability. *Mol. Cell* **4**, 915–924.

Hagedorn, M., Balke, M., Schmidt, A., Bloch, W., Kurz, H., Javerzat, S., Rousseau, B., Wilting, J., and Bikfalvi, A. (2004). VEGF coordinates interaction of pericytes and endothelial cells during vasculogenesis and experimental angiogenesis. *Dev. Dyn.* **230**, 23–33.

Han, Z., Ni, J., Smits, P., Underhill, C. B., Zie, B., Chen, Y., Liu, N., Tylzanowski, P., Parmelee, D., Feng, P., Ding, I., Gao, F., *et al.* (2001). Extracellular matrix protein 1 (ECM1) has angiogenic properties and is expressed by breast tumor cells. *FASEB J.* **15**, 988–994.

Hood, J. D., Frausto, R., Kiosses, W. B., Schwartz, M. A., and Cheresh, D. A. (2003). Differential alphav integrin-mediated Ras-ERK signaling during two pathways of angiogenesis. *J. Cell Biol.* **162**, 933–943.

Melkonian, G., Munoz, N., Chung, J., Tong, C., Marr, R., and Talbot, P. (2002). Capillary plexus development in the day five to day six chick chorioallantoic membrane is inhibited by cytochalasin D and suramin. *J. Exp. Zool.* **292**, 241–254.

Miller, W. J., Kayton, M. L., Patton, A., O'Connor, S., He, M., Vu, H., Baibakov, G., Lorang, D., Knezevic, V., Kohn, E., Alexander, H. R., Stirling, D., *et al.* (2004). A novel technique for quantifying changes in vascular density, endothelial cell proliferation and protein expression in response to modulators of angiogenesis using the chick chorioallantoic membrane (CAM) assay. *J. Transl. Med.* **2,** 4.

Murphy, J. B. (1913). Transplantability of tissues to the embryo of foreign species. Its bearing on questions of tissue specificity and tumor immunity. *J. Exp. Med.* **17,** 482–493.

Murugesan, S., Mousa, S. A., O'Connor, J., Lincoln, D. W., 2nd, and Linhardt, R. J. (2007). Carbon inhibits vascular endothelial growth factor- and fibroblast growth factor-promoted angiogenesis. *FEBS Lett.* **581,** 1157–1160.

Nguyen, M., Shing, Y., and Folkman, J. (1994). Quantitation of angiogenesis and anti-angiogenesis in the chick embryo chorioallantoic membrane. *Microvasc. Res.* **47,** 31–40.

Reizis, A., Hammel, I., and Ar, A. (2005). Regional and developmental variations of blood vessel morphometry in the chick embryo chorioallantoic membrane. *J. Exp. Biol.* **208,** 2483–2488.

Ribatti, D., Nico, B., Vacca, A., and Presta, M. (2006). The gelatin sponge–chorioallantoic membrane assay. *Nat. Protoc.* **1,** 85–91.

Ribatti, D., Nico, B., Vacca, A., Roncali, L., Burri, P. H., and Djonov, V. (2001). Chorioallantoic membrane capillary bed: A useful target for studying angiogenesis and anti-angiogenesis *in vivo. Anat. Rec.* **264,** 317–324.

Ribatti, D., Urbinati, C., Nico, B., Rusnati, M., Roncali, L., and Presta, M. (1995). Endogenous basic fibroblast growth factor is implicated in the vascularization of the chick embryo chorioallantoic membrane. *Dev. Biol.* **170,** 39–49.

Ribatti, D., Vacca, A., Roncali, L., and Dammacco, F. (2000). The chick embryo chorioallantoic membrane as a model for *in vivo* research on anti-angiogenesis. *Curr. Pharm. Biotechnol.* **1,** 73–82.

Riboldi, E., Musso, T., Moroni, E., Urbinati, C., Bernasconi, S., Rusnati, M., Adorini, L., Presta, M., and Sozzani, S. (2005). Cutting edge: Proangiogenic properties of alternatively activated dendritic cells. *J. Immunol.* **175,** 2788–2792.

Romanoff, A. L. (1960). "The Avian Embryo." New York: Macmillan, New York.

Sahni, A., Khorana, A. A., Baggs, R. B., Peng, H., and Francis, C. W. (2006). FGF-2 binding to fibrin(ogen) is required for augmented angiogenesis. *Blood* **107,** 126–131.

Seandel, M., Noack-Kunnmann, K., Zhu, D., Aimes, R. T., and Quigley, J. P. (2001). Growth factor-induced angiogenesis *in vivo* requires specific cleavage of fibrillar type I collagen. *Blood* **97,** 2323–2332.

Struman, I., Bentzien, F., Lee, H., Mainfroid, V., D'Angelo, G., Goffin, V., Weiner, R. I., and Martial, J. A. (1999). Opposing actions of intact and N-terminal fragments of the human prolactin/growth hormone family members on angiogenesis: an efficient mechanism for the regulation of angiogenesis. *Proc. Natl. Acad. Sci. USA* **96,** 1246–1251.

Tufan, A. C., and Satiroglu-Tufan, N. L. (2005). The chick embryo chorioallantoic membrane as a model system for the study of tumor angiogenesis, invasion and development of anti-angiogenic agents. *Curr. Cancer Drug Targets* **5,** 249–266.

Vazquez, F., Hastings, G., Ortega, M. A., Lane, T. F., Oikemus, S., Lombardo, M., and Iruela-Arispe, M. L. (1999). METH-1, a human ortholog of ADAMTS-1, and METH-2 are members of a new family of proteins with angio-inhibitory activity. *J. Biol. Chem.* **274,** 23349–23357.

Watanabe, K., Hasegawa, Y., Yamashita, H., Shimizu, K., Ding, Y., Abe, M., Ohta, H., Imagawa, K., Hojo, K., Maki, H., Sonoda, H., and Sato, Y. (2004). Vasohibin as an endothelium-derived negative feedback regulator of angiogenesis. *J. Clin. Invest.* **114,** 898–907.

Yang, E. Y., and Moses, H. L. (1990). Transforming growth factor beta 1–induced changes in cell migration, proliferation, and angiogenesis in the chicken chorioallantoic membrane. *J. Cell Biol.* **111,** 731–741.

Zijlstra, A., Aimes, R. T., Zhu, D., Regazzoni, K., Kupriyanova, T., Seandel, M., Deryugina, E. I., and Quigley, J. P. (2004). Collagenolysis-dependent angiogenesis mediated by matrix metalloproteinase-13 (collagenase-3). *J. Biol. Chem.* **279,** 27633–27645.

Zijlstra, A., Seandel, M., Kupriyanova, T. A., Partridge, J. J., Madsen, M. A., Hahn-Dantona, E. A., Quigley, J. P., and Deryugina, E. I. (2006). Proangiogenic role of neutrophil-like inflammatory heterophils during neovascularization induced by growth factors and human tumor cells. *Blood* **107,** 317–327.

CHAPTER THREE

THE ADENOVIRAL VECTOR ANGIOGENESIS/ LYMPHANGIOGENESIS ASSAY

Janice A. Nagy, Shou-Ching Shih, Wendy H. Wong, Ann M. Dvorak, *and* Harold F. Dvorak

Contents

Abstract

Adenoviral vectors expressing vascular permeability factor/vascular endothelial growth factor (VPF/VEGF, VEGF-A^{164}) offer a powerful method for elucidating the mechanisms of pathological angiogenesis and lymphangiogenesis and for

Department of Pathology, Beth Israel Deaconess Medical Center, Boston, Massachusetts
Department of Pathology, Harvard Medical School, Boston, Massachusetts

Methods in Enzymology, Volume 444
ISSN 0076-6879, DOI: 10.1016/S0076-6879(08)02803-6

evaluating the effectiveness of pro- and anti-angiogenesis therapies. When injected into any of a variety of tissues in nude mice or rats, adenoviral vectors expressing VEGF-A^{164} (Ad-VEGF-A^{164}) induce the formation of six structurally and functionally distinct types of new blood vessels: mother vessels (MV), capillaries, glomeruloid microvascular proliferations (GMP), vascular malformations (VM), feeding arteries (FA), and draining veins (DV). Each of these abnormal vessel types may be found in tumors and in other examples of pathological angiogenesis. In addition, Ad-VEGF-A^{164} induces the formation of highly abnormal and poorly functional "giant" lymphatics. The Ad-VEGF-A^{164} assay has provided a means of elucidating the steps and mechanisms by which each type of new blood and lymphatic vessel forms, and for generating at defined times and in large numbers each of these different types of vessels for molecular study. Ear injection sites are advantageous in that the angiogenic and lymphangiogenic responses can be followed visually over time in intact animals, thus providing a convenient, inexpensive global screening assay for assessing the efficacy and toxicity of anti- or pro-angiogenic therapies. The assay can be readily extended to the study of the new blood vessels/lymphatics induced by adenoviral vectors expressing other growth factors and cytokines.

1. INTRODUCTION

The normal adult vasculature arises from the twin processes of vasculogenesis and physiological angiogenesis and is comprised of evenly spaced blood vessels that are organized into a hierarchy of elastic and muscular arteries, arterioles, capillaries, postcapillary venules, and small and large veins. Each of these vessel types has a size and structure that allows it to perform specialized functions optimally. In contrast, pathological angiogenesis, which differs strikingly from physiological angiogenesis in both process and outcome, is a characteristic property of cancer, wound healing, macular degeneration, and a number of diseases that are associated with chronic inflammation such as rheumatoid arthritis and psoriasis. The new vessels that form in pathological angiogenesis are structurally heterogeneous, nonuniformly distributed, branch irregularly, form arteriovenous shunts, and do not conform to a clear hierarchal organization of arterioles, capillaries, and venules (reviewed in Dvorak, 2003, 2007; Fu et al., 2007; Nagy et al., 2007). On average, the blood vessels of pathological angiogenesis are larger than normal blood vessels, overexpress VEGF-A receptors, are lined by actively dividing endothelial cells, and are often deficient in pericytes. Recently we have been able to classify these abnormal blood vessels into six structurally and functionally distinct types: mother vessels (MV), normal–appearing capillaries, glomeruloid microvascular proliferations (GMP), vascular malformations (VM), feeding arteries (FA), and draining veins (DV) (Dvorak, 2007; Fu et al., 2007; Nagy et al., 2007). The first four of these are derived from pre-existing venules, whereas the latter two arise from pre-existing arteries and veins (Nagy *et al.*, 2002b, 2003) and are examples of abnormal arteriogenesis and venogenesis. Studies to define the distinct properties of each of these vessel types have begun but are not as yet complete.

Given the observed differences in vascular structure, organization, and function, it is not surprising that physiological and pathological angiogenesis are generated by distinctly different mechanisms. Although not as yet well understood, the normal vasculature results from the balanced and coordinated activities of numerous cytokines, growth factors, inhibitors, and modulators, each expressed in appropriate amounts and in ordered sequence (Beck and D'Amore, 1997; Folkman, 1997; Gale and Yancopoulos, 1999). By contrast, the mechanisms of pathological angiogenesis are simpler and therefore are much better understood. Pathological angiogenesis results from the unbalanced expression of one or a few individual angiogenic cytokines, particularly the 164–amino-acid isoform of VEGF-A, VEGF-A^{164} (the corresponding human version, VEGF-A^{165}, has 165 amino acids). Each of the six types of highly abnormal vessels that we have identified in tumors and other examples of pathological angiogenesis can be induced by expressing VEGF-A^{164} in normal mouse tissues with an adenoviral vector (Dvorak, 2003; Nagy *et al.*, 2007; Pettersson *et al.*, 2000; Sundberg *et al.*, 2001).

In addition, Ad-VEGF-A^{164} induces abnormal lymphangiogenesis, the formation of greatly enlarged, "giant" lymphatic vessels of the type found in lymphangiomas and Crohn's disease (Nagy *et al.*, 2002a,b; Wirzenius *et al.*, 2007). For this reason we have confined the present discussion to the new blood and lymphatic vessels that are induced by Ad-VEGF-A^{164}, recognizing that the same general approach can be used to study the angiogenic response induced by other VEGF-A isoforms or unrelated protein angiogenic factors.

A number of important diseases are expected to benefit from treatments that inhibit pathological angiogenesis, including cancer, wet form of macular degeneration, and chronic inflammatory diseases. On the other hand, other equally important diseases would benefit from therapies that enhance high-quality angiogenesis, including myocardial infarctions, strokes, wound healing in diabetics, and so on. Evaluating the effectiveness of such therapies, as well as elucidating the mechanisms of normal and pathological angiogenesis and lymphangiogenesis, requires a robust, high-throughput assay that can reproducibly mimic angiogenesis as it occurs *in vivo*. Nonreplicating adenoviral vectors expressing VEGF-A^{164} or other growth factors provide just such an assay. Adenoviral vectors infect both dividing and nondividing cells via the CAR receptor (Meier and Greber, 2004) and give strong cytokine mRNA and protein expression with virtually no background. Cytokine expression levels are readily modulated by adjusting viral dose, no foreign matrix need be introduced, and combinations of cytokines can be efficiently studied, together or in sequence.

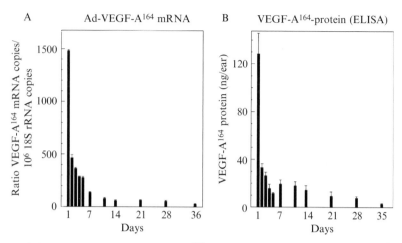

Figure 3.1 Expression levels of VEGF-A^{164} mRNA (A) and protein (B) in mouse ears at various times after injection of 2×10^8 PFU Ad-VEGF-A^{164}. VEGF-A^{164} mRNA is expressed as the ratio of copies of VEGF-A^{164} mRNA per 10^6 copies of 18S rRNA. (From Nagy, J. A., Feng, D., Vasile, E., Wong, W. H., Shih, S. C., Dvorak, A. M., and Dvorak, H. F. (2006). Permeability properties of tumor surrogate blood vessels induced by VEGF-A. *Lab. Invest.* **86**, 767–780.)

Genes introduced by adenoviral vectors are not incorporated into the host genome and therefore expression is transient (Fig. 3.1). Ear injection sites offer an important advantage in that angiogenesis and lymphangiogenesis can be followed visually over time in living animals, thus providing a rapid and relatively inexpensive screening assay. More detailed studies can be performed on ear or other injection sites to elucidate mechanisms.

2. Methods

2.1. Adenoviral vectors and animals

Adenoviral vectors are prepared by inserting the coding sequences of murine VEGF-A^{164}, or of other of its isoforms, PlGF, bFGF, LacZ, green fluorescent protein (gfp), and so on into the pMDM transcriptional cassette, which consists of a complete immediate early cytomegalovirus promoter and intron and poly A–containing sequences derived from the human beta-globin gene (Ory *et al.*, 1996). These viruses are engineered using standard procedures (Hardy *et al.*, 1997) and are purified by double cesium-chloride banding, concentrated to high titer ($\sim$1 to 2×10^{11} PFU/ml), and then dialyzed into 10 mM Tris HCl pH 8.0, 2 mM MgCl$_2$, 4% sucrose storage buffer. Aliquots are stable at $-80\,^{\circ}$C for 1 to 2 years. Immediately prior to injection, the virus is diluted directly into PBS-3% glycerol. Adenoviral vectors in volumes of 10 to 50 μl containing 10^6 to 5×10^8 PFU are injected with a Hamilton syringe (10 μl) fitted with a fixed 26-g needle intradermally into the dorsal ear skin or with an insulin syringe (0.3 ml) into any of a variety of other tissues such as flank skin, skeletal muscle of the hind limb, peritoneal cavity, or myocardium. Immunoincompetent nude mice (BALB/c ByJ Hfh11nu or athymic NCr Nu/Nu) or rats (Tac:N:NIH-rnufDF) of either sex are used to avoid an immune response against the viral vector. All studies are performed under hospital IACUC-approved protocols.

2.2. Quantitative real-time PCR

Total RNA is prepared using the RNeasy RNA extraction kit with DNase-I treatment following the manufacturer's protocol (Qiagen). To generate cDNA, total RNA (100 ng) from each of triplicate samples is mixed and converted into cDNA using random primers and SuperscriptIII reverse transcriptase (Invitrogen, Carlsbad, CA). All cDNA samples are aliquoted and stored at −80 °C. Primers are designed using the Primer Express oligo design software (Applied BioSystems, Foster City, CA) and are synthesized by Integrated DNA Technologies (Coralville, IA). All primer sets are subjected to rigorous database searches to identify potential conflicting transcript matches to pseudogenes or homologous domains within related genes. Amplicons generated from the primer set are analyzed for melting point temperatures using the first derivative primer melting curve software supplied by Applied BioSystems. As an example, the real-time PCR primer sequences for murine VEGF-A follow: forward primer, 5'-AACAAAGC-CAGAAAATCACTGTGA-3'; reverse primer, 5'-CGGATCTTGGA-CAAACAAATGC-3'. The SYBR Green I assay and the ABI Prism 7500 Sequence Detection System are used for detecting real-time PCR products from the reverse-transcribed cDNA samples, as described previously (Shih and Smith, 2005). 18S rRNA, a uniformly expressed housekeeping gene (Aerts *et al.*, 2004), is used as the normalizer. PCR reactions for each sample are performed in duplicate, and copy numbers are measured as described previously (Shih and Smith, 2005). The level of target gene expression is normalized against the 18S rRNA expression in each sample, and data are reported as mRNA copies per 10^6 18S rRNA copies (Fig. 3.1, left panel).

2.3. Quantitative analysis of VEGF-A protein by ELISA

The amount of growth factor protein generated in adenoviral vector–injected tissues is quantified by ELISA. For Ad-VEGF-A[164], a murine VEGF-A-specific ELISA (Quantikine Mouse VEGF Immunoassay Kit; R&D Systems, Minneapolis, MN, USA) is used. At time points of interest following Ad-VEGF-A[164] injection, animals are euthanized and 8-mm punch biopsy samples of injected ears, or equivalent amounts of other tissue injection sites, are homogenized in 1 ml of T-PER (tissue protein extraction reagent) (Pierce, Rockford, IL) containing 10 μl of each of the following: HALT protease inhibitor cocktail (Pierce), Phosphatase Inhibitor Cocktails I and II (Sigma), and 0.5 M EDTA. Lysates are centrifuged at 14,000 rpm for 30 min at 4 °C to pellet insoluble material, and the supernatant is removed and stored at −20 °C prior to use. Tissue lysates are thawed, diluted in lysis buffer, and analyzed according to the manufacturer's instructions. Four to six independent samples are analyzed at each time point, and each sample is analyzed in duplicate. The VEGF-A concentration in each sample is corrected for protein recovery using a correction factor that is determined as follows. A known volume of purified mouse VEGF-A protein is injected into the ear skin of control animals, and the ear tissue is processed as described above for ELISA. The correction factor for recovery is calculated by comparing the amount of purified VEGF-A detected in the ear specimen by ELISA with the amount of VEGF-A determined by ELISA when the same volume of purified VEGF-A was injected directly into the well of the ELISA microtiter plate. The mean corrected amount of VEGF-A present in the Ad-VEGF-A[164] ears is reported as nanograms per ear ± standard error of the mean (SEM) (Fig. 3.1, right panel).

2.4. Qualitative analysis of angiogenesis and vascular permeability using Evans blue dye as tracer

As noted earlier, an important advantage of injecting Ad-VEGF-A[164] or other adenoviral vectors into the ears of nude mice is that the angiogenic response can be followed visually in the intact animal (Fig. 3.2, upper panel). Thus, the effects of pro- or anti-angiogenic treatments can be evaluated inexpensively. Mice are anesthetized with Avertin (tribromoethanol, 200 mg/kg) and ears are photographed using a Wild M-400 Photomacroscope. Vascular hyperpermeability, a property of some angiogenic blood vessels (MV and GMP), can also be followed readily. Mice are injected intravenously with 100 μl 0.5% Evans Blue (Sigma, St. Louis, MO) in saline, and ears are photographed 30 min later (Fig. 3.2, lower panel).

2.5. Qualitative analysis of lymphangiogenesis using colloidal carbon as tracer

Intravital perfusion of lymphatics can be used to evaluate Ad-VEGF-A[164]-induced lymphangiogenesis in mouse ears. At various times after the intradermal injection of the adenovirus, mice are anesthetized with Avertin (tribromoethanol, 200 mg/kg) and cradled in a transparent acrylic resin mold (Syndicate Sales, Inc.). Ears are mounted flat on the resin support and held in place by silicone vacuum grease and viewed in a Wild M400

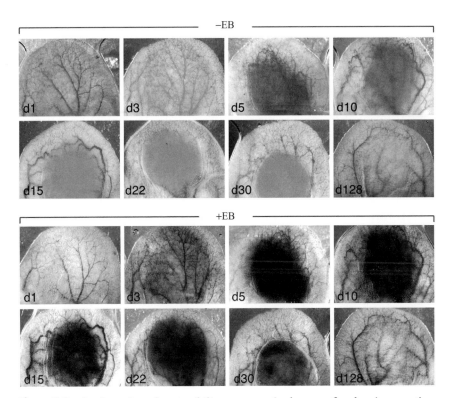

Figure 3.2 Angiogenic and permeability responses in the ears of nude mice over time from 1 to 128 days after injection of 1×10^8 PFU of Ad-VEGF-A[164]. Ears were photographed (–EB, top panel). Evans blue dye was then injected intravenously and 30 min later ears were rephotographed to assess plasma protein leakage (+EB, bottom panel). Note that ears at later times (30 and 128 days) exhibit reduced or no leakage of EB. (From Nagy, J. A., Feng, D., Vasile, E., Wong, W. H., Shih, S. C., Dvorak, A. M., and Dvorak, H. F. (2006). Permeability properties of tumor surrogate blood vessels induced by VEGF-A. *Lab. Invest.* **86**, 767–780.) (See color insert.)

Photomacroscope. Colloidal carbon (Higgins nonwaterproof drawing ink, Sanford) is diluted 1:1 in Tyrode's buffer and injected through a 10-μl prepulled thin-wall borosilicate glass micropipette (World Precision Instruments, Sarasota, FL) attached via flexible silicone tubing (Cole Parmer) to a 500-μl threaded Hamilton syringe fitted with a blunt 20-g needle. The micropipette is mounted on a joystick-controlled micromanipulator (World Precision Instruments) and repeatedly injected into the dorsal surface of the peripheral ear until a lymphatic capillary is entered. Additional carbon tracer (5 to 20 μl) is then slowly injected into the lymphatic lumen under the control of the threaded plunger to avoid overfilling and lymphatic vessel damage. For photography, ears are flooded with immersion oil, a coverslip applied, and photographs taken with a SPOT Insight Digital camera.

Colloidal carbon–filled lymphatics are illustrated in Fig. 3.3. In control ears, carbon delineates fine, distinct, normal lymphatics that extend radially from the ear periphery to its base and are punctuated by periodic bulbous swellings that mark valves (Fig. 3.3, left panel). Although joined at intervals by interconnecting side branches, intact valves largely prevent lateral back filling. Therefore, injected carbon remains largely within one or two radially oriented lymphatics. Therefore, multiple injections into different lymphatics are required to delineate fully the lymphatics of normal ears. Very different results are obtained when colloidal carbon is injected into the terminal ear lymphatics of mice previously injected with Ad-VEGF-A[164] (Fig. 3.3, right panel). By day 7, both radial and interconnecting lymphatics are back-filled, and the bulbous swellings that mark the location of valves in normal lymphatics are no longer seen. By days 14 to 21, a new, highly abnormal lymphatic vascular plexus is visible, and this persists for at least 1 year (Nagy *et al.*, 2002a).

2.6. Quantitative analysis of vascular volume and permeability

A dual-tracer method (Dvorak *et al.*, 1984; Graham and Evans, 1991; Nagy *et al.*, 2006) is used to quantify the leakage rate (LR) of albumin from the circulation into the interstitial space of nude mouse ears (or other injected tissue sites) at various times after local injection of adenoviral vectors. This method also determines the intravascular plasma volume at virus injection sites, providing a quantitative measure of the extent of the angiogenic response. Operationally, radioiodine-labeled bovine serum albumins (^{125}I-BSA, specific

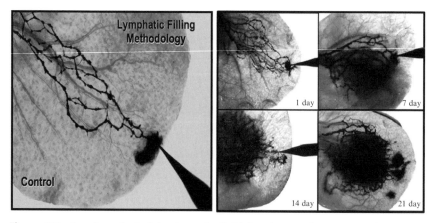

Figure 3.3 Intralymphatic injection of colloidal carbon to demonstrate the normal lymphatics of control ears (left panel) and of the poorly functional, back-filling giant lymphatics induced at varying times after ear injections of Ad-VEGF-A[164] (right panel).

activity 3.7 $\mu Ci/\mu g$ and ^{131}I-BSA, specific activity 72.5 $\mu Ci/\mu g$, PerkinElmer Life Sciences, Boston, MA) are diluted in Hanks' balanced salt solution supplemented with 1 mg/ml unlabeled carrier BSA. Free iodine is less than 5% as determined by trichloroacetic acid precipitation. Five μCi of ^{125}I-BSA (0.1 ml) are injected intravenously via the tail vein into nude mice that have been anesthetized with Avertin (tribromoethanol, 200 mg/kg). Twenty-five minutes later, animals receive a second intravenous injection of 5 μCi of ^{131}I-BSA (0.1 ml) to provide a measure of intravascular tracer. Five minutes after the second injection, an aliquot of blood (0.5 ml) is taken in heparin by cardiac puncture and animals are immediately euthanized by CO_2 narcosis. Ears (or other tissues) are removed by cauterization to prevent blood leakage and are weighed. Plasma and tissue samples are subjected to gamma counting in a Wallac-LKB Model 1470 Wizard gamma counter that has been normalized to account for spillover of ^{131}I counts into the ^{125}I window and vice versa. Counts per gram of tissue for the ears and counts per μl of plasma are used to calculate plasma volume, and albumin LR ($\mu l/min/g$), using the following equations (Nagy *et al.*, 2006), modified from Graham and Evans (1991).

$$A_{131} = C_{p131}{}^* V_p + LR^* C_{p131}{}^* 5$$

$$A_{125} = C_{p125}{}^* V_p + LR^* C_{p125}{}^* 30$$

$$LR = 1/25^* \left(A_{125}/C_{p125} - A_{131}/C_{p131} \right)$$

$$Vp = 1/5^* \left(6^* A_{131}/C_{p131} - A_{125}/C_{p125} \right)$$

$$A = C_p{}^* Vp + LR^* C_p{}^* t$$

where A is total tissue radioactivity ($\mu Ci/g$) of ^{125}I-albumin or ^{131}I-albumin; Cp is concentration of radioactive tracer in plasma ($\mu Ci/\mu l$); Vp is volume of plasma in tissue ($\mu l/g$); LR, equivalent to the permeability-surface area product, is the leakage rate from plasma into tissue expressed as $\mu l/min/g$; and t is time elapsed after injection of tracer (min).

An underlying assumption of this method is that even in the case of highly leaky blood vessels, only negligible amounts of ^{131}I-albumin will have had time to extravasate from the blood into tissues at 5 min after injection. Therefore, the ^{131}I-albumin value at 5 min provides a measure of intravascular volume, whereas the ^{125}I-albumin value provides a measure of the sum of both intravascular and extravascular albumin. Extravasated albumin (i.e., the volume of plasma extravasated in 25 min) can then be determined by subtracting the 5-min value from the 30-min value. This method has the disadvantages of using a strong, short-lived gamma emitter (^{131}I), and of not permitting visual inspection of tracer leakage as is provided when Evans' blue dye is used as tracer. To circumvent these limitations we recently modified the method by substituting Evans' blue dye (hence plasma albumin) for the first tracer (30-min time point) and using ^{125}I-albumin for the second (5-min time point).

2.7. Morphological studies

In order to investigate additional features of the angiogenic or lymphangio-genic responses, tissues may be taken for histology. Animals are euthanized by CO_2 narcosis and tissue is harvested at times of interest and processed in different ways. Tissues are fixed in 10% formalin for preparation of routine, paraffin-embedded, H&E-stained slides. For in situ hybridization and immunohistochemistry, tissues are fixed in RNAse-free 4% paraformalde-hyde in phosphate buffered saline, pH 7.4 (PBS) for 2 to 4 h at 4 °C. They are then transferred to 30% sucrose in PBS overnight at 4 °C, frozen in OCT compound (Miles Diagnostics, Elkhart, IN), and stored at −70 °C for subsequent preparation of cryostat sections and immunohistochemistry or in situ hybridization (Brown *et al.*, 1998). Other properties of the angiogenic response can be evaluated by confocal microscopy. For example, to evaluate vascular permeability in ear injection sites, mice are injected intravenously with 100 μl of saline containing 1 mg of lysine-fixable FITC-labeled dextran (FITC-D, 2×10^6 MW, Molecular Probes, Eugene, OR). Ears are then excised and fixed in cold 4% paraformaldehyde for 1 h. The dorsal half of each ear is then gently separated from the underlying cartilage which remains attached to the ventral half. Both halves are washed in PBS containing 2% sucrose for 30 min, mounted in imaging chambers in Vectashield (Vector Laboratories, Burlingame, CA). Whole mounts are visualized by confocal microscopy (Nagy *et al.*, 2006).

For 1-μm Epon sections and for electron microscopy, as well as for immunohistochemistry at the electron microscopic level, tissues are fixed in paraformaldehyde-glutaraldehyde, and processed as previously described, using either of two postfixation protocols: osmium-collidine uranyl en bloc (OCUB) or potassium ferrocyanide–reduced osmium (OPF) (Dvorak, 1987; Feng *et al.*, 1996, 1998, 2000a, 2004). For evaluating the pathways by which macromolecules extravasate from leaky blood vessels, anionic ferritin (cadmium-free, from horse spleen, Sigma Chemical Co., St. Louis, MO) is injected intravenously at a dose of 1 mg/g of body weight 20 min prior to euthanasia. Ferritin is an iron-rich protein that is normally present in plasma at low concentrations and that provides a convenient tracer to follow macromo-lecular transport because individual ferritin molecules can be visualized directly by electron microscopy (Feng *et al.*, 1996). Colloidal carbon can also be used as a tracer (Feng *et al.*, 1997).

3. Blood Vessels Induced by Ad-VEGF-A¹⁶⁴

As noted above, Ad-VEGF-A^{164} induces the formation of four types of new angiogenic blood vessels from pre-existing venules (reviewed in Dvorak, 2003; Nagy *et al.*, 2006, 2007) (Fig. 3.4). MV form first and evolve over time into capillaries, GMP, and VM. In addition, Ad-VEGF-A^{164} induces the formation of large FA and DV as well as abnormal giant lymphatics. Each of these vessel types is found in tumors and other forms of pathological angiogenesis (Fu *et al.*, 2007; Nagy *et al.*, 1995, 2002b; Ren *et al.*, 2002).

3.1. Mother vessels

Mother vessels (Figs. 3.4 to 3.6) are the first new blood vessels to develop in response to Ad-VEGF-A^{164}. Derived from preexisting venules, they are greatly enlarged, thin-walled, serpentine, pericyte-poor, hyperpermeable, and strongly VEGFR-positive sinusoids (Figs. 3.5A and B and 3.6C and D). MV begin to appear within 24 h after Ad-VEGF-A^{164} administration, and develop for at least 2 days without significant endothelial cell division; they achieve maximal numbers and size at ~5 days, at which time they comprise

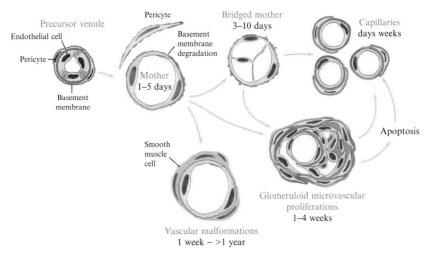

Figure 3.4 Schematic diagram of the angiogenic response induced by Ad-VEGF-A^{164} in nude mouse ears and other tissues. (Pettersson, A., Nagy, J. A., Brown, L. F., Sundberg, C., Morgan, E., Jungles, S., Carter, R., Krieger, J. E., Manseau, E. J., Harvey, V. S., Eckelhoefer, I. A., Feng, D., Dvorak, A. M., Mulligan, R. C., and Dvorak, H. F. (2000). Heterogeneity of the angiogenic response induced in different normal adult tissues by vascular permeability factor/vascular endothelial growth factor. *Lab. Invest.* **80**, 99–115.)

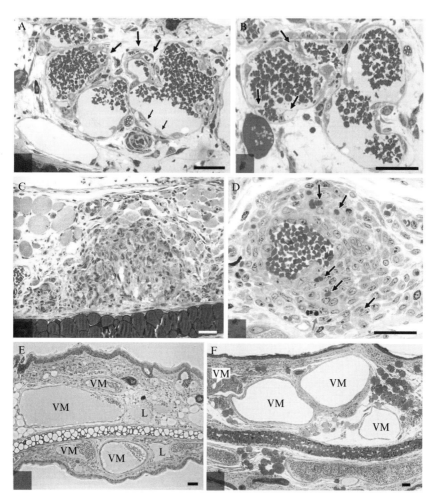

Figure 3.5 Illustration of mother vessels (A, B), glomeruloid microvascular proliferations (GMP) (C, D), and vascular malformations (VM) (E, F) induced in tissues injected with Ad-VEGF-A[164]. The arrows in panels (A) and (B) indicate intraluminal bridging by endothelial cells that divide MV into smaller channels. The black arrows in panel (D) indicate small residual vascular lumens in GMP; the red arrow indicates apoptotic cells. In panel (E), L identifies two lymphatic channels (the upper one is perfused with colloidal carbon). Giemsa-stained, 1-micron Epon sections. Scale bars, 50 μm. (From Nagy, J. A., Dvorak, H. F., and Dvorak, A. M. (2007). VEGF-A and the induction of pathological angiogenesis. *Annu. Rev. Pathol. Mech. Dis.* **2**, 251–275.)

the bulk of the vascular mass at Ad-VEGF-A[164] injection sites. MV form by a three-step process of basement membrane degradation, pericyte detachment, and extensive vessel lumen enlargement. Basement membrane degradation is an essential early step because basement membranes are noncompliant (nonelastic) structures that do not allow microvessels to

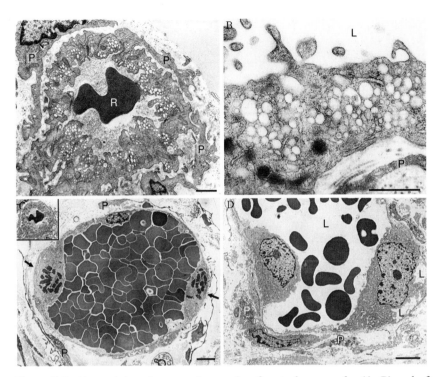

Figure 3.6 Transmission electron micrographs of control ear venules (A, B) and of mother vessels (C, D) 3 days after local injection of Ad-VEGF-A[164]. (A, B) Typical normal venules lined by cuboidal endothelium. The cytoplasm contains prominent vesiculo-vacuolar organelles (VVOs) and is enveloped by a complete coating of pericytes (P). R, red blood cell. (C, D) Typical mother vessels are greatly enlarged vascular structures, characterized by extensive endothelial cell thinning, striking reduction in VVOs and other cytoplasmic vesicles, enlarged nuclei that project into the vascular lumen, frequent mitotic figures (arrows, c), endothelial cell bridging with the formation of multiple lumens (L, d), and decreased pericyte (P) coverage. Note that the mother vessel lumen (c) is packed with red blood cells, indicative of extensive plasma extravasation. (Inset, c) The normal venule depicted in (A) is reproduced in (C) at the same magnification as the mother vessel to illustrate differences in relative size of normal venules and mother vessels. OCUB processing. Scale bars: A, B, 1 μm; C, D, 5 μm. (From Nagy, J. A., Feng, D., Vasile, E., Wong, W. H., Shih, S. C., Dvorak, A. M., and Dvorak, H. F. (2006). Permeability properties of tumor surrogate blood vessels induced by VEGF-A. *Lab. Invest.* **86**, 767–780.)

expand their cross-sectional area by more than ~30% (Swayne *et al.*, 1989), that is, far less than the three- to five-fold cross-sectional enlargement characteristic of MV. Whether pericytes detach from MV by an active process or simply fall off as the result of basement membrane degradation is not known. Vascular enlargement requires a rapid increase in plasma membrane, and this is accomplished in part by transfer of the membranes of a cytoplasmic organelle, the vesiculo-vacuolar organelle (VVO), to the cell surface (Fig. 3.6A and B). VVOs are clusters of interconnected,

uncoated vesicles, and vacuoles that span the cytoplasm of venular endothelial cells from lumen to albumen (Fig. 3.6) (Dvorak *et al.*, 1996). VVOs
provide a VEGF-A–regulated, transcellular pathway for macromolecule
extravasation (Feng *et al.*, 1996, 1997, 2000a,b; Kohn *et al.*, 1992). They
also provide an abundant intracellular membrane store, one that corresponds to more than twice that of the plasma membrane of normal venular
endothelium; this membrane store contributes importantly to the expanded
plasma membrane of MV (Nagy *et al.*, 2006).

Mother vessels are common in both animal and human tumors, where
they were first described (Paku and Paweletz, 1991), and are also abundant
in other types of pathological angiogenesis including skin wounds, myocardial infarcts, and inflammation (Brown *et al.*, 1992; Ren *et al.*, 2002).
They require the continuing presence of exogenous VEGF-A for their
maintenance, and, likely, are the vessel subpopulation most susceptible
to anti–VEGF-A therapy. Over time, mother vessels evolve into several
different types of daughter vessels (Fig. 3.4).

3.2. Evolution of MV into capillaries

MV evolve into capillaries by intraluminal bridging, a process that we
originally described in tumor blood vessels (Nagy *et al.*, 1995, 2007)
(Figs. 3.4 and 3.5). Intraluminal bridging has subsequently been found by
others in wound healing (Ren *et al.*, 2002) and in capillaries that have been
subjected to chronic dilatation (Zhou *et al.*, 1998). Endothelial cells extend
cytoplasmic processes into and across mother vessel lumens to form transluminal bridges that divide blood flow into multiple smaller-sized channels. These
smaller channels then separate from each other to form individual, normal
appearing capillaries. The mechanisms by which transluminal bridging takes
place is not fully understood, but is favored by two processes: intraluminal
clotting and sluggish blood flow. Whether induced by Ad-VEGF-A[164] or by
tumors, MV often undergo partial or complete thrombosis. This is not
surprising in that VEGF-A upregulates endothelial cell expression of the
procoagulant tissue factor (reviewed in Dvorak and Rickles, 2006). Vascular
clotting may facilitate transluminal bridge formation as fibrin provides a
favorable matrix for endothelial cell migration. Bridge formation is also
facilitated in MV because these vessels are highly permeable to plasma, leaving
behind compacted red blood cells and sluggish blood flow.

3.3. Evolution of MV into glomeruloid microvascular proliferations

Glomeruloid microvascular proliferations, also referred to as glomeruloid
bodies, are poorly organized vascular structures that resemble renal glomeruli at the macroscopic level (hence the name) (Pettersson *et al.*, 2000;

Sundberg *et al.*, 2001) (Figs. 3.4 and 3.5C and D). GMP are especially numerous in glioblastoma multiforme, but are also common in cancers of the stomach, breast, etc. where they have been associated with an unfavorable prognosis (Foulkes *et al.*, 2004). All of the human tumors known to form GMP express VEGF-A and tumors such as glioblastoma multiforme that make unusually large amounts of VEGF-A are among those that most commonly induce GMP. GMP are permeable to plasma and plasma proteins, but, because they are poorly perfused, are thought to account for much less plasma extravasation than MV (Nagy *et al.*, 2006). GMP begin to develop from MV at $\sim$7 days after Ad-VEGF-A^{164} administration, starting as focal accumulations of immature, CD-31- and VEGFR-2–positive cells in the endothelial lining (Sundberg *et al.*, 2001). The source of these cells, whether from local vascular endothelium or from circulating endothelial progenitor cells, is not known. Whatever their source, they proliferate rapidly and extend inwardly into MV lumens, and also outwardly into the surrounding extravascular matrix. In this manner, they replace the MV from which they arose, eventually dividing single large MV lumens into much smaller channels that barely admit the passage of single red blood cells. As GMP develop, pericytes accumulate around endothelial cell–lined channels. Macrophages, identified by markers (F4/80, HLA DR2) and by characteristic ultrastructure, also accumulate peripherally. An additional prominent feature is deposition of abnormal multilayered basal lamina. Like MV, GMP apparently require the continued presence of exogenous VEGF-A^{164} for their maintenance because, as adenoviral vector-derived VEGF-A^{164} expression declines, GMP undergo apoptosis and progressively devolve into smaller, normal-appearing capillaries (Nagy *et al.*, 2007; Sundberg *et al.*, 2001).

3.4. Evolution of MV into vascular malformations

Vascular malformations develop from MV by acquiring an irregular coat of smooth muscle cells (Dvorak, 2003; Nagy *et al.*, 2007; Pettersson *et al.*, 2000) (Figs. 3.4 and 3.5E and F). VM are readily distinguished from normal arteries and veins by their inappropriately large size (for their location) and by their thinner and often asymmetric muscular coat. Vessels of this description closely resemble the nonmalignant vascular malformations that occur in skin, brain, and other tissues. As their structure implies, VM are not permeable to plasma proteins (Fig. 3.2, lower panel, 30 and 128 days) (Nagy *et al.*, 2006). Also unlike MV and GMP, VM persist indefinitely (for more than a year), long after adenoviral vector-induced VEGF-A^{164} expression has ceased. Thus, VM have attained independence from exogenous Ad-VEGF-A^{164}, although it is possible that they depend on VEGF-A secreted locally by their enveloping smooth muscle cells. This independence from exogenous VEGF-A has important implications because these vessels would not be expected to be susceptible to anti–VEGF-A therapy.

3.5. FA and DV

Ad-VEGF-A[164] also induces the formation of large FA and DV, in addition
to MV, GMP, capillaries, and VM (Fu *et al.*, 2007; Nagy *et al.*, 2007)
(Fig. 3.7). Unlike angiogenic vessels, FA and DV develop not from venules
but by processes of arteriogenesis and venogenesis from preexisting arteries
and veins. FA and DV form with kinetics that slightly lag MV formation. The
factors that cause FA and DV to develop are not known but are certainly
related to the events taking place, vascular and otherwise, at sites of tumor
growth or Ad-VEGF-A[164] injection. FA and DV, like VM, persist indefinitely in mouse tissues, long after VEGF-A[164] expression levels have become
undetectable, indicating that these vessels, like VM, no longer require exogenous VEGF-A for their survival. Part of the reason seems to be that FA and
DV undergo structural changes in their walls including collagen deposition
that make reversal difficult (unpublished data). FA and DV would seem to be
ideal therapeutic targets in that they are relatively few in number and reside to
a large extent outside of the tumor mass or Ad-VEGF-A[164] injection site.
They supply and drain the tumor vasculature and that of the immediately
surrounding normal tissue. Hence, if occluded, they would be expected to
compromise, not only the tumor center, but also the peripheral tumor rim
that has often proved to be resistant to antiangiogenic therapy.

4. GIANT LYMPHATICS INDUCED BY AD-VEGF-A[164]

In addition to generating new blood vessels, VEGF-A[164] also induces
the formation of abnormal "giant" lymphatic vessels from preexisting
normal lymphatics (Nagy *et al.*, 2002a) (Fig. 3.8). Like the formation of
mother blood vessels, giant lymphatics result from extensive lymphatic
endothelial cell replication with relatively little sprouting (Nagy *et al.*,
2002a). Ad-VEGF-C also induces the formation of enlarged lymphatics
but these exhibit more sprouting than those induced by Ad-VEGF-A[164]
(Wirzenius *et al.*, 2007). The basement membrane enveloping lymphatics is
much less extensive than that surrounding venules and therefore provides
less of a barrier to expansion. The giant lymphatics induced by Ad-VEGF-
A[164] function poorly with sluggish flow and delayed lymph clearance. In
part this is because with their great enlargement the lymphatic valves are not
able to function properly. Also, because VEGF-A[164] induces vascular permeability, fibrinogen, and other clotting proteins escape and the clotting
system is triggered; as a result, the giant lymphatics are often filled with
fibrin clots that obstruct lymphatic flow. Once formed, the giant lymphatics
induced by Ad-VEGF-A[164] become VEGF-A independent and persist
indefinitely, long after VEGF-A expression has ceased.

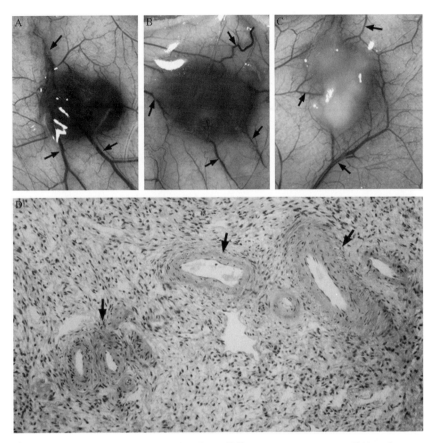

Figure 3.7   FA and DV induced by three different mouse tumors—B16 melanoma (A), TA3/St mammary carcinoma (B), and MOT ovarian cancer (C), and by a human bladder cancer (D) (arrows). Ad-VEGF-A[164] induces identical FA and DV. (From Fu, Y., Nagy, J., Dvorak, A., and Dvorak, H. (2007). Tumor blood vessels: Structure and function. In Teicher, B., and Ellis, L., eds., "Cancer Drug Discovery and Development: Antiangiogenic Agents in Cancer Therapy." Humana Press Inc., Totowa, NJ, pp. 205–224.) (See color insert.)

5. USE OF EAR ANGIOGENESIS ASSAY TO EVALUATE ANTIANGIOGENESIS DRUGS

After some years of disappointment, anti-angiogenesis with a humanized monoclonal antibody to VEGF-A (Avastin, bevacizumab) has shown promise in the treatment of some malignant tumors and also of macular degeneration (Dafer *et al.*, 2007; Hurwitz *et al.*, 2004; Jain *et al.*, 2006; Quiroz-Mercado *et al.*, 2007) (Fig. 3.9). There is reason to expect even greater success as Avastin therapy is optimized and new vascular targets are

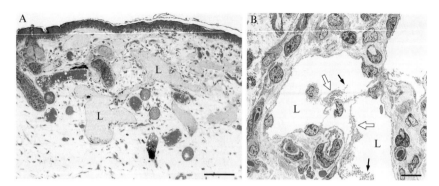

Figure 3.8   Lymphangiogenic response induced at 7 (A) and 14 (B) days after ear injection of Ad-VEGF-A[164]. (A) Giemsa-stained 1-micron Epon section demonstrating giant lymphatics (L) defined by intralymphatic injection of colloidal carbon. (B) Electron micrograph illustrating a giant lymphatic with intraluminal fibrin deposition (solid arrows) and transluminal bridging by lymphatic endothelial cells (outlined arrows). (Modified from Nagy, J. A., Vasile, E., Feng, D., Sundberg, C., Brown, L. F., Detmar, M. J., Lawitts, J. A., Benjamin, L., Tan, X., Manseau, E. J., Dvorak, A. M., and Dvorak, H. F. (2002a). Vascular permeability factor/vascular endothelial growth factor induces lymphangiogenesis as well as angiogenesis. *J. Exp. Med.* **196,** 1497–1506.)

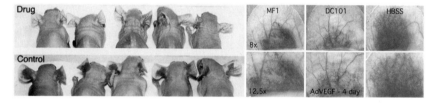

Figure 3.9   Effect of a receptor tyrosine kinase inhibitor targeting VEGFR-1 and -2 suppresses angiogenesis induced by Ad-VEGF-A[164] (left panel). Mice in top panel were treated with drug, bottom panel with vehicle (right panel). Imclone antibodies that block either VEGFR-1 (MF1) or VEGF-2 (DC101) both suppressed Ad-VEGF-A[164]–induced angiogenesis but DC101 had a larger effect than MF1.

discovered. At present, many antiangiogenic agents are in clinical trials (see Angiogenesis Foundation database, www.angio.org). The ear angiogenesis assay provides a powerful, convenient preclinical method for evaluating the effectiveness of antiangiogensis therapies and for elucidating the steps in angiogenesis and the signaling mechanisms that are affected.

6. COMMENTARY AND FUTURE PERSPECTIVES

Adenoviral vectors expressing VEGF-A[164] provide a convenient means of reproducing pathological angiogenesis and lymphangiogenesis in any of a variety of tissues in immunodeficient mice and rats. Unlike *in vitro* assays that measure only one or another property of angiogenesis such as endothelial cell proliferation or migration, adenoviral vectors provide a global *in vivo* assay that measures all aspects of the angiogenic response. This assay has demonstrated that the neovascular response induced by VEGF-A[164] is heterogeneous and leads to the formation of at least six distinctly different types of new blood vessels as well as highly abnormal lymphatics. Moreover, it has provided a means for elucidating the steps and mechanisms by which each type of new vessel forms and for generating at defined times and in large numbers, each of these different types of vessels for molecular study. The adenoviral vector assay avoids some of the complications associated with studying tumor vessels directly, such as necrosis, inflammatory cell infiltration, the relative paucity of blood vessels of each type in tumors, and so on. When used in the mouse ear, it provides a powerful screening test for evaluating both the effectiveness and toxicity of new treatments and for determining whether all or only a subset of the several types of newly formed blood vessels are being targeted. Further, it can be extended to the study of the new blood vessels that are induced by VEGF-A isoforms other than 164/5 and by other angiogenic growth factors such as PlGF (Luttun *et al.*, 2002) and bFGF. Finally, as physiological angiogenesis becomes better understood, the assay may come to have value in assessing the value of new treatments that enhance not only the quantity but also the quality of the angiogenic response as would be desirable in diseases such as myocardial infarction, stroke, and impaired wound healing.

ACKNOWLEDGMENTS

This work was supported by U.S. Public Health Service grants HL-64402 and P01 CA92644 (HFD), and by a contract from the National Foundation for Cancer Research (HFD).

REFERENCES

Aerts, J. L., Gonzales, M. I., and Topalian, S.L (2004). Selection of appropriate control genes to assess expression of tumor antigens using real-time RT-PCR. *Biotechniques* **36,** 84–86, 88, 90–91.

Beck, L., Jr., and D'Amore, P. A. (1997). Vascular development: Cellular and molecular regulation. *FASEB J.* **11,** 365–373.

Brown, L. F., Guidi, A. J., Tognazzi, K., and Dvorak, H. F. (1998). Vascular permeability factor/vascular endothelial growth factor and vascular stroma formation in neoplasia. Insights from *in situ* hybridization studies. *J. Histochem. Cytochem.* **46,** 569–575.

Brown, L. F., Yeo, K. T., Berse, B., Yeo, T. K., Senger, D. R., Dvorak, H. F., and van de Water, L. (1992). Expression of vascular permeability factor (vascular endothelial growth factor) by epidermal keratinocytes during wound healing. *J. Exp. Med.* **176,** 1375–1379.

Dafer, R. M., Schneck, M., Friberg, T. R., and Jay, W. M. (2007). Intravitreal ranibizumab and bevacizumab: A review of risk. *Semin. Ophthalmol.* **22,** 201–204.

Dvorak, A. (1987). Procedural guide to specimen handling for the ultrastructural pathology service laboratory. *J. Electron Microsc. Tech.* **6,** 255–301.

Dvorak, H. F., Harvey, V. S., and McDonagh, J. (1984). Quantitation of fibrinogen influx and fibrin deposition and turnover in line 1 and line 10 guinea pig carcinomas. *Cancer Res.* **44,** 3348–3354.

Dvorak, A. M., Kohn, S., Morgan, E. S., Fox, P., Nagy, J. A., and Dvorak, H. F. (1996). The vesiculo-vacuolar organelle (VVO): A distinct endothelial cell structure that provides a transcellular pathway for macromolecular extravasation. *J. Leukoc. Biol.Leukoc. Biol.* **59,** 100–115.

Dvorak, H. F. (2003). Rous–Whipple Award Lecture. How tumors make bad blood vessels and stroma. *Am. J. Pathol.* **162,** 1747–1757.

Dvorak, H. (2007). Tumor blood vessels. *In* "Endothelial Biomedicine." (Aird, W., ed.), pp. 1457–1470. Cambridge University Press, New York.

Dvorak, H. F., and Rickles, F. R. (2006). Malignancy and hemostasis. "Hemostasis and Thrombosis: Basic Principles and Clinical Practice." (R. W. Colman, V. J. Marder, A. W. Clowes, J. N. George, and S. Z. Goldhaber, eds.), 5th ed. pp. 851–873. Lippincott Williams & Wilkins, Philadelphia.

Feng, D., Nagy, J., Brekken, R., Pettersson, A., Manseau, E., Pyne, K., Mulligan, R., Thorpe, P., Dvorak, H., and Dvorak, A. (2000a). Ultrastructural localization of the vascular permeability factor/vascular endothelial growth factor (VPF/VEGF) receptor-2 (FLK-1, KDR) in normal mouse kidney and in the hyperpermeable vessels induced by VPF/VEGF-expressing tumors and adenoviral vectors. *J. Histochem. Cytochem.* **48,** 545–555.

Feng, D., Nagy, J. A., Dvorak, A. M., and Dvorak, H. F. (2000b). Different pathways of macromolecule extravasation from hyperpermeable tumor vessels. *Microvasc. Res.* **59,** 24–37.

Feng, D., Nagy, J. A., Hipp, J., Dvorak, H. F., and Dvorak, A. M. (1996). Vesiculo–vacuolar organelles and the regulation of venule permeability to macromolecules by vascular permeability factor, histamine, and serotonin. *J. Exp. Med.* **183,** 1981–1986.

Feng, D., Nagy, J. A., Hipp, J., Pyne, K., Dvorak, H. F., and Dvorak, A. M. (1997). Reinterpretation of endothelial cell gaps induced by vasoactive mediators in guinea-pig, mouse and rat: Many are transcellular pores. *J. Physiol.* **504,** 747–761.

Feng, D., Nagy, J. A., Pyne, K., Dvorak, H. F., and Dvorak, A. M. (1998). Neutrophils emigrate from venules by a transendothelial cell pathway in response to FMLP. *J. Exp. Med.* **187,** 903–915.

Feng, D., Nagy, J. A., Pyne, K., Dvorak, H. F., and Dvorak, A. M. (2004). Ultrastructural localization of platelet endothelial cell adhesion molecule (PECAM-1, CD31) in vascular endothelium. *J. Histochem. Cytochem.* **52,** 87–101.

Folkman, J. (1997). Angiogenesis and angiogenesis inhibition: An overview. *EXS* **79,** 1–8.

Foulkes, W. D., Brunet, J. S., Stefansson, I. M., Straume, O., Chappuis, P. O., Begin, L. R., Hamel, N., Goffin, J. R., Wong, N., Trudel, M., Kapusta, L., Porter, P., *et al.* (2004). The prognostic implication of the basal-like (cyclin E high/p27 low/p53+/glomeruloid-microvascular-proliferation+) phenotype of BRCA1-related breast cancer. *Cancer Res.* **64,** 830–835.

Fu, Y., Nagy, J., Dvorak, A., and Dvorak, H. (2007). Tumor blood vessels: Structure and function. "Cancer Drug Discovery and Development: Antiangiogenic Agents in Cancer Therapy." (B. Teicher and L. Ellis, eds.), pp. 205–224. Humana Press Inc., Totowa, NJ.

Gale, N. W., and Yancopoulos, G. D. (1999). Growth factors acting via endothelial cell-specific receptor tyrosine kinases: VEGFs, angiopoietins, and ephrins in vascular development. *Genes Dev.* **13,** 1055–1066.

Graham, M. M., and Evans, M. L. (1991). A simple, dual tracer method for the measurement of transvascular flux of albumin into the lung. *Microvasc. Res.* **42**, 266–279.

Hardy, S., Kitamura, M., Harris-Stansil, T., Dai, Y., and Phipps, M. L. (1997). Construction of adenovirus vectors through Cre-lox recombination. *J. Virol.* **71**, 1842–1849.

Hurwitz, H., Fehrenbacher, L., Novotny, W., Cartwright, T., Hainsworth, J., Heim, W., Berlin, J., Baron, A., Griffing, S., Holmgren, E., Ferrara, N., Fyfe, G., *et al.* (2004). Bevacizumab plus irinotecan, fluorouracil, and leucovorin for metastatic colorectal cancer. *N. Engl. J. Med.* **350**, 2335–2342.

Jain, R. K., Duda, D. G., Clark, J. W., and Loeffler, J. S. (2006). Lessons from phase III clinical trials on anti-VEGF therapy for cancer. *Nat. Clin. Pract. Oncol.* **3**, 24–40.

Kohn, S., Nagy, J. A., Dvorak, H. F., and Dvorak, A. M. (1992). Pathways of macromolecular tracer transport across venules and small veins. Structural basis for the hyperpermeability of tumor blood vessels. *Lab. Invest.* **67**, 596–607.

Luttun, A., Tjwa, M., Moons, L., Wu, Y., Angelillo-Scherrer, A., Liao, F., Nagy, J. A., Hooper, A., Priller, J., De Klerck, B., Compernolle, V., Daci, E., *et al.* (2002). Revascularization of ischemic tissues by PlGF treatment, and inhibition of tumor angiogenesis, arthritis and atherosclerosis by anti-Flt1. *Nat. Med.* **8**, 831–840.

Meier, O., and Greber, U.F (2004). Adenovirus endocytosis. *J. Gene Med.* **6**(Suppl 1), S152–S163.

Nagy, J. A., Dvorak, A. M., and Dvorak, H. F. (2003). VEGF-A(164/165) and PlGF: Roles in angiogenesis and arteriogenesis. *Trends Cardiovasc. Med.* **13**, 169–175.

Nagy, J. A., Dvorak, H. F., and Dvorak, A. M. (2007). VEGF-A and the induction of pathological angiogenesis. *Annu. Rev. Pathol. Mech. Dis.* **2**, 251–275.

Nagy, J. A., Feng, D., Vasile, E., Wong, W. H., Shih, S. C., Dvorak, A. M., and Dvorak, H. F. (2006). Permeability properties of tumor surrogate blood vessels induced by VEGF-A. *Lab. Invest.* **86**, 767–780.

Nagy, J. A., Morgan, E. S., Herzberg, K. T., Manseau, E. J., Dvorak, A. M., and Dvorak, H. F. (1995). Pathogenesis of ascites tumor growth: Angiogenesis, vascular remodeling, and stroma formation in the peritoneal lining. *Cancer Res.* **55**, 376–385.

Nagy, J. A., Vasile, E., Feng, D., Sundberg, C., Brown, L. F., Detmar, M. J., Lawitts, J. A., Benjamin, L., Tan, X., Manseau, E. J., Dvorak, A. M., and Dvorak, H. F. (2002a). Vascular permeability factor/vascular endothelial growth factor induces lymphangiogenesis as well as angiogenesis. *J. Exp. Med.* **196**, 1497–1506.

Nagy, J. A., Vasile, E., Feng, D., Sundberg, C., Brown, L. F., Manseau, E. J., Dvorak, A. M., and Dvorak, H. F. (2002b). VEGF-A induces angiogenesis, arteriogenesis, lymphangiogenesis, and vascular malformations. *Cold Spring Harb. Symp. Quant. Biol.* **67**, 227–237.

Ory, D. S., Neugeboren, B. A., and Mulligan, R. C. (1996). A stable human-derived packaging cell line for production of high titer retrovirus/vesicular stomatitis virus G pseudotypes. *Proc. Natl. Acad. Sci. USA* **93**, 11400–11406.

Paku, S., and Paweletz, N. (1991). First steps of tumor-related angiogenesis. *Lab. Invest.* **65**, 334–346.

Pettersson, A., Nagy, J. A., Brown, L. F., Sundberg, C., Morgan, E., Jungles, S., Carter, R., Krieger, J. E., Manseau, E. J., Harvey, V. S., Eckelhoefer, I. A., Feng, D., *et al.* (2000). Heterogeneity of the angiogenic response induced in different normal adult tissues by vascular permeability factor/vascular endothelial growth factor. *Lab. Invest.* **80**, 99–115.

Quiroz-Mercado, H., Ustariz-Gonzalez, O., Martinez-Castellanos, M. A., Covarrubias, P., Dominguez, F., and Sanchez-Huerta, V. (2007). Our experience after 1765 intravitreal injections of bevacizumab: The importance of being part of a developing story. *Semin. Ophthalmol.* **22**, 109–125.

Ren, G., Michael, L. H., Entman, M. L., and Frangogiannis, N. G. (2002). Morphological characteristics of the microvasculature in healing myocardial infarcts. *J. Histochem. Cytochem.* **50**, 71–79.

Shih, S. C., and Smith, L. E. (2005). Quantitative multi-gene transcriptional profiling using real-time PCR with a master template. *Exp. Mol. Pathol.* **79,** 14–22.

Sundberg, C., Nagy, J. A., Brown, L. F., Feng, D., Eckelhoefer, I. A., Manseau, E. J., Dvorak, A. M., and Dvorak, H. F. (2001). Glomeruloid microvascular proliferation follows adenoviral vascular permeability factor/vascular endothelial growth factor-164 gene delivery. *Am. J. Pathol.* **158,** 1145–1160.

Swayne, G. T., Smaje, L. H., and Bergel, D. H. (1989). Distensibility of single capillaries and venules in the rat and frog mesentery. *Int. J. Microcirc. Clin. Exp.* **8,** 25–42.

Wirzenius, M., Tammela, T., Uutela, M., He, Y., Odorisio, T., Zambruno, G., Nagy, J. A., Dvorak, H. F., Yla-Herttuala, S., Shibuya, M., and Alitalo, K. (2007). Distinct vascular endothelial growth factor signals for lymphatic vessel enlargement and sprouting. *J. Exp. Med.* **204,** 1431–1440.

Zhou, A., Egginton, S., Hudlicka, O., and Brown, M. D. (1998). Internal division of capillaries in rat skeletal muscle in response to chronic vasodilator treatment with alpha1-antagonist prazosin. *Cell Tissue Res.* **293,** 293–303.

CHAPTER FOUR

Using the Zebrafish to Study Vessel Formation

Mary C. McKinney *and* Brant M. Weinstein

Contents

Abstract

Danio rerio, commonly referred to as the zebrafish, is a powerful animal model for studying the formation of the vasculature. Zebrafish offer unique opportunities for *in vivo* analysis of blood and lymphatic vessels formation because of their accessibility to large-scale genetic and experimental analysis as well as the small size, optical clarity, and external development of zebrafish embryos and larvae. A wide variety of established techniques are available to study vessel formation in the zebrafish, from early endothelial cell differentiation to adult vessel patterning. In this chapter, we review methods used to functionally

Laboratory of Molecular Genetics, National Institute of Child Health and Human Development, Bethesda, Maryland

Methods in Enzymology, Volume 444
ISSN 0076-6879, DOI: 10.1016/S0076-6879(08)02804-8

manipulate and visualize the vasculature in the zebrafish and illustrate how these methods have helped further understanding of the genetic components regulating formation and patterning of developing vessels.

1. INTRODUCTION

The zebrafish has been studied as a model organism for development and disease since the 1970s, but only became a regular addition to animal model research beginning in the mid-1990s with the first large-scale mutagenic screens (Driever *et al.*, 1996; Haffter *et al.*, 1996). The basic mechanisms and gene pathways controlling many cellular functions are highly conserved through all vertebrate species, and zebrafish then lend themselves well to studying overarching conserved pathways in development such as angiogenesis. Zebrafish have become a popular model due to their many advantageous features. The small size (2 mm for a 2-day-old embryo) and optical clarity of zebrafish embryos, large clutch size with up to hundreds of embryos per pair of fish, rapid development, low cost, and increasing genetic knowledge of zebrafish have made them a popular model for many studies. In addition to angiogenesis, zebrafish have been used in cancer models (Amatruda *et al.*, 2002; Nicoli *et al.*, 2007), neurobiology (Key and Devine, 2003), infection and immunity studies (Traver *et al.*, 2003; van der Sar *et al.*, 2004), and drug discovery (MacRae and Peterson, 2003; Zon and Peterson, 2005).

Angiogenesis is a complex process with many intersecting gene pathways that control the growth and remodeling of vessels. Many model systems exist to study vasculogenesis and angiogenesis, but these models lack the *in vivo* accessibility of the zebrafish. The vascular system of zebrafish develops in similar fashion to other vertebrates, so comparison can be made across species (Isogai *et al.*, 2001). Because of its small size, the zebrafish embryo can absorb enough oxygen through passive diffusion from the water to survive for several days even in the complete absence of a functioning vasculature, facilitating determination of the vascular specificity of genetic or experimental defects. Because of their external development and transparency, zebrafish embryos and larvae can be manipulated and optically imaged at all stages of their development with unprecedented resolution. Blood flow can be easily observed in any vessel in the living animal, using even relatively low-power microscopes. As described in detail below, fluorescent transgenic fish permit collection of highly detailed images and time-lapse movies of vessel development.

The zebrafish has already provided many fundamental new insights into vascular development. Zebrafish studies have helped to dissect the molecular requirements for arterial–venous differentiation and the critical role played by hedgehog, vascular endothelial growth factor (VEGF), and notch signaling in this process (Fouquet *et al.*, 1997; Lawson and Weinstein, 2002a; Lawson *et al.*, 2001, 2002, 2003; Weinstein and Lawson, 2002; Weinstein *et al.*, 1995; Zhong *et al.*, 2000, 2001). Research in the zebrafish has also permitted dynamic imaging and observation of the assembly of vascular networks in living animals (Isogai *et al.*, 2003) and dissection of the molecular mechanisms regulating the stereotypic patterning of early vessels. These mechanisms include guidance by well-characterized factors important for repulsive guidance of neural axons and patterning of the developing brain such as semaphorins (Torres-Vazquez *et al.*, 2004) and regulation of vessel branching and vascular "tip cell" specification by notch signaling (Siekmann and Lawson, 2007). Zebrafish plexinD1 is an endothelial-specific receptor for type 3 semaphorins expressed in growing intersegmental vessels that navigate through semaphorin-free corridors between the somites (future muscle blocks) in the developing trunk. Loss of plexinD1 results in mispatterning and overbranching of the intersegmental vessels into regions of the somites from which they are normally excluded (Torres-Vazquez *et al.*, 2004) (Fig. 4.1). Loss of notch signaling from growing intersegmental vessel cells also results in increased branching and patterning defects, in this case because lack of notch signals permits too many endothelial cells to become tip cells, or leading cells in angiogenic sprouting and branching (Siekmann and Lawson, 2007). The zebrafish also provides a powerful new model system for studying the development of the lymphatic vascular system, a completely separate and distinct vascular system found in vertebrates that drains fluid "leaked" from blood vessels into tissues, returning it back to the blood circulation. In addition

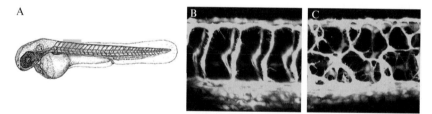

Figure 4.1 PlexinD1 is required for proper vessel patterning. (A) Camera lucida drawing of a 48-hpf embryo with area viewed in (B) and (C) highlighted. (B, C) Multiphoton images of the trunk of control (B) and plxnD1 (C) morpholino injected *Tg(fli1-EGFP)^{y1}* embryos at approximately 48 hpf. All endothelial cells in these transgenic animals are green fluorescent. The control morpholino–injected animal (B) displays regularly patterned trunk intersegmental vessels, while a plxnD1 morpholino–injected sibling embryo (C) exhibits loss of proper patterning and overbranching of these vessels. (Drawing modified from Kimmel, C. B., Ballard, W. W., Kimmel, S. R., Ullmann, B., and Schilling, T. F. (1995). Stages of embryonic development of the zebrafish. *Dev. Dyn.* **203**, 253–310. Images from Torres-Vazquez, J., Gitler, A. D., Fraser, S. D., Berk, J. D., Van, N. P., Fishman, M. C., Childs, S., Epstein, J. A., and Weinstein, B. M. (2004). Semaphorin-plexin signaling guides patterning of the developing vasculature. *Dev. Cell* **7**, 117–123.)

to its critical role in fluid homeostasis, a functional lymphatic system is also important for immune function and lipid absorption. Recent evidence suggests that lymphatic vessels are a major route of tumor metastasis, making these vessels important therapeutic targets. Recent studies in the zebrafish have shown conclusively that the zebrafish have a bona fide lymphatic system (Yaniv *et al.*, 2006). Using sophisticated, time-lapse two-photon imaging of transgenic zebrafish (described further below), these authors were also able to perform *in vivo* lineage tracing to demonstrate conclusively that early lymphatic endothelial cells arise from the endothelium of primitive veins, resolving a more than 100-year-old debate about the origins of these cells.

Clearly, the zebrafish provides an important tool for studying vasculogenesis and angiogenesis, and it is likely that further studies in the fish will continue to lead to important novel insights. In this review we discuss some of the methods used to manipulate and visualize the vasculature in zebrafish. In the first part of the review we briefly describe some of the methods available for functional manipulation of zebrafish, including genetic screens, injection of RNA, DNA, and antisense morpholine oligonucleotides ("morpholinos"), cell transplantation, and chemical treatment. In the second part of the review, we describe methods for visualizing the effects of these manipulations. We briefly describe methods used for fixed samples and then provide detailed protocols for *in vivo* methods including microangiography and confocal, two-photon, and dynamic time-lapse imaging of transgenic fish.

2. Methods for Functional Manipulation of Zebrafish

2.1. Genetic screening methods

Genetic screening is an important tool in the zebrafish that can be used to alter the normal patterns of angiogenesis and explore the factors regulating vessel formation (Patton and Zon, 2001). The ability to maintain large number of animals, obtain large numbers of progeny, and visualize the externally developing embryo make the zebrafish uniquely accessible to forward-genetic analysis of vertebrate development. Large-scale mutagenesis screens have already resulted in the identification of hundreds of genes involved in many different aspects of zebrafish development (Driever *et al.*, 1996; Haffter *et al.*, 1996). Chemical treatment with N-ethyl-N-nitrosourea (ENU) has been the most widely used method to create randomly mutated genomic DNA for forward genetic screening and details can be found in several reviews (Amsterdam and Hopkins, 2006; Bahary *et al.*, 2004; Wienholds *et al.*, 2002). The mutagenicity of ENU results in very high frequencies of mutations, making it a very efficient method for generating the largest number of mutants per fish screened. However, positional cloning of ENU-induced point mutants is still a time-consuming challenge. Screens performed with ENU chemical mutagenesis have yielded novel mutations with vascular phenotypes, including cloche, gridlock, plcg1, violet beauregard, out of bounds, and etsrp (Childs *et al.*, 2002; Lawson *et al.*, 2003; Pham *et al.*, 2007; Roman *et al.*, 2002; Stainier *et al.*, 1995; Torres-Vazquez *et al.*, 2004; Weinstein *et al.*, 1995). Cloning and identification of the genes involved in these mutations have broadened our understanding of vascular genetic pathways not only in the zebrafish but in all vertebrates. Mutagenesis can also be performed be generating gene-disrupting insertions in the genome using transposons or retroviruses. Although the mutagenic efficiency of insertion methods is much lower than chemical mutagenesis, the insertions provide a "tag" greatly facilitating molecular cloning of the disrupted gene. This method has been successful in *Drosophila*, *C. elegans*, mouse, plants, and zebrafish (Amsterdam *et al.*, 1999; Ballinger and Benzer, 1989; Das and Martienssen, 1995; Zambrowicz *et al.*, 1998; Zwaal *et al.*, 1993). The frequency of insertion is higher with retroviral vectors rather than transposon methods but both methods have been used (Wienholds *et al.*, 2002).

Although the zebrafish is best known for its strength in forward-genetic analysis, reverse-genetic methods can also be employed. Technology for targeted disruption in embryonic stem cells is currently unavailable in the zebrafish, but mutations in specific genes of interest can be uncovered by screening randomly mutated pools of fish. Targeted induced local lesions in genomes ("TILLING") is a method originally developed in Arabidopsis, where point mutations were found using enzyme-mediated mismatch recognition (McCallum *et al.*, 2000). Since its original development, this method has been applied to other plants, flies, and zebrafish (Sood *et al.*, 2006; Till *et al.*, 2006; Wienholds *et al.*, 2002). Briefly, nested PCR is performed on genomic DNA from offspring of the mutated fish. PCR products are pooled, melted, and annealed. The PCR products are then digested with CEL1 endonuclease, which cleaves mismatched DNA hybrids (from mutations or polymorphisms). The fragments can be detected on a gel and traced back to the mutant animal. Hundreds of fish can be analyzed at once and with a high throughput sequencing system the mutations can be found relatively quickly, although these enzyme-based methods have a substantial error and failure rate. Mutations can be detected more reliably by simply massively resequencing specific PCR products of a gene of interest from a large library of mutagenized animals. However, this is time consuming and costly when searching through many fish.

2.2. DNA and RNA injection

In the absence of a genetic mutant in a gene of interest, a variety of additional powerful and useful methods for functional manipulation are available. RNA injections can be used to express a gene of interest.

Dominant-negative or constitutive-active forms of many genes can also be generated and injected as RNA, facilitating understanding of the functional roles of these genes. However, RNA injection methods are limited by a lack of spatial and temporal control. Injected RNAs are translated ubiquitously throughout the entire animal after activation of zygotic gene expression at the mid-blastula transition, and injected RNA is rapidly degraded in the embryo, making this method useful for functional manipulation only for approximately the first day of development. By coupling a gene of interest to a tissue-specific promoter and injecting this DNA construct, spatially, and/or temporally restricted and somewhat longer-lasting expression of the gene can be accomplished. Unlike after RNA injections, expression from injected DNA constructs is mosaic. That is, only a fraction of the cells in the embryo that are potentially able to activate the promoter employed actually receive the DNA construct and express the target gene, although recent methods allow mosaicism of up to 50% or higher. This mosaicism can be disadvantageous or advantageous depending on the experimental paradigm being used. The generation of germline transgenic animals from DNA constructs eliminates mosaicism, but methods for generating germline transgenic animals are not discussed here. Loss of function for specific gene products can be accomplished by injection of morpholino oligonucleotides, as discussed in the next section. The protocols for RNA, DNA and morpholino injections are similar (Westerfield, 2000; Xu, 1999), and a rough outline is provided below. Embryos should preferentially be injected at the one-cell stage in order to ensure uniform and ubiquitous distribution of injected materials throughout the animal.

2.2.1. Materials needed

- Fish
- Breeder tanks
- Net to collect the eggs
- Glass Pasteur pipettes for handling embryos
- Petri dishes (Falcon)
- Ramp molds (Adaptive Science Tools)
- Fine forceps (Dumont #55)
- Blunted forceps (Dumont #5)
- Microloader pipette tips (Eppendorf)
- Glass capillaries, 1-mm outer diameter (World Precision Instruments, Inc.)
- Micropipette puller (Sutter Instruments)
- Pneumatic pump with micromanipulator and micropipette holder (World Precision Instruments)
- Dissecting microscope

2.2.2. The day before injection

1. Place male and female fish in one breeding tank separated by a barrier. Depending on the health and virility of the fish, three pairs for one tank may be sufficient. However, more breeding tanks should be prepared as backup.
2. Melt 2.5% agarose in egg water (Westerfield, 2000) and let cool in a flask until viscous. Prepare approximately 35 ml per ramp.
3. Pour molten agarose into lid of 10-cm Petri dish until half full.
4. Submerge ramp mold in distilled water, remove, and tap on paper to remove excess water.
5. Drop mold into molten agarose from a height of approximately 4". V-shaped or square molds are both sufficient for injection, and are a matter of personal preference. If a large number of bubbles are surrounding the mold, pull the mold out and rinse and brush bubbles to the side and try again.
6. Let the agarose cool completely, and then remove the mold with dull forceps by prying from one end.
7. Pour a small amount of egg water on top of the injection ramp and cover with the bottom of the Petri dish. Ramps can be stored at 4 °C and reused until they dry out or are damaged.
8. Pull 1-mm filamented or nonfilamented microneedles for injection using approximate settings of 500 heat, 60 pull, 60 velocity, 80 time, and 200 pressure on Sutter Instruments micropipette puller. The resulting needle should resemble Fig. 4.2A or B.

2.2.3. Day of injection

1. Remove barrier between male and female fish to permit mating to occur.
2. Remove injection ramps from 4 °C storage and place either at room temperature or 28 °C.
3. Prepare the RNA or DNA solution in the desired concentration in filtered 30% Danieu's Solution (100% Danieu's: 58 mM NaCl, 0.7 mM KCl, 0.4 mM MgSO$_4$, 0.6 mM Ca(NO$_3$)$_2$, 5 mM Hepes, pH 7.6). For visualization of the injection, phenol red can be added to 0.1%.
4. Back-fill needles with 5 to 10 μl of injection solution.
5. Clip end of needle to an opening of 5 μm or less using fine forceps.
6. Calibrate the needle by the following method:
   - Mark needle at 1-mm intervals along barrel before loading.
   - Back-fill with injection solution.
   - Mount the needle into the holder attached to the pneumatic pump.
   - With the pneumatic pump set to constant pressure, measure the amount of time in milliseconds that it takes for the solution to flow 1 mm (dispense 440 nl of liquid).
   - Calculate the milliseconds needed to release 1 nl of liquid with the following formula: ms/nl = 100 * time/(440 nl).

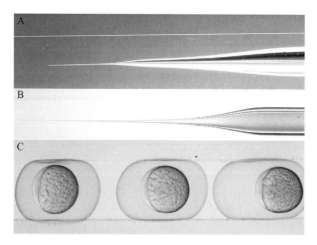

Figure 4.2 Injection into one-cell–stage embryos. (A) Desired needle shape for angiography. (B) Desired needle shape for nucleotide injection. (C) Orientation of embryos in ramp for needle entering on the right. Embryo chorion (shell) is approximately 1 mm in diameter.

7. If flow though the needle is faster than 2 nl/ms or slower than 0.2 nl/ms, it will be physically difficult to inject with this needle. If the needle becomes clogged by debris from the solution, spin the solution in a microcentrifuge at 13,000 rpm and load a new needle.

8. Collect the embryos from the breeder tank by pouring over a net and transfer to a Petri dish.

9. Using the microscope, remove healthy embryos from debris and place on ramp. Using blunted forceps, push embryos into channels of the ramp and orient embryos with yolk facing needle and the cell facing away (Fig. 4.2C).

10. Insert needle through chorion and yolk into the cytoplasm. It is easiest if the needle is approximately 45 degrees from horizontal. Inject up to 4 nl of solution into each embryo.

11. For suitable statistics, at least 100 injected embryos should be scored, so more than this should be injected to allow for unfertilized or damaged embryos.

12. Use the blunt forceps and glass pipette to remove embryos from ramp and store at 28 °C in a Petri dish with fresh egg water.

2.2.4. After injection

1. Clean the injected embryos a few hours after injection by removing unfertilized embryos and replacing the water. Also, begin examining for effects at this time.

2. Embryos can be fixed and stained or imaged alive as desired to document phenotypes. Score embryos by counting the number of embryos exhibiting the phenotype, documenting the severity as well if there is a range, the number dead, and the number of wildtype animals.

To visualize the effects of injections on the developing vasculature, injections can be performed into transgenic animals such as *Tg(fli1-EGFP)ʸ¹* that expresses the fluorescent GFP protein in all endothelial cells. This allows for quick scoring and high-resolution imaging of vascular defects (Torres-Vazquez *et al.*, 2004; Yaniv *et al.*, 2006). Alternatively, *in situ* hybridization or alkaline phosphatase staining of the embryos at various ages can be performed to assess effects (Kamei *et al.*, 2004; Thisse *et al.*, 2004). When injecting RNA, care must be taken to keep the environment and equipment as RNAse free as possible. All tubes and needles should only be handled with gloves. The amount of RNA to be injected will vary depending on the gene and desired effects, but can be up to approximately 1 μg. Ideally, the highest concentration should be used that is not lethal and does not result in severe off-target effects, but that permits most embryos to exhibit gene-specific phenotypes. For DNA injections, the gene of interest must be cloned into a vector downstream from a promoter permitting expression in zebrafish. Depending on the promoter chosen, the expression can be regulated spatially and/or temporally in the embryo but as noted above it will have a mosaic expression in the injected animals. Zebrafish promoters are available for many tissue types including endothelium, blood, and muscle (Cross *et al.*, 2003; Du *et al.*, 2003; Lawson and Weinstein, 2002b; Zhu *et al.*, 2005). DNA is usually injected as a linear construct in amounts up to 200 to 300 ng per embryo. Methods for production of transgenic animals by DNA injection have been reviewed elsewhere (Lin, 2000; Stuart *et al.*, 1988; Udvadia and Linney, 2003; Xu, 1999). DNA injection is relatively inefficient for generating germline transgenics, with a transgenesis rate of as little as 1 to 10% (Stuart *et al.*, 1988, 1990). Recently developed transposon-based systems offer higher efficiencies. Sleeping beauty transposons can give up to a 30% transgenesis rate (Davidson *et al.*, 2003), while transgenesis with Tol2 vectors can be up to 50% (Fisher *et al.*, 2006; Kawakami, 2007; Kawakami *et al.*, 2004).

2.3. Morpholinos

Loss-of-function analysis of genes of interest in the zebrafish has been greatly facilitated by the application of antisense oligonucleotide technologies to the zebrafish. Although RNA interference methods used extensively in cell culture studies have not yet shown promising results in zebrafish (Chen and Ekker, 2004; Ekker, 2000; Oates *et al.*, 2000), highly stable and specific morpholine oligonucleotides or "morpholinos" have proven to be very effective for reducing the translation or splicing of target genes (Corey and Abrams, 2001; Ekker, 2000). Morpholinos are oligonucleotides with normal nucleic acid bases, but a phosphorodiamidate linkage between bases and a morpholine moiety instead of a ribose along the backbone.

Morpholinos bind tightly and specifically to mRNA, but are otherwise chemically inactive in the cell. They can be designed to block translation by binding at the beginning of the start codon (Nasevicius and Ekker, 2000), or to block splicing by binding across an exon/intron boundary (Draper et al., 2001). Because of their stability, morpholinos injected into one-cell stage embryos can persist and effectively block their targets for up to 5 days (Ekker, 2000). Morpholinos have already been extensively used to study vessel formation in zebrafish (e.g., Alvarez et al., 2007; Buchner et al., 2007; Covassin et al., 2006; Lawson et al., 2001, 2002, 2003; Nasevicius et al., 2000; Pham et al., 2007; Roman et al., 2002; Siekmann and Lawson, 2007; Torres-Vazquez et al., 2004; Yaniv et al., 2006; Zhong et al., 2000). Many morpholinos exhibit off-target effects and nonspecific toxicity, so multiple morpholinos targeting different parts of the gene should be tested at a range of different concentrations each, to see if they give the same specific phenotype. Additional controls that should be attempted include (1) co-inject the correct RNA sequence with the morpholino to rescue the phenotype, (2) compare the morpholino phenotype to a known mutant in the same gene, and (3) inject a mismatched control morpholino.

The protocol for injection of the morpholino is essentially the same as RNA or DNA injection. Morpholino solutions are prepared according to the manufacturer's instructions and can be stored at −20 °C for several months. The concentration of the morpholino to be used must be empirically determined and widely varies according to the particular morpholino used. The injected concentration should be titrated in small increments until most of the embryos show the phenotype but injecting slightly more morpholino would prove lethal to most of the embryos and/or cause substantial nonspecific morphological defects. The margin between minor phenotype and lethality can be less than twofold. The unused portion of morpholino after injection can be saved for future injections. Scoring for the effects of morpholinos is similar to that for RNA injections although as noted above controls must also be performed to ensure specific gene targeting.

2.4. Late-stage functional manipulation

A complication to injecting RNA, DNA or morpholinos is that perturbation of many genes at the one cell-stage or later is lethal to the embryo before angiogenesis begins. A number of methods have recently been tested to perform functional manipulation of animals specifically at later stages of development or even during adulthood. Several transgenic tools are available that can potentially be useful for late-stage regulation of gene expression, including a heat-shock promoter to select a time point to start universally expressing a gene (Pyati *et al.*, 2005), a cre/lox system using a chemical treatment to start the expression of a gene (Langenau *et al.*, 2003), or a heat-shock promoter driving gal4 production crossed with a UAS promoter line driving the gene of interest (Le *et al.*, 2007). Each of these temporally and spatially regulated expression schemes occasionally misexpress target genes. In addition, electroporation has been used to deliver DNA, RNA, or morpholinos into zebrafish embryos by applying current to an area of the body with electrodes (Cerda *et al.*, 2006). Electroporating into an animal provides effective temporal control, but relatively crude control over spatial expression compared to use of specific promoters in transgenic schemes.

2.5. Transplant assays

Blastomere transplantation has been widely used by zebrafish researchers for generating chimeric embryos for experimental analysis. Most commonly, transplants are performed on shield-stage or younger embryos (Mizuno *et al.*, 1999), but transplants and xenotransplants have been performed even in juvenile fish. Transplanting cells at later stages carries a risk of rejection and lethality to the host embryo. Highly inbred animals have extremely homogeneous genetics that might result in easier transplants, but these lines are difficult to create and maintain and are not currently widely available for zebrafish (Davidson and Zon, 2004). Genetic clones can also be made from a line of homozygous diploid zebrafish that provide less response to the transplanted tissue between the clones (Mizgireuv and Revskoy, 2006). For early transplantation genetically compatible transplants are not necessary, since the immune system of the zebrafish embryo is not fully developed for several days after fertilization (Nicoli and Presta, 2007). Implantation of cultured cells or embryo-to-embryo transplants can be performed relatively easily (Mizuno *et al.*, 1999; Westerfield, 2000). Recently, there have been several papers describing xenografts of cancerous cell lines into zebrafish embryos and juvenile fish to study angiogenesis into the transplanted tissue (Lee *et al.*, 2005; Nicoli *et al.*, 2007; Stoletov *et al.*, 2007), but the potential has yet to be fully explored. For juvenile fish greater than 1 month old, immune system chemical inhibitors (e.g., dexamethasone or Cyclosporin A) (Stoletov *et al.*, 2007; Zhang *et al.*, 2003) may need to be employed to reduce the chance of rejection. The progress of angiogenesis after the transplant can be monitored by several methods, from fixing and staining with antibodies or *in situ* hybridization to live *in vivo* imaging. A drawback to the zebrafish model for transplant assays is that only roughly 1000 to 2000 cells may be transplanted into even a several day-old embryo because of its small size (Nicoli and Presta, 2007).

2.6. Chemical treatment

Treatment with specific chemical inhibitors has also been used in the zebrafish to target specific genes or pathways (Chan and Serluca, 2004; Chan et al., 2002; Lawson et al., 2002; Peterson et al., 2004). These chemical treatments are usually quite easy to perform on zebrafish embryos and larvae since the chemicals can be added directly to the water, they are immersed in and absorbed into the small body of the embryo. Embryos can tolerate up to 1% DMSO in the egg water, so even chemicals that are only slightly soluble in water can be used (Chan and Serluca, 2004). Many if not most vascular genes are conserved between vertebrates, so many drugs that bind to proteins in other animal models can be used on zebrafish. Chemicals that have been used for vascular studies in zebrafish include PTK787/ZK222584, SU5416 and SU1498, VEGF receptor inhibitors (Bayliss et al., 2006; Chan et al., 2002); mycophenolic acid, an immunosuppressive drug (Wu et al., 2006); and cyclopamine, a hedgehog signaling inhibitor (Lawson et al., 2002).

Zebrafish also make a particularly useful model for drug screening. Three embryos can survive in as little as 100 to 200 μl of water during a treatment period, thus using less chemical (Chan and Serluca, 2004; Peterson et al., 2004), and many animals can be screened for chemical effects together in high-density-format microtiter plates. Fish thus provide an effective model for medium- to high-throughput small chemical screens (Murphey and Zon, 2006; Peterson et al., 2004). In one study, gridlock mutant embryos were treated with 5000 different compounds to screen for compounds capable of rescuing the vascular defect. Two compounds were identified that rescued the defect by increasing VEGF receptor signaling. Both have comparable effects in mammalian cells (Hong et al., 2006; Peterson et al., 2004). These results demonstrate the potential of the zebrafish for unbiased large-scale chemical screening in an intact vertebrate model, an approach that may prove highly useful for identification of new therapeutic agents for human disease.

3. METHODS FOR VISUALIZING VESSELS IN ZEBRAFISH

The functional tools available in the fish are powerful, but the accessibility of zebrafish embryos and larvae to high-resolution optical imaging has been perhaps the most useful feature of this model. The optical advantages of the fish have been exploited in many studies of vascular development, and a variety of useful methods have been developed to visualize vessels, some of which we describe in more detail below.

A number of methods are available for visualizing vessels in fixed specimens, including endogenous alkaline phosphatase (AP) staining, immuno-histochemistry, and in situ hybridization. These methods are all relatively rapid and can provide data on a large number of embryos at the same time. Each of these methods has been well documented elsewhere (Kamei *et al.*, 2004; Macdonald, 1999; Serbedzija *et al.*, 1999; Thisse *et al.*, 2004; Westerfield, 2000). AP staining relies on the embryos' high level of endogenous alkaline phosphatase activity to permit reasonably specific staining of blood vessels and is very simple and easy to perform. However, AP staining is not completely specific for endothelial cells and has a very high background after approximately 3 dpf (Chan and Serluca, 2004). *In situ* hybridization and immunohistochemistry generally provide better and more specific staining of vessels. Some commonly used probes for visualizing vessels in the zebrafish by *in situ* hybridization can be found in Table 4.1. Whole-mount staining methods can be carried out on animals up to approximately 7 dpf, after which penetration by probes or antibodies becomes very difficult. Whole-mount staining (Fig. 4.3) is useful for showing the spatial relation of marked tissues relative to the animal as a whole (Thisse *et al.*, 2004). Later-stage staining can be performed on sectioned tissue, which also provides a higher-resolution closer view at earlier stages (Fouquet *et al.*, 1997). Multiple *in situ* hybridization probes or antibodies can also be used together for comparison of expression patterns (Jowett, 2001; Thisse *et al.*, 2004). Fluorescent *in situ* hybridization has also been used to visualize multiple RNA probes at once. Protocols are described elsewhere, but results to date have been of variable quality (Clay and Ramakrishnan, 2005; Trinh le *et al.*, 2007).

Although staining methods on fixed specimens provide very useful information, particularly those methods that allow visualization of the expression of particular genes, the optical clarity of zebrafish embryos and larvae is probably most usefully harnessed for fluorescence imaging of the

Table 4.1 Selected *in situ* probes commonly used in vascular research

Marker genes	Expression pattern	Reference
fli1	Pan-endothelial	Brown *et al.*, 2000
tie2	Pan-endothelial	Lyons *et al.*, 1998
scl	Hematopoietic	Gering *et al.*, 1998
VE-cadherin (cdh5)	Pan-endothelial	Larson *et al.*, 2004
flk1 (kdr)	Pan-endothelial	Sumoy *et al.*, 1997
Flt4	Initially pan-endothelial, later restricted to vein only	Thompson *et al.*, 1998
efnb2	Artery only	Zhong *et al.*, 2000
grl	Artery only	Zhong *et al.*, 2000
notch5	Artery only	Lawson *et al.*, 2001
ephb4	Vein only	Lawson *et al.*, 2001

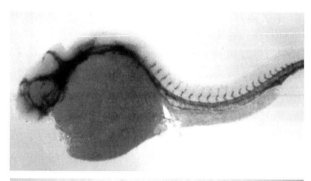

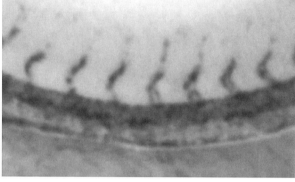

Figure 4.3 *In situ* hybridization. Whole-mount *in situ* hybridization performed on a 24-hpf embryo using a probe for the fli1 gene. (A) Lower magnification later view of a whole animal. (B) Close-up of the trunk. (From Lawson, N. D., and Weinstein, B. M. (2002b). *In vivo* imaging of embryonic vascular development using transgenic zebrafish. *Dev. Biol.* 248, 307–318.)

vasculature in living animals. In the rest of this review, we describe methods for confocal and two-photon imaging of animals whose vessels or endothelial cells have been fluorescently "tagged" by microangiographic injections or transgene expression.

3.1. Preparing zebrafish for fluorescence imaging

In order to image living zebrafish embryos and larvae, they must be prepared and then held immobile for imaging on a microscope stage without killing or damaging them. Pigment autofluorescence can hinder fluorescent imaging. 1-phenyl-2-thiourea (PTU) 0.3 mg/ml can be added to the egg water to reduce the pigment in the embryo, but one must ensure that it does not cause developmental delay or sickness in the embryo (Karlsson *et al.*, 2001; Westerfield, 2000). If the area to be imaged is unobstructed by pigment, or the embryo is to be grown to adulthood, PTU may be omitted or a reduced concentration should be considered.

In addition, a number of pigment-reduced lines of fish can potentially be used (albino [Streisinger *et al.*, 1986], roy [Ren *et al.*, 2002], and nacre [Lister *et al.*, 1999]).

The embryo to be imaged should be removed from its chorion (eggshell) if it has not yet hatched and anesthetized. The embryo can be anesthetized in 3-amino benzoic acid ethylester (tricaine, 0.003%) in egg water so that it will remain still for imaging. This works well for short-term imaging, but for long time-lapse imaging runs, it is best to avoid the use of tricaine, which will eventually slow the heart and can lead to death. As an alternative, chrna1 mutant animals with a deficiency of the nicotinic receptor can be used (Westerfield *et al.*, 1990). Embryos homozygous for the mutation have no skeletal muscle movement, but the heart and circulation are normal and the mutants can be imaged without the use of tricaine, although they do not survive longer than approximately a week.

Animals can be mounted with methylcellulose for short-term imaging at lower magnifications, but for time-lapse imaging or very high magnification imaging, mounting in agarose is recommended.

3.1.1. Protocol for methycellulose mounting

1. Prepare a solution of 6 to 9% methylcellulose in 30% Danieu solution (or egg water). This can be stored at 4 °C for many months.
2. Drop a small amount of methylcellulose, approximately 200 μl, onto a depression slide.
3. Draw the embryo into a pipette and place on the pool of methylcellulose.
4. The embryo can be immersed using a fine brush in the desired orientation and gently pushed to the bottom of the pool.
5. Dispense a few hundred microliters of tricaine–egg water on top of the methylcellulose. As long as the methylcellulose stays moist, the embryo will survive this mounting for several hours.
6. Removal of the embryo can be achieved by soaking the mounting in embryo medium to soften it before removing with a brush and/or pipette.

3.1.2. Protocol for agarose mounting

1. Prepare 1 to 2% low-melting temperature agarose in 30% Danieu solution or egg water.
2. Let the hot agarose cool, but while still molten, fill the depression in a thick depression slide.
3. Two methods can be used to mount the embryo:
   Method 1: The embryo can be implanted into the agarose using a pipette directly and then quickly oriented with a brush and left to cool.
   Method 2: The heat from the agarose could damage a sensitive embryo; alternatively, the agarose can be left to cool completely first.

- Cut a trench in the agarose just larger than the embryo in the orientation the embryo will eventually set using forceps (Dumont #5 or #55).
- Place the embryo into the trench and remove excess water.
- With a pipette, gently drop a thin layer of cool yet molten agarose over the embryo.

4. In any agarose mounting, the excess agar over the area to be imaged must be removed for clear fluorescence images. This can be achieved with forceps or a fine brush by gently flaking off bits of agarose.

5. The embryo's health will be improved during imaging by removing as much agar as possible, leaving the animal held by only small bridges or wedges of agar (Fig. 4.4).

6. Pipette more egg water with tricaine over the embryo for imaging. This mounting can be used up to a few days as long as fresh water is available to the embryo.

7. The embryo can be dismounted by carving away sections of the agarose with an eyelash brush, forceps, or other delicate tool, being careful not to let the shards of agarose crush the embryo as they dislodge.

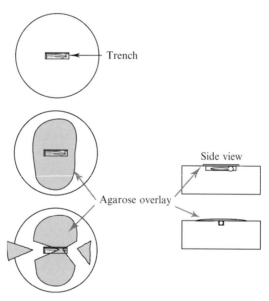

Figure 4.4 Mounting an embryo in agarose. A well in a depression slide or confocal imaging chamber is filled with molten low-melt agarose. A trench is cut large enough to accommodate the embryo. Additional molten agarose is dropped over the embryo and left to cool, and then the overlay agarose is cut away leaving only a bridge across the embryo. (Modified from Kamei, M., Isogai, S., and Weinstein, B. (2004). Imaging blood vessels in the zebrafish. *Methods Cell Biol.* **76,** 51–74.)

Most effective high-resolution images of fluorescent animals will be carried out using a confocal (or preferentially two-photon, if available) microscope. A number of factors must be considered in carrying out confocal imaging.

3.1.3. Considerations for confocal imaging

- The maximum penetration depth is roughly 250 μm.
- Magnification of 10 to 100× can be used.
- The spacing between planes can be from 0.5 μm to 10 μm for reasonable quality images.
- If the images are to be rotated afterward, imaging software performs better with closer spaces between planes; however, this will increase the time the embryo is under the laser and the probability of photo-bleaching or photo-damage.
- If possible, adjust laser power during stack collection with higher power for deep image planes and lower power for shallow planes.
- Images may be frame-averaged up to about five times, depending on the brightness of the fluorescence, quality of the optics, detectors, and lasers to reduce noise.
- Commercial confocal microscopes come with software to analyze the data once imaging is complete, but this software can sometimes be limiting. There are many image-processing software packages available to suitably analyze the data (MetaMorph, ImageJ, Imaris software, Adobe Photoshop).

3.2. Microangiography

Dye injection has been extensively used to visualize lumenized blood vessels in vertebrates since the early 20th century (Evans, 1910; Sabin, 1917). A confocal microangiography method developed for the zebrafish has coupled this useful classical method with modern, high-resolution optical imaging technologies (Weinstein *et al.*, 1995). A number of fluorescent particles are available for imaging patent blood vessels. Small fluorescent dyes such as the AlexaFluor® probes or fluorescent dextran (Invitrogen) can be used but leak quickly from the embryonic zebrafish bloodstream and images must be collected very quickly after injection. Alternatively, quantum dots® or fluorescent microspheres in various colors and sizes (0.02 to 0.04 μm are useful) can be obtained. Whatever dye or particle used, keep in mind that the red blood cells are approximately 7 μm, so any particle must be at least significantly smaller to avoid the risk of blocking the heart or vessels. Depending on the imaging capabilities and the application, blue, red, orange, or yellow-green fluorescence can be chosen. Fish older than 1 dpf can be used in angiography experiments. Once in the blood vessels, fluorescent particles will remain there for approximately 40 min before either leaking out of the vessels or being phagocytosed into cells lining the vessels (Westerfield, 2000). Animals are imaged using either confocal or two-photon microscopy, and the collected image stacks can be digitally reconstructed into very high-resolution, three-dimensional representations of patent vascular spaces of lumenized vessels anywhere in the small and transparent developing zebrafish. Confocal microangiography was used to create a vascular atlas of the zebrafish (Fig. 4.5) (Isogai et al., 2001). Recently, a similar method was also developed to study lymphatic vessels (Yaniv et al., 2006).

3.2.1. Materials needed

- Glass capillaries, 1-mm outer diameter (World Precision Instruments)
- Microloader pipette tips (Eppendorf)
- Micropipette puller (Stutter instruments)
- Pneumatic pump with needle holder and micromanipulator (World Precision instruments)
- Dye or beads to be injected (Invitrogen)
- Dissecting microscope with epifluorescence capabilities with the appropriate excitation light and filters for your chosen injection solution

3.2.2. Protocol

1. Prepare needles from capillaries with or without internal filaments using a glass pipette puller (either vertical or horizontal) similar to what is used for DNA injection (Fig. 4.2A or B).
2. Break needle tip with forceps (Dumont #55) to make an opening of less than 5 μm.
3. Examine the needle for a beveled shape on the tip after breaking with forceps. Blunt-ended needles will not easily insert into the embryo, causing damage to the vessels and poor angiography. A needle beveler and microforge can also be used instead of simply clipping the glass for a smooth, narrow tip.

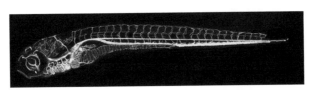

Figure 4.5 Microangiography reveals functional zebrafish vasculature. Confocal image of a 7-dpf larva that has been injected intravascularly with fluorescent microspheres. (From Isogai, S., Horiguchi, M., and Weinstein, B. M. (2001). The vascular anatomy of the developing zebrafish: An atlas of embryonic and early larval development. *Dev. Biol.* **230**, 278–301.)

4. Suspend fluorescent molecules or spheres in 1× Danieu solution and back-fill the glass needle with 5 to10 μl with a microloader pipette tip. Sonicate the solution before filling for fewer aggregates

5. Insert the needle into the pipette holder attached to a pneumatic pump, and calibrate the needle following the RNA injection protocols.

6. Place the embryo in tricaine egg water. Once anesthetized, mount the embryo with the area to be injected facing upward in 0.5% low-melt agarose or methylcellulose. If mounted in agarose, clear the area to be injected of agarose or it may clog the needle when inserted.

7. Place mounted embryo under a dissecting microscope oriented such that the needle is pointed along the axis of the vessel to be injected.

8. To inject the solution, insert microneedle obliquely into the sinus venosus, cardiac atrium, otic vesicle or dorsal aorta depending on the application. It is essential to use a shallow angle of needle insertion to not puncture the vessel.

9. Using the pump, deliver 1-nl dye or bead suspension into the embryo in three or four doses over the course of a minute. Delivery of large volumes at once can cause cardiac arrest.

10. Carefully remove the embryo from the mount and place it in tricaine-free, warmed media to recover for a few minutes before mounting in methylcellulose or low-melt agarose for imaging. During the recovery period, the dye will spread uniformly through the embryo and the injection site in the skin will close.

11. Injected animals can be examined on a microscope using the appropriate excitation light and filters for the fluorescent particles.

For 1- to 3-dpf embryos, the sinus venosus is the most accessible region for angiography. For older embryos, injection directly into the heart or caudal vein may be the easiest. The injection site will often have fluorescent particles extravasated near it so do not inject directly into the area you wish to image. Fluorescent particles, which initially will be well distributed throughout the vasculature, slowly begin to be phagocytosed by venous vessels. Within an hour, the vasculature of injected animals can no longer be clearly imaged. The animal can be imaged and photographed using a dissecting microscope, compound microscope, or confocal or two-photon microscope. Red fluorescent quantum dots are useful for two-photon imaging in conjunction with green fluorescent transgenic animals since the dots can be excited with the same two-photon laser line.

3.3. Cell tracing

Lineage tracing has been used extensively in many model systems to track the movement of a set of cells through development (Kissa *et al.*, 2008; Mills *et al.*, 1999; Stern and Fraser, 2001; Woo *et al.*, 1995). A subset of cells can be marked by (1) injecting fluorescent dextran into just a few cells; (2) injecting an embryo with photoconvertible fluorescent molecules at the one-cell stage and photoactivating a small subset of cells at a later stage of development; and (3) using a transgenic animal expressing a photoconvertible fluorescent protein, photoactivating a small subset of cells at the desired stage. Injected fluorescent dextran has been used to make a fate map of the zebrafish nervous system from the shield stage onward (Woo *et al.*, 1995). Caged-fluorescent molecules have been used to study the developing lateral plate mesoderm and the rise of the intersegmental vessels (Childs *et al.*, 2002; Zhong *et al.*, 2001). Transgenic embryos expressing a photoconvertible protein like kaede or dronpa (Hatta *et al.*, 2006) have great potential for studying endothelial migration in zebrafish, but have not yet been fully explored. For photoactivation of molecules and proteins, a high-power microscope must be used and the embryo mounted in methylcellulose or agarose to achieve precise spatial control. The power from a mercury lamp on an epifluorescence microscope may be enough to convert some of the superficial molecules, but for reliable conversion, a high-power laser light is necessary. The imaging is most often performed on a one or two-photon confocal microscope and can be continued as long as the dye can be visualized.

3.4. Imaging transgenic zebrafish

Confocal and two-photon imaging use all of the advantages of the zebrafish, providing very clear pictures of the developing vascular system. A fluorescent protein can be localized very specifically to the cell linage of interest by expressing it under the control of a specific tissue promoter such as the fli promoter for endothelial cells. While angiography only permits visualization of already lumenized vessels, transgenic animals allow visualization of progenitor cells, isolated or migrating cells, and nonlumenized vessels long before the vessels begin to carry blood flow. As noted above, the two techniques can also be used together (Kamei *et al.*, 2006). Furthermore, several different transgenes can be combined to generate fish with several tissue types expressing different colors of fluorescent proteins. Although transgenic lines require a great deal and effort and time to generate, there are fortunately a number of established lines of well-studied transgenic fish expressing fluorescent proteins already available for vascular studies (Table 4.2). The most widely used line is the *Tg(fli1-EGFP)^{y1}* expressing eGFP in all endothelial cells and some neural crest–derived cells (Lawson and Weinstein, 2002b). The *Tg(fli1-EGFP)^{y1}* line has been used to examine blood vessels in both embryonic and adult fish in a variety of studies including vascular development (Fig. 4.6) (Lawson *et al.*, 2002; Lawson and Weinstein, 2002b), lymphanogenesis (Yaniv *et al.*, 2006),

Table 4.2 Some useful transgenic fish lines for vascular research

Name of line	Expression pattern	Reference
$Tg(fli1\text{-}eGFP)^{y1}$	Pan-endothelial plus neural crest	Lawson et al., 2002
$Tg(kdr\text{-}eGFP)$	Pan-endothelial	Cross et al., 2003
$Tg(kdr\text{-}RFP)$	Pan-endothelial	Huang et al., 2005
$Tg(gata1:GFP)^{la781}$	Red blood cells	Sumanas et al., 2005
$Tg(gata1:dsRed)^{sd2}$	Red blood cells	Traver et al., 2003
$Tg(Tie2:EGFP)^{s849}$	Pan-endothelial, weak eGFP expression	Motoike et al., 2000
$Tg(lmo2:dsRed)$ and $Tg(lmo2:eGFP)$	Hematopoietic stem cell	Zhu et al., 2005
$Tg(fli1:GFP\text{-}cdc42wt)^{y48}$	Pan-endothelial plus neural crest, membrane localized expression	Kamei et al., 2005
$Tg(fli1:negfp)^{y7}$	Pan-endothelial plus neural crest, nucleus localized fluorescence	Roman et al., 2002

cell signaling and differentiation (Siekmann and Lawson, 2007), tumor angiogenesis (Stoletov et al., 2007), and chemical mutagenesis screens (Buchner et al., 2007; Peterson et al., 2004; Torres-Vazquez et al., 2004).

3.5. Time-lapse imaging

One of the most important advantages of using living, transgenic animals to image blood vessels, instead of fixed and stained animals, is that the use of transgenics permits repeated imaging of the same animal at different time points. The ability to observe the dynamics of cell and tissue behavior in a living animal is an extremely powerful tool for understanding developmental processes, and one that has been used to great effect in the optically clear zebrafish. Time-lapse two-photon imaging has been used to examine how vascular networks assemble in the developing zebrafish trunk (Isogai et al., 2003) and how the patterning of these networks is directed (Torres-Vazquez et al., 2004). Time-lapse imaging has also been used to "lineage trace" the origins of cells contributing to the formation of the lymphatic vascular system in the zebrafish (Yaniv et al., 2006) and to examine the regulation of endothelial cell behavior by notch signaling (Siekmann and Lawson, 2007). Two-photon imaging is preferred for time-lapse imaging, particularly for very-long-term imaging experiments. The benefits of multiphoton confocal imaging have been reviewed elsewhere (Kamei et al., 2004). Briefly, two-photon imaging results in less photobleaching of

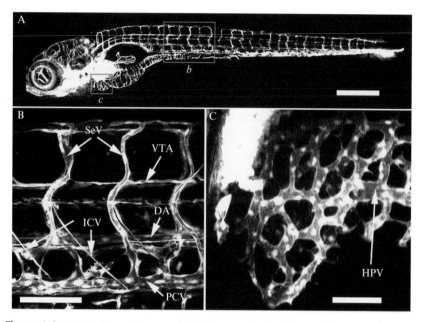

Figure 4.6 Multiphoton images of eGFP expression in live $Tg(fli1\text{-}EGFP)^{y1}$, albino transgenic fish. (A) A 7-dpf larva, compiled from five separate image reconstructions. Boxed areas labeled "B" and "C" indicate regions shown in (B) and (C), respectively. (B) Image of trunk vessels at 7 dpf showing patent vessels such as the dorsal aorta (DA), posterior cardinal vein (PCV), and intersegmental vessels (SeV), as well as vessels that do not carry blood flow at this time such as the vertebral arteries (VTA) and intercostal vessels (ICV). (C) Hepatic (liver) vasculature in a 5-dpf larva. The hepatic portal vein (HPV) is noted (arrow). Scale bar: (A) 500 μm; (B) 100 μm; (C) 50 μm. (From Lawson, N. D., and Weinstein, B. M. (2002b). *In vivo* imaging of embryonic vascular development using transgenic zebrafish. *Dev. Biol.* **248**, 307–318.)

fluorescent proteins with less photodamage to embryonic tissues, and also generally gives better spatial resolution in the z-axis. Two disadvantages to multiphoton imaging of developing zebrafish are that certain pigment cells reflect or fluoresce at two-photon wavelengths and that many standard microscopes and optics are not optimized for the near-infrared, two-photon laser light. In the sections below, we describe preparation of a flow chamber for time-lapse imaging, mounting of the animal, and procedures for actually performing the imaging.

3.5.1. Time-lapse imaging chamber

We have previously reported methods for construction of a flow chamber for long-term time-lapse imaging (Kamei and Weinstein, 2005). The embryo can be mounted in this chamber for long-term confocal imaging over many days. The chamber allows for heated, aerated water to flow over the embryo, continuously providing a hospitable environment for the developing embryo while it is being imaged (Fig. 4.7). We provide a somewhat abbreviated version of these methods below.

Materials needed

- 30-mm angled cell culture flask (Nunc)
- Luer bulkheads (3/32", barbed), silicone O-rings (1/4" ID × 3/8" OD × 1/16" wide), and locking nuts (World Precision Instruments)

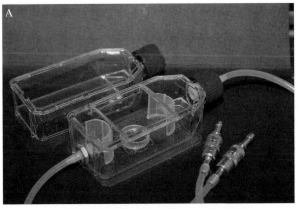

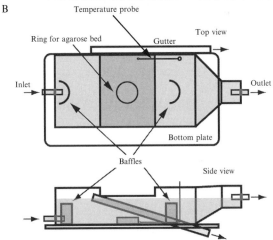

Figure 4.7 Imaging chamber construction. (A) Tissue culture flask before and after modification pictured with silicone tubing. (B) Diagram of top and side views of flask with water level indicated by shading. (From Kamei, M., and Weinstein, B. M. (2005). Long-term time-lapse fluorescence imaging of developing zebrafish. *Zebrafish* 2, 113–123.)

- 15-ml cell culture tube (Falcon)
- 5-min epoxy resin
- Peristaltic pump. Either two medium-speed, variable-speed pumps (Fisher Scientific), or one multichannel pump (World Precision Instruments)
- Silicon tubing (3/32" ID × 5/32" OD; Cole-Palmer)
- Variable temperature water bath large enough to hold a 1-l bottle
- Aquarium air pump (pet and aquarium stores)
- Digital thermometer with wire probes (Fisher Scientific)
- Plastruct 1-mm clear plastic sheet cut to the size of the depression in the confocal microscope stage
- Dremel tool, or other small electric drill and saw (hardware store).
- Upright confocal microscope—this design is for an upright, not inverted, microscope.

Construction of chamber

1. The top of the chamber must be removed for the microscope objective to easily fit over the embryo. Cut an approximately 1.5" square from the top of the flask using either a Dremel tool or a saw.
2. Drill an outlet hole for the water in the flask with an electric drill to 3/32" diameter in the side of the flask. Cut an inlet hole into the cap of the flask the same way.
3. Fit O-rings over the female luers, and then insert into these holes such that the O-ring is snugly between the flask and the bulkhead of the luer.
4. The luer can then be secured with a nut on the outside of the flask.
5. A small amount of epoxy can be applied to the joint between the luer and the flask if the chamber leaks.
6. To prevent agarose or other debris from clogging hoses, cut a small amount of netting and secure over the outlet drain with a locking nut.
7. For emergency overflows, install a drain on the flask. Cut one side of the chamber at the top shorter than the other side by approximately 4 mm.
8. Secure a lengthwise cut plastic straw to this side of the flask to direct overflow water into a container below the microscope.
9. Alternatively, if a multichannel pump is being used, a third port can be drilled into the side of the chamber near the top using luers and O-rings similar to the other ports. The pump can be set to constantly draw from this port, and in the event of water climbing to the top, two outlets will then be available.
10. Cut the 15-ml culture tube in cross-sections, creating an approximately 15-mm ring using either a razor blade or saw. This will be the holding chamber for the embryo.
11. Cut the cap of the falcon tube to make two semicircular baffles.
12. Apply epoxy to the falcon tube pieces and secure inside the flask. Allow the chamber to air dry for several hours to overnight.

13. Test for leaks in the chamber before preparing the embryo for imaging.
14. If the flask does not fit securely in the stage of the microscope, plastruct, a portion of a pipette tip box, or another type of sheet plastic can be cut to fit the microscope stage with a razor blade or an electric saw and then fastened to the flask with epoxy. A custom stage mount can also be made by a machine shop to fit the flask and inlet/outlet hoses.

3.5.2. Mounting the embryo

Embryos or larvae must be mounted in agarose for time-lapse imaging. The mounting procedure is essentially the same as that described in Section 3.1.2 and Fig. 4.4.

Protocol

1. Prepare 2% low-melting-temperature agarose in 30% Danieu solution or egg water.
2. Fill the polypropylene ring with the agarose to the top.
3. For time-lapse imaging, a trench must be cut in the agarose using forceps (Dumont #5 or #55) to hold the embryo. The trench should allow enough room for the development of the embryo for the duration of the time-lapse images.
4. It should be cut carefully to hold the embryo in the orientation for imaging as closely as possible with the head facing the inflow of water.
5. Place the embryo into the trench using a pipette, and then remove excess water.
6. Using 1.5% molten, low-melt agarose, carefully drop agarose over the embryo, filling in the trench and covering the embryo to a shallow depth.
7. An eyelash brush can be used to make small adjustments to the orientation of the embryo before the agarose sets.
8. After it has cooled completely, remove as much agarose as possible over the head and tail of the embryo, leaving just a bridge covering the trunk.

There must be ample free space around the embryo for it to grow during the time-lapse imaging. Failure to remove enough agarose can lead to not only distortions in the morphology, but necrosis or death. However, failure to provide a secure bridge or wedges holding the embryo in place may result in the embryo dislodging from the mount. For longer time-lapse runs, the embryo may need to be remounted, as it will grow too large for the trench and/or agar holding it in place. Also, beyond 5 days, the embryo may need to be removed briefly and allowed to rest in fresh water to swim and/or eat to stay healthy.

3.5.3. Imaging

Once an animal is properly mounted in the imaging chamber, the rest of the preparations prior to imaging are relatively straightforward.

Protocol

1. Warm approximately 500 ml of egg water (with tricaine and/or PTU as needed) in a water bath. Methylene blue will stain the tubing.
2. Place the chamber with mounted embryo onto the stage carefully without dislodging the embryo.
3. Once the chamber is placed onto the microscope stage, connect the tubing to the inlet and outlets of the flask, through the pump and into the bottle of egg water.
4. Adjust the outflow of water to be slightly faster than the inflow of water to avoid overfilling the chamber.
5. Allow the pump to fill the flask and measure the temperature inside the flask near the embryo. Adjust the water bath so that the temperature near the embryo is roughly 28 °C. Depending on the ambient temperature in the room and speed of water flow, the water bath temperature may need to be raised up to as high as 50 °C and foam or cloth insulation can be wrapped around the tubing.
6. Place a tube from the aquarium filter into the bottle of egg water to keep it aerated.
7. Cover the top of the bottle and tubes with parafilm to prevent evaporation.
8. The bottle of water can be changed as often as needed but will last for a few days.

Two-photon time-lapse imaging is performed much the same as one-photon confocal imaging, but a few additional considerations need to be taken.

- The wavelength of maximum two-photon excitation is twice the standard one-photon excitation maximum for a given fluorophore.
- Microscope objectives must be capable of transmitting long wavelength light.
- If possible, an adjustable laser power setting is desirable to use higher power for deep images, and lower power for shallow images.
- The spacing between time points can range from 1 min to 1 h, depending on the application and the speed at which the cells move.
- The time lapse can continue for up to a week but the embryo should be checked periodically for shifting, photobleaching, damage, or sickness. Major movements, sickness, or morphological problems as a result of the mounting may require a new embryo to be mounted.

- The stage can be adjusted during the experiment to correct for shifting, or the data can be digitally shifted at the end if the movement is minor.
- To obtain images with less background noise, up to five frames may be averaged.
- A motorized stage (ProScanII, Prior Scientific) accessible through the microscope software will permit imaging of multiple spatial points in the x–y plane during the time lapse. Therefore, multiple embryos or different areas on a single embryo can be imaged in a single experiment.
- Keep in mind that long time-lapse experiments with large Z-stacks at each time point generate a very large amount of data. Make sure that your imaging system has adequate RAM and hard drive space to permit collection and storage of the necessary data.

4. CONCLUSION

The zebrafish model has come to play an increasingly important role in understanding mechanisms of vasculogenesis and angiogenesis during development. The tools available for functional manipulation and high-resolution imaging of the vasculature in this vertebrate organism make it uniquely well suited for studying this diffuse and complex tissue. Studies carried out to date have revealed extensive conservation of anatomical form and molecular mechanisms between the vasculature in the zebrafish and other vertebrates. Many of these studies have already had a far-reaching impact in angiogenesis research, uncovering underlying mechanisms of endothelial specification, arterial–venous and lymphatic differentiation, and patterning and remodeling of the vasculature. Since many of the molecules that play key roles in developing vessels carry out analogous functions during postnatal angiogenesis, it is probable that the zebrafish will continue to yield insights important in both basic and clinical vascular research.

ACKNOWLEDGMENT

This work was supported by the intramural program of the National Institute of Child Health and Human Development.

REFERENCES

Alvarez, Y., Cederlund, M. L., Cottell, D. C., Bill, B. R., Ekker, S. C., Torres-Vazquez, J., Weinstein, B. M., Hyde, D. R., Vihtelic, T. S., and Kennedy, B. N. (2007). Genetic determinants of hyaloid and retinal vasculature in zebrafish. *BMC Dev. Biol.* **7,** 114.
Amatruda, J. F., Shepard, J. L., Stern, H. M., and Zon, L. I. (2002). Zebrafish as a cancer model system. *Cancer Cell* **1,** 229–231.

Amsterdam, A., Burgess, S., Golling, G., Chen, W., Sun, Z., Townsend, K., Farrington, S., Haldi, M., and Hopkins, N. (1999). A large-scale insertional mutagenesis screen in zebrafish. *Genes Dev.* **13**, 2713–2724.

Amsterdam, A., and Hopkins, N. (2006). Mutagenesis strategies in zebrafish for identifying genes involved in development and disease. *Trends Genet.* **22**, 473–478.

Bahary, N., Davidson, A., Ransom, D., Shepard, J., Stern, H., Trede, N., Zhou, Y., Barut, B., and Zon, L. I. (2004). The Zon laboratory guide to positional cloning in zebrafish. *Methods Cell Biol.* **77**, 305–329.

Ballinger, D. G., and Benzer, S. (1989). Targeted gene mutations in Drosophila. *Proc. Natl. Acad. Sci. U.S.A.* **86**, 9402–9406.

Bayliss, P. E., Bellavance, K. L., Whitehead, G. G., Abrams, J. M., Aegerter, S., Robbins, H. S., Cowan, D. B., Keating, M. T., O'Reilly, T., Wood, J. M., Roberts, T. M., and Chan, J. (2006). Chemical modulation of receptor signaling inhibits regenerative angiogenesis in adult zebrafish. *Nat. Chem. Biol.* **2**, 265–273.

Brown, L. A., Rodaway, A. R., Schilling, T. F., Jowett, T., Ingham, P. W., Patient, R. K., and Sharrocks, A. D. (2000). Insights into early vasculogenesis revealed by expression of the ETS-domain transcription factor Fli-1 in wild-type and mutant zebrafish embryos. *Mech. Dev.* **90**, 237–252.

Buchner, D. A., Su, F., Yamaoka, J. S., Kamei, M., Shavit, J. A., Barthel, L. K., McGee, B., Amigo, J. D., Kim, S., Hanosh, A. W., Jagadeeswaran, P., Goldman, D., *et al.* (2007). pak2a mutations cause cerebral hemorrhage in redhead zebrafish. *Proc. Natl. Acad. Sci. U.S.A.* **104**, 13996–14001.

Cerda, G. A., Thomas, J. E., Allende, M. L., Karlstrom, R. O., and Palma, V. (2006). Electroporation of DNA, RNA, and morpholinos into zebrafish embryos. *Methods* **39**, 207–211.

Chan, J., Bayliss, P. E., Wood, J. M., and Roberts, T. M. (2002). Dissection of angiogenic signaling in zebrafish using a chemical genetic approach. *Cancer Cell* **1**, 257–267.

Chan, J., and Serluca, F. C. (2004). Chemical approaches to angiogenesis. *Methods Cell Biol.* **76**, 475–487.

Chen, E., and Ekker, S. C. (2004). Zebrafish as a genomics research model. *Curr. Pharm. Biotechnol.* **5**, 409–413.

Childs, S., Chen, J. N., Garrity, D. M., and Fishman, M. C. (2002). Patterning of angiogenesis in the zebrafish embryo. *Development* **129**, 973–982.

Clay, H., and Ramakrishnan, L. (2005). Multiplex fluorescent *in situ* hybridization in zebrafish embryos using tyramide signal amplification. *Zebrafish* **2**, 105–111.

Corey, D. R., and Abrams, J. M. (2001). Morpholino antisense oligonucleotides: Tools for investigating vertebrate development. *Genome Biol.* **2**, reviews 1015.1–1015.3.

Covassin, L. D., Villefranc, J. A., Kacergis, M. C., Weinstein, B. M., and Lawson, N. D. (2006). Distinct genetic interactions between multiple Vegf receptors are required for development of different blood vessel types in zebrafish. *Proc. Natl. Acad. Sci. U.S.A.* **103**, 6554–6559.

Cross, L. M., Cook, M. A., Lin, S., Chen, J. N., and Rubinstein, A. L. (2003). Rapid analysis of angiogenesis drugs in a live fluorescent zebrafish assay. *Arterioscler. Thromb. Vasc. Biol.* **23**, 911–912.

Das, L., and Martienssen, R. (1995). Site-selected transposon mutagenesis at the hcf106 locus in maize. *Plant Cell* **7**, 287–294.

Davidson, A. E., Balciunas, D., Mohn, D., Shaffer, J., Hermanson, S., Sivasubbu, S., Cliff, M. P., Hackett, P. B., and Ekker, S. C. (2003). Efficient gene delivery and gene expression in zebrafish using the Sleeping Beauty transposon. *Dev. Biol.* **263**, 191–202.

Davidson, A. J., and Zon, L. I. (2004). The "definitive" (and "primitive") guide to zebrafish hematopoiesis. *Oncogene* **23**, 7233–7246.

Draper, B. W., Morcos, P. A., and Kimmel, C. B. (2001). Inhibition of zebrafish fgf8 pre-mRNA splicing with morpholino oligos: A quantifiable method for gene knock-down. *Genesis* **30,** 154–156.

Driever, W., Solnica-Krezel, L., Schier, A. F., Neuhauss, S. C., Malicki, J., Stemple, D. L., Stainier, D. Y., Zwartkruis, F., Abdelilah, S., Rangini, Z., Belak, J., and Boggs, C. (1996). A genetic screen for mutations affecting embryogenesis in zebrafish. *Development* **123,** 37–46.

Du, S. J., Gao, J., and Anyangwe, V. (2003). Muscle-specific expression of myogenin in zebrafish embryos is controlled by multiple regulatory elements in the promoter. *Comp. Biochem. Physiol. B Biochem. Mol. Biol.* **134,** 123–134.

Ekker, S. C. (2000). Morphants: A new systematic vertebrate functional genomics approach. *Yeast* **17,** 302–306.

Evans, H. (1910). "Manual of Human Embryology." J. B. Lippincott & Co, Philadelphia.

Fisher, S., Grice, E. A., Vinton, R. M., Bessling, S. L., Urasaki, A., Kawakami, K., and McCallion, A. S. (2006). Evaluating the biological relevance of putative enhancers using Tol2 transposon-mediated transgenesis in zebrafish. *Nat. Protoc.* **1,** 1297–1305.

Fouquet, B., Weinstein, B. M., Serluca, F. C., and Fishman, M. C. (1997). Vessel patterning in the embryo of the zebrafish: Guidance by notochord. *Dev. Biol.* **183,** 37–48.

Gering, M., Rodaway, A. R., Gottgens, B., Patient, R. K., and Green, A. R. (1998). The SCL gene specifies haemangioblast development from early mesoderm. *EMBO J.* **17,** 4029–4045.

Haffter, P., Granato, M., Brand, M., Mullins, M. C., Hammerschmidt, M., Kane, D. A., Odenthal, J., van Eeden, F. J., Jiang, Y. J., Heisenberg, C. P., Kelsh, R. N., Furutani-Seiki, M., *et al.* (1996). The identification of genes with unique and essential functions in the development of the zebrafish, *Danio rerio. Development* **123,** 1–36.

Hatta, K., Tsujii, H., and Omura, T. (2006). Cell tracking using a photoconvertible fluorescent protein. *Nat. Protoc.* **1,** 960–967.

Hong, C. C., Peterson, Q. P., Hong, J. Y., and Peterson, R. T. (2006). Artery/vein specification is governed by opposing phosphatidylinositol-3 kinase and MAP kinase/ERK signaling. *Curr. Biol.* **16,** 1366–1372.

Huang, H., Zhang, B., Hartenstein, P. A., Chen, J. N., and Lin, S. (2005). NXT2 is required for embryonic heart development in zebrafish. *BMC Dev. Biol.* **5,** 7 (abstract).

Isogai, S., Horiguchi, M., and Weinstein, B. M. (2001). The vascular anatomy of the developing zebrafish: An atlas of embryonic and early larval development. *Dev. Biol.* **230,** 278–301.

Isogai, S., Lawson, N. D., Torrealday, S., Horiguchi, M., and Weinstein, B. M. (2003). Angiogenic network formation in the developing vertebrate trunk. *Development* **130,** 5281–5290.

Jowett, T. (2001). Double *in situ* hybridization techniques in zebrafish. *Methods* **23,** 345–358.

Kamei, M., Isogai, S., and Weinstein, B. (2004). Imaging blood vessels in the zebrafish. *Methods Cell Biol.* **76,** 51–74.

Kamei, M., Saunders, W. B., Bayless, K. J., Dye, L., Davis, G. E., and Weinstein, B. M. (2006). Endothelial tubes assemble from intracellular vacuoles *in vivo. Nature* **442,** 453–456.

Kamei, M., and Weinstein, B. M. (2005). Long-term time-lapse fluorescence imaging of developing zebrafish. *Zebrafish* **2,** 113–123.

Karlsson, J., von Hofsten, J., and Olsson, P. E. (2001). Generating transparent zebrafish: A refined method to improve detection of gene expression during embryonic development. *Mar. Biotechnol. (N. Y.)* **3,** 522–527.

Kawakami, K. (2007). Tol2: A versatile gene transfer vector in vertebrates. *Genome Biol.* **8,** (Suppl 1), S7.

Kawakami, K., Takeda, H., Kawakami, N., Kobayashi, M., Matsuda, N., and Mishina, M. (2004). A transposon-mediated gene trap approach identifies developmentally regulated genes in zebrafish. *Dev. Cell* **7,** 133–144.

Key, B., and Devine, C. A. (2003). Zebrafish as an experimental model: Strategies for developmental and molecular neurobiology studies. *Methods Cell Sci.* **25,** 1–6.

Kimmel, C. B., Ballard, W. W., Kimmel, S. R., Ullmann, B., and Schilling, T. F. (1995). Stages of embryonic development of the zebrafish. *Dev. Dyn.* **203,** 253–310.

Kissa, K., Murayama, E., Zapata, A., Cortes, A., Perret, E., Machu, C., and Herbomel, P. (2008). Live imaging of emerging hematopoietic stem cells and early thymus colonization. *Blood* **111,** 1147–1156.

Langenau, D. M., Traver, D., Ferrando, A. A., Kutok, J. L., Aster, J. C., Kanki, J. P., Lin, S., Prochownik, E., Trede, N. S., Zon, L. I., and Look, A. T. (2003). Myc-induced T cell leukemia in transgenic zebrafish. *Science* **299,** 887–890.

Larson, J. D., Wadman, S. A., Chen, E., Kerley, L., Clark, K. J., Eide, M., Lippert, S., Nasevicius, A., Ekker, S. C., Hackett, P. B., and Essner, J. J. (2004). Expression of VE-cadherin in zebrafish embryos: A new tool to evaluate vascular development. *Dev. Dyn.* **231,** 204–213.

Lawson, N. D., Scheer, N., Pham, V. N., Kim, C. H., Chitnis, A. B., Campos-Ortega, J. A., and Weinstein, B. M. (2001). Notch signaling is required for arterial-venous differentiation during embryonic vascular development. *Development* **128,** 3675–3683.

Lawson, N. D., Vogel, A. M., and Weinstein, B. M. (2002). Sonic hedgehog and vascular endothelial growth factor act upstream of the Notch pathway during arterial endothelial differentiation. *Dev. Cell* **3,** 127–136.

Lawson, N. D., Mugford, J. W., Diamond, B. A., and Weinstein, B. M. (2003). Phospholipase C gamma-1 is required downstream of vascular endothelial growth factor during arterial development. *Genes Dev.* **17,** 1346–1351.

Lawson, N. D., and Weinstein, B. M. (2002a). Arteries and veins: Making a difference with zebrafish. *Nat. Rev. Genet.* **3,** 674–682.

Lawson, N. D., and Weinstein, B. M. (2002b). *In vivo* imaging of embryonic vascular development using transgenic zebrafish. *Dev. Biol.* **248,** 307–318.

Le, X., Langenau, D. M., Keefe, M. D., Kutok, J. L., Neuberg, D. S., and Zon, L. I. (2007). Heat shock-inducible Cre/Lox approaches to induce diverse types of tumors and hyperplasia in transgenic zebrafish. *Proc. Natl. Acad. Sci. U.S.A.* **104,** 9410–9415.

Lee, L. M., Seftor, E. A., Bonde, G., Cornell, R. A., and Hendrix, M. J. (2005). The fate of human malignant melanoma cells transplanted into zebrafish embryos: Assessment of migration and cell division in the absence of tumor formation. *Dev. Dyn.* **233,** 1560–1570.

Lin, S. (2000). Transgenic zebrafish. *Methods Mol. Biol.* **136,** 375–383.

Lister, J. A., Robertson, C. P., Lepage, T., Johnson, S. L., and Raible, D. W. (1999). Nacre encodes a zebrafish microphthalmia-related protein that regulates neural-crest–derived pigment cell fate. *Development* **126,** 3757–3767.

Lyons, M. S., Bell, B., Stainier, D., and Peters, K. G. (1998). Isolation of the zebrafish homologues for the tie-1 and tie-2 endothelium-specific receptor tyrosine kinases. *Dev. Dyn.* **212,** 133–140.

Macdonald, R. (1999). Zebrafish immunohistochemistry. *Methods Mol. Biol.* **127,** 77–88.

MacRae, C. A., and Peterson, R. T. (2003). Zebrafish-based small molecule discovery. *Chem. Biol.* **10,** 901–908.

McCallum, C. M., Comai, L., Greene, E. A., and Henikoff, S. (2000). Targeted screening for induced mutations. *Nat. Biotechnol.* **18,** 455–457.

Mills, K. R., Kruep, D., and Saha, M. S. (1999). Elucidating the origins of the vascular system: A fate map of the vascular endothelial and red blood cell lineages in *Xenopus laevis*. *Dev. Biol.* **209,** 352–368.

Mizgireuv, I. V., and Revskoy, S. Y. (2006). Transplantable tumor lines generated in clonal zebrafish. *Cancer Res.* **66,** 3120–3125.

Mizuno, T., Shinya, M., and Takeda, H. (1999). Cell and tissue transplantation in zebrafish embryos. *Methods Mol. Biol.* **127,** 15–28.

Motoike, T., Loughna, S., Perens, E., Roman, B. L., Liao, W., Chau, T. C., Richardson, C. D., Kawate, T., Kuno, J., Weinstein, B. M., Stainier, D. Y., and Sato, T. N. (2000). Universal GFP reporter for the study of vascular development. *Genesis* **28,** 75–81.

Murphey, R. D., and Zon, L. I. (2006). Small molecule screening in the zebrafish. *Methods* **39,** 255–261.

Nasevicius, A., and Ekker, S. C. (2000). Effective targeted gene "knockdown" in zebrafish. *Nat. Genet* **26,** 216–220.

Nasevicius, A., Larson, J., and Ekker, S. C. (2000). Distinct requirements for zebrafish angiogenesis revealed by a VEGF-A morphant. *Yeast* **17,** 294–301.

Nicoli, S., and Presta, M. (2007). The zebrafish/tumor xenograft angiogenesis assay. *Nat. Protoc. Nat. Protoc.* **2,** 2918–2923.

Nicoli, S., Ribatti, D., Cotelli, F., and Presta, M. (2007). Mammalian tumor xenografts induce neovascularization in zebrafish embryos. *Cancer Res.* **67,** 2927–2931.

Oates, A. C., Bruce, A. E., and Ho, R. K. (2000). Too much interference: Injection of double-stranded RNA has nonspecific effects in the zebrafish embryo. *Dev. Biol.* **224,** 20–28.

Patton, E. E., and Zon, L. I. (2001). The art and design of genetic screens: Zebrafish. *Nat. Rev. Genet.* **2,** 956–966.

Peterson, R. T., Shaw, S. Y., Peterson, T. A., Milan, D. J., Zhong, T. P., Schreiber, S. L., MacRae, C. A., and Fishman, M. C. (2004). Chemical suppression of a genetic mutation in a zebrafish model of aortic coarctation. *Nat. Biotechnol.* **22,** 595–599.

Pham, V. N., Lawson, N. D., Mugford, J. W., Dye, L., Castranova, D., Lo, B., and Weinstein, B. M. (2007). Combinatorial function of ETS transcription factors in the developing vasculature. *Dev. Biol.* **303,** 772–783.

Pyati, U. J., Webb, A. E., and Kimelman, D. (2005). Transgenic zebrafish reveal stage-specific roles for Bmp signaling in ventral and posterior mesoderm development. *Development* **132,** 2333–2343.

Ren, J. Q., McCarthy, W. R., Zhang, H., Adolph, A. R., and Li, L. (2002). Behavioral visual responses of wild-type and hypopigmented zebrafish. *Vision Res.* **42,** 293–299.

Roman, B. L., Pham, V. N., Lawson, N. D., Kulik, M., Childs, S., Lekven, A. C., Garrity, D. M., Moon, R. T., Fishman, M. C., Lechleider, R. J., and Weinstein, B. M. (2002). Disruption of acvrl1 increases endothelial cell number in zebrafish cranial vessels. *Development* **129,** 3009–3019.

Sabin, F. R. (1917). Origin and development of the primitive vessels of the chick and of the pig. *Carnegie Contrib. Embryol.* **6,** 61–124.

Serbedzija, G. N., Flynn, E., and Willett, C. E. (1999). Zebrafish angiogenesis: A new model for drug screening. *Angiogenesis* **3,** 353–359.

Siekmann, A. F., and Lawson, N. D. (2007). Notch signalling limits angiogenic cell behaviour in developing zebrafish arteries. *Nature* **445,** 781–784.

Sood, R., English, M. A., Jones, M., Mullikin, J., Wang, D. M., Anderson, M., Wu, D., Chandrasekharappa, S. C., Yu, J., Zhang, J., and Paul Liu, P. (2006). Methods for reverse genetic screening in zebrafish by resequencing and TILLING. *Methods* **39,** 220–227.

Stainier, D. Y., Weinstein, B. M., Detrich, H. W., 3rd., Zon, L. I., and Fishman, M. C. (1995). Cloche, an early acting zebrafish gene, is required by both the endothelial and hematopoietic lineages. *Development* **121,** 3141–3150.

Stern, C. D., and Fraser, S. E. (2001). Tracing the lineage of tracing cell lineages. *Nat. Cell Biol.* **3,** E216–E218.

Stoletov, K., Montel, V., Lester, R. D., Gonias, S. L., and Klemke, R. (2007). High-resolution imaging of the dynamic tumor cell vascular interface in transparent zebrafish. *Proc. Natl. Acad. Sci. U.S.A.* **104,** 17406–17411.

Streisinger, G., Singer, F., Walker, C., Knauber, D., and Dower, N. (1986). Segregation analyses and gene-centromere distances in zebrafish. *Genetics* **112**, 311–319.

Stuart, G. W., McMurray, J. V., and Westerfield, M. (1988). Replication, integration and stable germ-line transmission of foreign sequences injected into early zebrafish embryos. *Development* **103**, 403–412.

Stuart, G. W., Vielkind, J. R., McMurray, J. V., and Westerfield, M. (1990). Stable lines of transgenic zebrafish exhibit reproducible patterns of transgene expression. *Development* **109**, 577–584.

Sumanas, S., Jorniak, T., and Lin, S. (2005). Identification of novel vascular endothelial-specific genes by the microarray analysis of the zebrafish cloche mutants. *Blood* **106**, 534–541.

Sumoy, L., Keasey, J. B., Dittman, T. D., and Kimelman, D. (1997). A role for notochord in axial vascular development revealed by analysis of phenotype and the expression of VEGR-2 in zebrafish flh and ntl mutant embryos. *Mech. Dev.* **63**, 15–27.

Thisse, B., Heyer, V., Lux, A., Alunni, V., Degrave, A., Seiliez, I., Kirchner, J., Parkhill, J. P., and Thisse, C. (2004). Spatial and temporal expression of the zebrafish genome by large-scale in situ hybridization screening. *Methods Cell Biol.* **77**, 505–519.

Thompson, M. A., Ransom, D. G., Pratt, S. J., MacLennan, H., Kieran, M. W., Detrich, H. W., 3rd., Vail, B., Huber, T. L., Paw, B., Brownlie, A. J., Oates, A. C., Fritz, A., *et al.* (1998). The cloche and spadetail genes differentially affect hematopoiesis and vasculogenesis. *Dev. Biol.* **197**, 248–269.

Till, B. J., Zerr, T., Comai, L., and Henikoff, S. (2006). A protocol for TILLING and Ecotilling in plants and animals. *Nat. Protoc.* **1**, 2465–2477.

Torres-Vazquez, J., Gitler, A. D., Fraser, S. D., Berk, J. D., Van, N. P., Fishman, M. C., Childs, S., Epstein, J. A., and Weinstein, B. M. (2004). Semaphorin-plexin signaling guides patterning of the developing vasculature. *Dev. Cell* **7**, 117–123.

Traver, D., Herbomel, P., Patton, E. E., Murphey, R. D., Yoder, J. A., Litman, G. W., Catic, A., Amemiya, C. T., Zon, L. I., and Trede, N. S. (2003). The zebrafish as a model organism to study development of the immune system. *Adv. Immunol.* **81**, 253–330.

Trinh le, A., McCutchen, M. D., Bonner-Fraser, M., Fraser, S. E., Bumm, L. A., and McCauley, D. W. (2007). Fluorescent *in situ* hybridization employing the conventional NBT/BCIP chromogenic stain. *Biotechniques* **42**, 756–759.

Udvadia, A. J., and Linney, E. (2003). Windows into development: Historic, current, and future perspectives on transgenic zebrafish. *Dev. Biol.* **256**, 1–17.

van der Sar, A. M., Appelmelk, B. J., Vandenbroucke-Grauls, C. M., and Bitter, W. (2004). A star with stripes: Zebrafish as an infection model. *Trends Microbiol.* **12**, 451–457.

Weinstein, B. M., and Lawson, N. D. (2002). Arteries, veins, Notch, and VEGF. *Cold Spring Harb. Symp. Quant. Biol.* **67**, 155–162.

Weinstein, B. M., Stemple, D. L., Driever, W., and Fishman, M. C. (1995). Gridlock, a localized heritable vascular patterning defect in the zebrafish. *Nat. Med.* **1**, 1143–1147.

Westerfield, M., Liu, D. W., Kimmel, C. B., and Walker, C. (1990). Pathfinding and synapse formation in a zebrafish mutant lacking functional acetylcholine receptors. *Neuron* **4**, 867–874.

Westerfield, M. (2000). "The Zebrafish Book. A Guide for the Laboratory Use of Zebrafish (*Danio rerio*)." University of Oregon Press, Eugene.

Wienholds, E., Schulte-Merker, S., Walderich, B., and Plasterk, R. H. (2002). Target-selected inactivation of the zebrafish rag1 gene. *Science* **297**, 99–102.

Woo, K., Shih, J., and Fraser, S. E. (1995). Fate maps of the zebrafish embryo. *Curr. Opin. Genet. Dev.* **5**, 439–443.

Wu, X., Zhong, H., Song, J., Damoiseaux, R., Yang, Z., and Lin, S. (2006). Mycophenolic acid is a potent inhibitor of angiogenesis. *Arterioscler. Thromb. Vasc. Biol.* **26**, 2414–2416.

Xu, Q. (1999). Microinjection into zebrafish embryos. *Methods Mol. Biol.* **127**, 125–132.

Yaniv, K., Isogai, S., Castranova, D., Dye, L., Hitomi, J., and Weinstein, B. M. (2006). Live imaging of lymphatic development in the zebrafish. *Nat. Med.* **12,** 711–716.

Zambrowicz, B. P., Friedrich, G. A., Buxton, E. C., Lilleberg, S. L., Person, C., and Sands, A. T. (1998). Disruption and sequence identification of 2,000 genes in mouse embryonic stem cells. *Nature* **392,** 608–611.

Zhang, C., Willett, C. E., and Fremgen, T. (2003). "Zebrafish: An Animal Model for Toxicological Studies." John Wiley & Sons, New York.

Zhong, T. P., Rosenberg, M., Mohideen, M. A., Weinstein, B., and Fishman, M. C. (2000). Gridlock, an HLH gene required for assembly of the aorta in zebrafish. *Science* **287,** 1820–1824.

Zhong, T. P., Childs, S., Leu, J. P., and Fishman, M. C. (2001). Gridlock signalling pathway fashions the first embryonic artery. *Nature* **414,** 216–220.

Zhu, H., Traver, D., Davidson, A. J., Dibiase, A., Thisse, C., Thisse, B., Nimer, S., and Zon, L. I. (2005). Regulation of the lmo2 promoter during hematopoietic and vascular development in zebrafish. *Dev. Biol.* **281,** 256–269.

Zon, L. I., and Peterson, R. T. (2005). *In vivo* drug discovery in the zebrafish. *Nat. Rev. Drug Discov.* **4,** 35–44.

Zwaal, R. R., Broeks, A., van Meurs, J., Groenen, J. T., and Plasterk, R. H. (1993). Target-selected gene inactivation in *Caenorhabditis elegans* by using a frozen transposon insertion mutant bank. *Proc. Natl. Acad. Sci. U.S.A.* **90,** 7431–7435.

CHAPTER FIVE

Evaluating Vascular Leak *In Vivo*

Sara M. Weis

Contents

Abstract

Vascular permeability during normal physiology is necessary, reversible, and highly regulated. At the site of an injury, hyperpermeable blood vessels allow deposition of fibrin which can form a barrier to contain the wound and prevent infection. Likewise, ongoing extravasation or transendothelial migration of immune cells is necessary for surveillance of tissues. In contrast, uncontrolled vascular leak occurs as a side effect in a number of pathologies including age-related macular degeneration, ischemic disease, cancer, and lung injuries. During the progression of disease, permeability is often dysregulated and results in ongoing accumulation of edema, which exacerbates disease and prevents recovery. An expanding number of mouse models to assess permeability *in vivo* have led to the design of new therapies and approaches to limit pathological hyperpermeability. This chapter will describe several mouse models which can be used to directly test the pro- or anti-permeability properties of novel compounds, as well as mouse models of ischemic disease in which hyperpermeability plays a key role. These methods can be used to evaluate the efficacy of potential new antileak therapies and to dissect the molecular mechanisms regulating permeability *in vivo*.

Moores Cancer Center, University of California, San Diego, La Jolla, California

Methods in Enzymology, Volume 444
ISSN 0076-6879, DOI: 10.1016/S0076-6879(08)02805-X

1. INTRODUCTION

Vascular permeability occurs in response to a number of permeability agents, including vascular endothelial growth factor (VEGF), histamine, or thrombin. These agents elicit a rapid and reversible permeability response when applied to blood vessels *in vivo*. The molecular pathways regulating the response to each permeability factor consist of downstream mediators that are shared by permeability factors in general, or which may be unique to a certain permeability agent. Identification of antileak agents has led to therapeutic advances and the alleviation of hyperpermeability in a number of models. Understanding how and why permeability occurs during pathological situations will be critical for the design and optimization of antileak therapies.

2. MECHANISMS OF VASCULAR LEAK

The molecular mechanisms contributing to vascular leak *in vivo* are complicated, since vascular leak is a side effect of some pathologies while a contributing factor to others. Permeability *in vivo* is a function of many factors including blood flow, shear stress, pressure differentials, and concentration gradients. Particles or solutes of varying sizes can exit the circulation through pathways within individual endothelial cells (transcellular permeability) or between adjacent endothelial cells (paracellular permeability). The signaling pathways regulating paracellular or transcellular permeability do overlap to some degree, and some permeability agents induce responses that involve both types of leak.

Transcellular permeability can occur via caveolae, indentations in the luminal cell membrane that move through the cytoplasm and act as vesicular transport machines. Entire pathways of fused vesicles, termed vesiculo-vacuolar organelles (VVOs) (Feng *et al.*, 1996), can also act as a route for the transport of solutes from the luminal to abluminal surface of a single endothelial cell. Permeability factors such as VEGF also induce the formation of fenestrations, which are focal, thinned regions of endothelium in which the luminal and abluminal surfaces are separated by only a thin diaphragm (Feng *et al.*, 2000; Palade *et al.*, 1979). Paracellular permeability occurs at the cell-cell junctions between adjacent endothelial cells (see review by Mehta and Malik, 2006), through gap junctions (connexins), tight junctions (occludins, claudins, junctional adhesion molecules), and adherens junctions (cadherins). Adhesion molecules at tight and adherens junctions are connected by a series of intermediate proteins (ZO-1, catenins) to the actin cytoskeleton, which can generate contractile forces via MLCK and the Rho-family small GTPases—RhoA, Rac1, and Cdc42.

Cytoskeletal links to the cell–matrix interface via integrins, focal adhesion kinase (FAK), and p21-activated kinase (PAK), also impact the balance of cytoskeletal forces. All of these physical regulators of endothelial barrier function work together to maintain vascular homeostasis and to promote controlled and reversible leak at injury sites. Unfortunately, some of these signaling pathways are activated by tumor cells, inflammation, or hypoxia. The pathological hyperpermeability response is often dysregulated and leads to deleterious side effects during the progression of disease.

3. Consequences of Vascular Leak in Disease

Vascular leak has been a well-known hallmark of the leaky blood vessels associated with solid tumors (Dvorak, 1990). Tumor cells often release pro-leak and pro-angiogenic agents that ignite a series of vascular remodeling events to support tumor growth. Thus, antiangiogenic strategies have been somewhat successful in treating cancer. For example, VEGF-Trap and Avastin target the activity of VEGF, and were designed as antiangiogenic agents to rob the tumor of a blood supply and thus prevent tumor growth. However, it became clear that such anti-VEGF agents also act as vascular normalizers, which could reduce the tortuosity and leakiness of tumor-associated blood vessels. In this respect, vessel normalization could lead to better delivery of anticancer agents to a solid tumor.

Vascular leak syndrome (VLS) is a troublesome side effect in the majority of melanoma or renal cell carcinoma patients receiving interleukin 2 (IL-2) anticancer immunotherapy. Inhibition of angiopoetin-2 (Ang-2) (Gallagher et al., 2007) or endogenous hyaluronan (HA) (Guan et al., 2007) has proven to reduce VLS in mouse models, improving the efficacy and dose tolerance of IL-2 therapy.

Vascular leak and edema also occurs following ischemic injury, and exacerbates hypoxia-induced tissue damage by allowing accumulation of fluids and activated blood cells within sensitive tissue beds within the brain, heart, or limb muscles. Blockade of VEGF-induced leak following ischemic injury can effectively reduce the accumulation of edema and limit the expansion of the infarct zone (van Bruggen et al., 1999; Paul et al., 2001; Weis et al., 2004).

A number of pulmonary complications involve acute hyperpermeability, including asthma, acute lung injury (ALI), acute respiratory distress syndrome (ARDS), ventilator-induced lung injury (VILI), and transfusion-related acute lung injury (TRALI). The act of mechanical ventilation using a large tidal volume or high levels of positive pressure generates pulmonary edema, and involves both stretch-related and inflammatory-induced signaling pathways (Ricard et al., 2003). Blood transfusion also induces lung injury, a process that

involves passively transfused neutrophil and leukocyte antibodies and biologically active lipids (Looney and Matthay, 2006).

In the eye, uncontrolled vascular leak is a contributing factor to age-related
macular degeneration (AMD). Anti-VEGF strategies (such as pegaptanib
[Vinores, 2006] or Lucentis [Lowe *et al.*, 2007]) are approved to treat AMD,
and have shown great clinical promise in terms of reducing vascular leak and
preventing further vision loss. Macular edema is also a key component of
diabetic retinopathy, and involves crosstalk between growth factors, including
VEGF and the Ang-2/Tie-2 system, such that blocking both VEGF and Ang-2
pathways results in more than an additive response (Peters *et al.*, 2007).

4. MODELS TO STUDY VASCULAR LEAK *IN VIVO*

There are a wide array of models to study vascular leak *in vivo*, each
with its own unique strengths and challenges. The first section will address
models which can be used to test the direct effects of pro- or anti-
permeability agents on vascular leak or accumulation of edema in tissues.
The second section will review several mouse models of ischemic disease in
which permeability plays a critical role.

4.1. Evaluation of vascular leak in mice

The protocols detailed in this section use VEGF as a permeability agent, but
can easily be adapted to test the permeability induction by other mediators.
Likewise, these methods are well-suited for comparing how gene-targeted
mouse models, small molecule inhibitors, endogenous regulators of vascular
leak, and so on, may impact the normal response to a permeability factor.

4.1.1. The Miles Assay and adaptations

The Miles Assay was developed in 1952 to assess and quantify the vascular
leak response in the skin of guinea pigs (Miles and Miles, 1952), and has
since been adapted to assess leak in a number of different mouse organs. In
general, this assay measures leak of a systemically injected tracer (commonly
Evans blue dye) from the vascular compartment into a tissue bed by
homogenization of the tissue, extraction of the dye, and quantification
using spectrophotometry. The basic protocol to assess leak in the skin as
well as several variations are outlined below.

The basic protocol to evaluate leak in the skin follows (Fig. 5.1):

1. Shave flank skin areas for injection 2 to 3 days before experiment.
2. Inject tracer dye via tail vein (100 μl of 1% Evans blue dye in sterile saline).
3. After 10 min, inject vehicle control or permeability agent (e.g., 400 ng
   VEGF in 10 μl sterile saline) intradermally to the ventral flank.

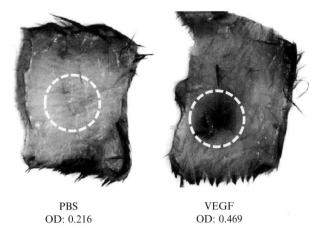

PBS   VEGF
OD: 0.216   OD: 0.469

Figure 5.1 Representative photographs from the Miles Assay to measure leak of Evans blue dye in the skin. Images show significant leak of blue dye near the injection of VEGF, but not the PBS control. Raw OD values are from skin punches that were extracted overnight in formamide and read on a spectrophotometer at 600 nm.

4. After 15 to 30 min, euthanize mice in a carbon dioxide chamber.
5. Perfuse each mouse with 10 ml saline via left ventricle (after venting the left atrium).
6. Remove skin patch with 8-mm biopsy punch, weigh, and store on ice.
7. Add 200 μl of formamide to extract the Evans blue dye.
8. Incubate at 56 °C for 24 h.
9. Read absorbance at 600 nm for 50-μl triplicates in a 96-well plate.
10. Normalize the absorbance values to each tissue weight.

An adaptation to assess the accumulation of leak within organs such as the lung follows (Liao *et al.*, 2002):

1. Inject 100 μl of 1% Evans blue dye containing 5 μg VEGF via tail vein.
2. After 15 min, euthanize mice in carbon dioxide chamber.
3. Slowly perfuse mouse with 10 ml saline via left ventricle (after venting left atrium).
4. Remove organs and rinse briefly in PBS.
5. Blot dry and then weigh immediately.
6. Photograph using digital camera.
7. Mince with razor blades on glass plates.
8. Homogenize in trichloroacetic acid:ethanol (1:1 v/v).
9. Read absorbance as above.

An adaptation to assess and visualize vascular leak in the ear over a longer time period by driving expression of VEGF using adenovirus follows (Nagy *et al.*, 2006):

1. Mice are anesthetized.
2. One ear is injected intradermally with 5 μl of AdVEGF165 ($\sim$1000 pfu/ml).
3. The opposite ear is injected intradermally with 5-μl vector control adenovirus.
4. After 5 days, Evans blue dye (1 mg/ml in 100 μl sterile saline) is injected intravenously via tail vein.
5. After 30 min, the ears are photographed, removed, and weighed.
6. Extract dye using formamide and read absorbance as above.

4.1.2. Measuring biochemical signaling events induced by permeability agents

In addition to assessing the physical accumulation of leak within tissues, it is also possible to measure the signaling response elicited by a permeability factor such as VEGF. To do this, VEGF is injected intravenously, tissues are removed at a range of time points, and the *in vivo* signaling response to VEGF is evaluated using immunoprecipitation and immunoblotting. This method typically yields several milligrams of protein per organ, which can be used for many immunoprecipitation and immunoblotting experiments. The protocol is detailed below:

1. Adult mice are injected i.v. with 2 to 5 μg VEGF diluted in 100 μl sterile saline.
2. After 2 to 60 min, tissues (e.g., heart, lung, brain) are rapidly excised, minced on a glass plate using razor blades, and homogenized in 3 ml RIPA buffer on ice.
3. Samples are rotated at 4 °C for 30 min, and then centrifuged at 4 °C for 10 min.
4. The supernatant is subject to immunoblotting or immunoprecipitation.
5. For example, immunoprecipitation can be performed to pull-down endothelial-specific complexes (e.g., containing VE-cadherin or VEGFR2) and subsequent Western blotting can detect phosphorylation events or co-association between proteins of interest.

4.1.3. Confocal microscopy approach to detect leak

Griffonia simplicifolia lectins have been used to identify endothelial cells for many years (Laitinen, 1987). Taking these a step further, McDonald and coworkers identified unique binding properties of individual lectins and applied this technology to the investigation of permeability *in vivo*. Lectins are injected intravenously, and peroxidase or fluorescence detection is used to visualize the binding to specific locations during the permeability response in tissue sections or whole-mount preparations (Thurston *et al.*, 1996). In particular, some lectins bind to all endothelial cells, while others bind only to inflamed endothelial cells (Thurston *et al.*, 1996). Extravasation of a particular tracer such as monastral blue can be used together with lectin staining to show sites of plasma leakage. These methods have been used to monitor the mechanisms of vascular leak during inflammation (McDonald *et al.*, 1999) and the leaky blood vessels associated with tumors (Hashizume *et al.*, 2000).

In addition to lectins that specifically bind endothelial cells, leaks can be visualized at the light microscope level as extravasation of labeled tracers. Intravenous injection of an endothelial-specific lectin along with a fluorescently labeled dextran allows visualization of both vessel perfusion and leak. Dextrans of different sizes can be used to evaluate the contribution of different pathways to the leak response (i.e., fenestrations, caveolae, VVOs, EC gaps). Together with electron microscopy (described below), these light microscopy techniques can offer valuable insights into the structural determinants of leak within pathological samples.

Figure 5.2 shows confocal images of fresh mouse tissues following injection of rhodamine lectin. Mice are injected i.v. via the tail vein with 100 μl of a 1-mg/ml solution of rhodamine lectin. After 10 min, organs are removed and placed into ice-cold PBS. Each tissue is placed on a glass

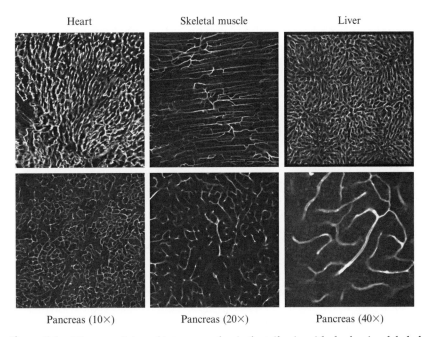

| Heart | Skeletal muscle | Liver |

| Pancreas (10×) | Pancreas (20×) | Pancreas (40×) |

Figure 5.2 Mice were injected intravenously via the tail vein with rhodamine-labeled lectin. Organs were removed, placed on glass coverslips, and imaged using confocal microscopy. Rhodamine lectin binding to the inner lumen of perfused blood vessels reveals the vascular networks within each organ.

coverslip and confocal microscopy is used to monitor blood vessel perfusion by the presence of lectin binding. FITC-dextran or other tracers can be co-injected with the lectin to visualize both blood vessel perfusion and areas of tracer accumulation (i.e., leak).

4.1.4. Electron microscopy approach to study leak at ultrastructural level

Electron microscopy (EM) has been used for years to study the ultrastructural changes to the endothelium during the induction of vascular leak. Palade and coworkers utilized EM to establish the structural aspects of the vascular endothelium (Simionescu et al., 1975, 1976a,b; Palade et al., 1979). Later work by Dvorak and colleagues provided a strong basis for using EM to study the vascular leak properties of VEGF and the leaky nature of tumor-associated blood vessels (Kohn et al., 1992; Dvorak et al., 1996). More recently, we have shown that heart or brain capillaries exposed to systemic VEGF respond with a series of events including gaps between adjacent endothelial cells, attraction of platelets to exposed basement membrane, and ultimately activation and aggregation of platelets forming microthrombi (Weis et al., 2004). Following a single injection of VEGF, this process occurs over 2 to 10 min, after which time vascular integrity returns to baseline. However, following ischemic injury or repeated VEGF injections over several hours, these processes create long-lasting effects such as damage to myocyte contractile machinery and mitochondria, interstitial edema, and regions of focal ischemia induced by platelet microthrombi (Weis et al., 2004). The following protocol details preparation of mouse heart tissue to study the effects of VEGF injection at the ultrastructural level.

1. Inject 2 μg VEGF in 100-μl sterile saline to tail vein.
2. After 2 to 30 min, euthanize mouse in carbon dioxide chamber.
3. Carefully excise heart, holding by aorta to preserve ventricular tissue.
4. On a glass plate, make each cut very gently with #11 scalpel, taking care not to touch tissue with forceps (do not poke, squeeze, compress, etc.).
5. Cut equatorial (horizontal) section one-third from base to apex.
6. Discard base of heart and retain remainder of left ventricle for analysis.
7. Tissue is immersion fixed in ice-cold 0.1 M of sodium cacodylate buffer (pH 7.3) containing 4% paraformaldehyde plus 1.5% glutaraldehyde for 2 h, transferred to 5% glutaraldehyde overnight, and then postfixed in 1% osmium tetroxide for 1 h at room temperature.
8. Blocks are washed, dehydrated, cleared in propylene oxide, infiltrated with Epon/Araldite, embedded in 100% resin, and allowed to polymerize.
9. Ultrathin sections are stained with uranyl acetate and lead citrate, and viewed using transmission electron microscopy (TEM).

To obtain quantitative data using EM, the incidence of certain events within blood vessels can be counted. Following myocardial infarction (MI), we graded the severity of damage to several cell types using the grading scale in Table 5.1. We used this method to correlate the improvements in infarct size, edema, and functional recovery with the barrier-preserving effects of treatment with antileak agents (i.e., Src kinase inhibitors) (Weis *et al.*, 2004). Although somewhat time consuming, this allows examination of a large number of variables that can change over time and vary by location within the heart following myocardial infarction. Similar methods can be used to study the vascular stabilizing effects of antileak agents in other tissues, in tumor-associated blood vessels, or in other ischemic models. Evidence at the ultrastructural level can lend insight into a variety of mechanisms by which an agent promotes or blocks vascular leak, and should certainly be considered as a valuable assay to evaluate the process of vascular leak.

4.2. Evaluation of leak and edema during the progression of ischemic disease

4.2.1. Mouse model of myocardial infarction

Studying the postischemia response in mouse models provides valuable insight into the deleterious effects of vascular leak during the early stages of the wound healing process in the heart (Figures 5.3 and 5.4). While the chronic mouse myocardial infarction preparation is difficult to perform experimentally (Michael *et al.*, 1995, 1999), many groups have published good success rates for creating chronic infarcts in mice. The ischemia/

Table 5.1 Ultrastructural tissue assessment following myocardial infarction

EC barrier dysfunction and adhesion	
Normal	Intact EC junctions
Abnormal	Gaps, fenestrations, platelet/RBC plugs, extravasated blood cells
Platelet activation	
Normal	Occasional, single granulated platelets
Abnormal	Increased platelet count, degranulated platelets, platelet clusters, increased neutrophil count
EC injury	
Normal	Normal EC with clear vessel lumen
Abnormal	Electron-lucent EC, sick/swollen EC, large EC vacuoles, occluded/collapsed vessel lumen
Cardiac damage	
Normal	Electron-dense mitochondria, well-ordered myofilaments with aligned I-band
Abnormal	Mitochondrial swelling/rupture, disordered cristae, intracellular edema, myofilament disintegration

Systemic VEGF injection induces leak of RBC in the heart

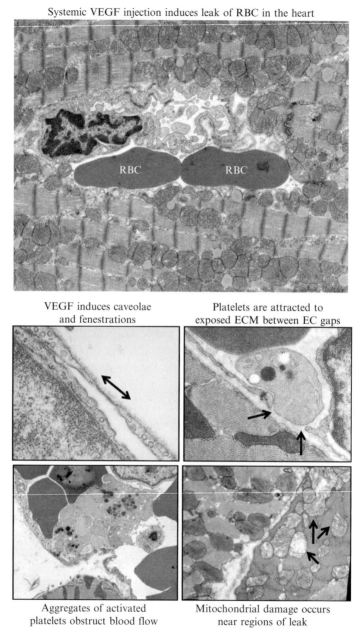

Figure 5.3 Myocardial infarction was induced in mice by occlusion of the left anterior descending coronary artery with suture, and hearts were processed 24 h later for TEM. Images show leak of red blood cells (RBC) from abnormal vessels at the perimeter of the infarct zone. Endothelial cells show fenestrations, increased caveolae and vesicles, and gaps between adjacent cells. Platelets bind to exposed basal lamina between gaps, and when activated aggregate to form microthrombi. Hypoxia, edema, and blood vessel damage ultimately leads to damage of myofilaments and cardiomyocyte mitochondria.

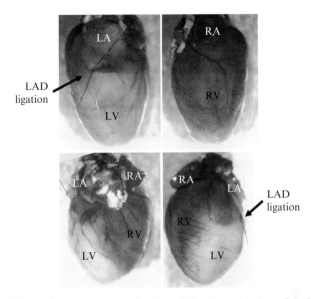

Figure 5.4   Photos show a representative heart following occlusion of the left anterior descending (LAD) coronary artery. After occlusion, mice were injected intravenously with Evans blue dye to demarcate the perfusion boundary. Orthogonal views of the same heart are shown, revealing the consequences of LAD ligation.

reperfusion model allows a more clinically relevant study in terms of a short period of ischemia followed by reperfusion as would be observed in patients receiving care in the emergency room. Detailed protocols to induce myocardial infarction in the mouse are described below:

Survival surgery   Animals are anesthetized with inhalant isoflurane (2.5 to 5% with O_2) and kept warm via a recirculating heated water pad. A tracheostomy or endotracheal intubation is performed, and the animal is ventilated throughout the surgical procedures with room air. A midline skin incision is made from the midsternal line, and the chest opened with a 1-cm lateral cut along the right side of the sternum, cutting between the ribs to expose the left ventricle of the heart. A coronary artery is ligated using 6-0 silk suture. For reperfusion studies, after 30 min of occlusion, the ligature is removed and reperfusion can be confirmed visually. After this time and for permanent occlusion studies, the chest is closed with one layer of suture through the chest wall and a second layer through skin and subcutaneous tissue. The entire surgery portion of the experiment takes no longer than 30 to 45 min.

Postoperative care   During recovery, 95% oxygen with 5% carbon dioxide is administered to the animal through a tube in the cage, and the animal is kept warm with a recirculating heated water pad. Buprenorphine is given before the ~30-min surgery as an analgesic. The animals are monitored continuously for several hours after surgery to verify recovery. Once the animals demonstrate normal habitus and have passed the initial "risk period," they are returned to their normal holding rooms, and cages are monitored daily until the completion of the experiment. Analgesic (buprenorphine) is administered twice daily for at least 3 days following the surgery.

Endpoints   Echocardiography can be performed to establish basal cardiac function and geometry before the induction of MI, and at several timepoints after MI to assess functional deficits or effects of potential treatments. At the end of the experiment, the heart is excised and processed for biochemistry, histology, infarct size determination, or ultrastructural analysis (using TEM).

4.2.2. Hindlimb ischemia and reperfusion injury in the mouse

Postischemia hindlimb reperfusion initiates both local limb and remote organ injury, including pulmonary edema. While VEGF expression may induce angiogenesis to revascularize injured muscle, it also induces edema, which can exacerbate injury and prolong recovery. Because the surgery is relatively easy to perform and amenable to large group sizes, this protocol is particularly well-suited for screening antileak compounds.

Induction of ischemia   There are three methods routinely used to induce hindlimb ischemia, and all mutually complement each other in elucidating the various components of the ischemia reperfusion injury.

Method 1: The femoral artery can be occluded using a microvascular clamp. A transverse oblique incision is made in the groin, and the femoral artery is isolated and clamped using a microvascular clamp.

Method 2: The femoral artery can be occluded noninvasively via a mouse limb tourniquet with the cuff inflated to 200 mmHg to occlude blood flow. After the period of ischemia, the cuff is deflated and the limb will be allowed to reperfuse.

Method 3: A flank incision can be made, and through a retroperitoneal approach, the lower abdominal aorta is accessed and a microvascular clamp is placed. Regardless of the method, the typical duration for ischemia is 2 to 6 h.

Restoration of flow (reperfusion)   After the relevant time of ischemia, the clamp/tourniquet is removed and flow is restored. Restoration of flow can be confirmed noninvasively with a Doppler probe while the animal is still anesthetized. The incision is closed with suture and the skin infiltrated with 0.5% lidocaine. Buprenorphine is administered as an analgesic during reperfusion. The reperfusion period for typical experiments is 1 to 6 h, with some additional samples at 12-, 24-, and 48-h time points.

Euthanasia and tissue collection   At the specified time point following femoral artery occlusion and reperfusion (via removal of the microvascular clips or release of the tourniquet), each animal is sacrificed and the limb skeletal muscle collected for measurement of infarct size (using standard triphenyltetrazolium chloride [TTC] methods), ultrastructural analysis (using TEM), and signal transduction (by homogenizing the tissue and using standard immunoprecipitation and western blotting techniques). Additionally, lung tissue should be harvested to determine potential pulmonary edema (by measuring wet-to-dry weight ratios or histology).

4.2.3. Mouse model of ischemic stroke

Similar to MI in the heart, ischemic stroke leads to VEGF expression, breakdown of the blood-brain barrier, and accumulation of edema. Thus, strategies to block VEGF-induced leak (e.g., anti-VEGF therapy [van Bruggen *et al.*, 1999] or Src kinase inhibition [Paul *et al.*, 2001]) can limit edema and improve long-term function following ischemic stroke. A mouse model of focal cerebral ischemia has been widely used to study the progression of stroke in mice (Nawashiro *et al.*, 1997; Paul *et al.*, 2001). Briefly, mice are anesthetized and an incision is made between the right ear and right eye, and the skull exposed. A small burr hole is drilled in the region over the middle cerebral artery (MCA). Meninges are removed, and the MCA is occluded by coagulation using microcauterization.

After 24 h, the brain is removed and infarct size measured by staining with triphenyltetrazolium chloride (TTC). Alternatively, mice can be injected i.v. with Evans blue dye to detect blood-brain barrier disruption using a method similar to the Miles Assay described above. An hour after Evans blue dye injection, mice are perfused with saline, and then the brain is removed and sliced into 1-mm-thick coronal sections. Areas of leak can be identified using confocal microscopy as patches of auto-fluorescent Evans Blue dye, either using unfixed cryosections or whole-mount preparations of the 1-mm-thick slices. Magnetic resonance imaging (MRI) can also be used to compare the extent of edema in the living mouse before and after the initiation of focal ischemia.

4.2.4. Using MRI to detect edema noninvasively

MRI has been widely used in preclinical and clinical studies to monitor the formation and progression of edema following ischemia. In particular, the T1/T2 relaxation times have been shown to provide good resolution of areas of edema in the heart or brain. To evaluate post-ischemic edema in mice, MRI can be performed before onset of ischemia to provide a baseline, and then 1 h and 24 h later to provide a timeline of edema progression. In the brain, we have used T2-weighted imaging to generate quantitative maps of apparent diffusion coefficient of water and T2 values (Paul *et al.*, 2001). Similarly, in the heart, T2-weighted imaging can be used to identify regions with T2 values greater that two standard deviations above the mean of normally perfused myocardium (Weis *et al.*, 2004). The MRI techniques correlate well with estimation of edema using wet-to-dry weight ratios, and also offer the valuable information regarding the locations of edema and progression of edema over time (Paul *et al.*, 2001; Weis *et al.*, 2004).

5. CONCLUDING REMARKS

The goal of this article was to provide selected methodologic approaches for the evaluation of vascular leak following direct VEGF injection, or the VEGF-induced leak that occurs during the progression of ischemic injury. Although VEGF and ischemic disease were used as examples, these *in vivo* techniques can be applied to study other inducers of vascular leak and to test agents that may preserve barrier function. It is important to note that mouse models of vascular permeability are particularly useful since the leak process *in vivo* involves the interaction of many cell types, physical forces, and geometries that cannot easily be recapitulated using *in vitro* models. Furthermore, the consequences of leak in organs such as the heart and brain lead to critical changes that impact survival. Thus, maintenance of the endothelial barrier under these circumstances is vastly more complex compared to a monolayer of cells growing in culture.

ACKNOWLEDGMENTS

The detailed methods described to study vascular permeability *in vivo* were developed in the laboratory of David Cheresh by a number of present and former postdoctoral fellows, including Robert Paul, Brian Eliceiri, Lisette Acevedo, Doinita Serban, and Joshua Greenberg. Malcolm Wood performed and helped to design the TEM studies, and Leo Barnes and Kimberly Lutu-Fuga provided excellent technical assistance.

REFERENCES

Dvorak, H. F. (1990). Leaky tumor vessels: Consequences for tumor stroma generation and for solid tumor therapy. *Prog. Clin. Biol. Res.* **354A,** 317–330.
Dvorak, A. M., Kohn, S., Morgan, E. S., Fox, P., Nagy, J. A., and Dvorak, H. F. (1996). The vesiculo-vacuolar organelle (VVO): A distinct endothelial cell structure that provides a transcellular pathway for macromolecular extravasation. *J. Leukoc. Biol.* **59,** 100–115.
Feng, D., Nagy, J. A., Hipp, J., Dvorak, H. F., and Dvorak, A. M. (1996). Vesiculo-vacuolar organelles and the regulation of venule permeability to macromolecules by vascular permeability factor, histamine, and serotonin. *J. Exp. Med.* **183,** 1981–1986.

Feng, D., Nagy, J. A., Dvorak, A. M., and Dvorak, H. F. (2000). Different pathways of macromolecule extravasation from hyperpermeable tumor vessels. *Microvasc. Res.* **59**, 24–37.

Gallagher, D. C., Bhatt, R. S., Parikh, S. M., Patel, P., Seery, V., McDermott, D. F., Atkins, M. G., and Sukhatme, V. P. (2007). Angiopoietin 2 is a potential mediator of high-dose interleukin 2-induced vascular leak. *Clin Cancer Res.* **13**, 2115–2120.

Guan, H., Nagarkatti, P. S., and Nagarkatti, M. (2007). Blockade of hyaluronan inhibits IL-2–induced vascular leak syndrome and maintains effectiveness of IL-2 treatment for metastatic melanoma. *J. Immunol.* **179**, 3715–3723.

Hashizume, H., Baluk, P., Morikawa, S., McLean, J. W., Thurston, G., Roberge, S., Jain, R. K., and McDonald, D. M. (2000). Openings between defective endothelial cells explain tumor vessel leakiness. *Am. J. Pathol.* **156**, 1363–1380.

Kohn, S., Nagy, J. A., Dvorak, H. F., and Dvorak, A. M. (1992). Pathways of macromolecular tracer transport across venules and small veins. Structural basis for the hyperpermeability of tumor blood vessels. *Lab. Invest.* **67**, 596–607.

Laitinen, L. (1987). Griffonia simplicifolia lectins bind specifically to endothelial cells and some epithelial cells in mouse tissues. *Histochem. J.* **19**, 225–234.

Liao, F., Doody, J. F., Overholser, J., Finnerty, B., Bassi, R., Wu, Y., Dejana, E., Kussie, P., Bohlen, P., and Hicklin, D. J. (2002). Selective targeting of angiogenic tumor vasculature by vascular endothelial-cadherin antibody inhibits tumor growth without affecting vascular permeability. *Cancer Res.* **62**, 2567–2575.

Looney, M. R., and Matthay, M. A. (2006). Animal models of transfusion-related acute lung injury. *Crit. Care Med.* **34**, S132–S136.

Lowe, J., Araujo, J., Yang, J., Reich, M., Oldendorp, A., Shiu, V., Quarmby, V., Lowman, H., Lien, S., Gaudreault, J., and Maia, M. (2007). Ranibizumab inhibits multiple forms of biologically active vascular endothelial growth factor in vitro and in vivo. *Exp. Eye Res.* **13**, 781–789.

McDonald, D. M., Thurston, G., and Baluk, P. (1999). Endothelial gaps as sites for plasma leakage in inflammation. *Microcirculation* **6**, 7–22.

Mehta, D., and Malik, A. B. (2006). Signaling mechanisms regulating endothelial permeability. *Physiol. Rev.* **86**, 279–367.

Michael, L. H., Ballantyne, C. M., Zachariah, J. P., Gould, K. E., Pocius, J. S., Taffet, G. E., Hartley, C. J., Pham, T. T., Daniel, S. L., Funk, E., and Entman, M. L. (1999). Myocardial infarction and remodeling in mice: Effect of reperfusion. *Am. J. Physiol.* **277**, H660–H668.

Michael, L. H., Entman, M. L., Hartley, C. J., Youker, K. A., Zhu, J., Hall, S. R., Hawkins, H. K., Berens, K., and Ballantyne, C. M. (1995). Myocardial ischemia and reperfusion: A murine model. *Am. J. Physiol.* **269**, H2147–H2154.

Miles, A. A., and Miles, E. M. (1952). Vascular reactions to histamine, histamine-liberator and leukotaxine in the skin of guinea pigs. *J. Physiol.* **118**, 228–257.

Nagy, J. A., Feng, D., Vasile, E., Wong, W. H., Shih, S. C., Dvorak, A. M., and Dvorak, H. F. (2006). Permeability properties of tumor surrogate blood vessels induced by VEGF-A. *Lab. Invest.* **86**, 767–780.

Nawashiro, H., Tasaka, K., Ruetzler, C. A., and Hallenbeck, J. M. (1997). TNF-alpha pretreatment induces protective effects against focal cerebral ischemia in mice. *J. Cereb. Blood Flow Metab.* **17**, 483–490.

Palade, G. E., Simionescu, M., and Simionescu, N. (1979). Structural aspects of the permeability of the microvascular endothelium. *Acta Physiol. Scand. Suppl.* **463**, 11–32.

Paul, R., Zhang, Z. G., Eliceiri, B. P., Jiang, Q., Boccia, A. D., Zhang, R. L., Chopp, M., and Cheresh, D. A. (2001). Src deficiency or blockade of Src activity in mice provides cerebral protection following stroke. *Nat. Med.* **7**, 222–227.

Peters, S., Cree, I. A., Alexander, R., Turowski, P., Ockrim, Z., Patel, J., Boyd, S. R., Joussen, A. M., Ziemssen, F., Hykin, P. G., and Moss, S. E. (2007). Angiopoietin modulation of vascular endothelial growth factor: Effects on retinal endothelial cell permeability. *Cytokine* **40**(2), 144–150.

Ricard, J. D., Dreyfuss, D., and Saumon, G. (2003). Ventilator-induced lung injury. *Eur. Respir. J. Suppl.* **42**, 2s–9s.

Simionescu, M., Simionescu, N., and Palade, G. E. (1975). Segmental differentiations of cell junctions in the vascular endothelium: The microvasculature. *J. Cell Biol.* **67**, 863–885.

Simionescu, M., Simionescu, N., and Palade, G. E. (1976a). Characteristic endothelial junctions in different segments of the vascular system. *Thromb. Res.* **8**(2 suppl), 247–256.

Simionescu, M., Simionescu, N., and Palade, G. E. (1976b). Segmental differentiations of cell junctions in the vascular endothelium: Arteries and veins. *J. Cell Biol.* **68**, 705–723.

Thurston, G., Baluk, P., Hirata, A., and McDonald, D. M. (1996). Permeability-related changes revealed at endothelial cell borders in inflamed venules by lectin binding. *Am. J. Physiol.* **271**, H2547–H2562.

van Bruggen, N., Thibodeaux, H., Palmer, J. T., Lee, W. P., Fu, L., Cairns, B., Tumas, D., Gerlai, R., Williams, S. P., van Lookeren Campagne, M., and Ferrara, N. (1999). VEGF antagonism reduces edema formation and tissue damage after ischemia/reperfusion injury in the mouse brain. *J. Clin. Invest.* **104**, 1613–1620.

Vinores, S. A. (2006). Pegaptanib in the treatment of wet, age-related macular degeneration. *Int. J. Nanomed.* **1**, 263–268.

Weis, S., Shintani, S., Weber, A., Kirchmair, R., Wood, M., Cravens, A., McSharry, H., Iwakura, A., Yoon, Y. S., Himes, N., Burstein, D., Doukas, J., *et al.* (2004). Src blockade stabilizes a Flk/cadherin complex, reducing edema and tissue injury following myocardial infarction. *J. Clin. Invest.* **113**, 885–894.

CHAPTER SIX

OCULAR MODELS OF ANGIOGENESIS

Edith Aguilar, Michael I. Dorrell, David Friedlander,
Ruth A. Jacobson, Audra Johnson, Valentina Marchetti,
Stacey K. Moreno, Matthew R. Ritter, *and* Martin Friedlander

Contents

Department of Cell Biology, The Scripps Research Institute, La Jolla, California

Methods in Enzymology, Volume 444
ISSN 0076-6879, DOI: 10.1016/S0076-6879(08)02806-1

Abstract

During normal retinal vascular development, vascular endothelial cells proliferate and migrate through the extracellular matrix in response to a variety of cytokines, leading to the formation of new blood vessels in a highly ordered fashion. However, abnormal angiogenesis contributes to the vast majority of diseases that cause catastrophic loss of vision. During abnormal neovascularization of the iris, retina, or choroid, angiogenesis is unregulated and usually results in the formation of dysfunctional blood vessels. Multiple models of ocular angiogenesis exist which recapitulate particular aspects of both normal and pathological neovascularization. These experimental methods are useful for studying the mechanisms of normal developmental angiogenesis, as well as studying various aspects of pathological angiogenesis including ischemic retinopathies, vascular leak, and choroidal neovascularization. This chapter will outline several protocols used to study ocular angiogenesis, put the protocols into brief historical context, and describe some of the questions for which these protocols are commonly used.

1. INTRODUCTION

Ocular angiogenesis, the abnormal growth of blood vessels in the eye, is associated with the vast majority of eye diseases that cause a catastrophic loss of vision. One condition that is developmental in origin is retinopathy of prematurity (ROP), a condition of premature infants (Smith, 2002). Infants born early are often exposed to hyperoxia to reduce pulmonary distress. If retinal vascularization, which occurs during the third trimester in utero, has not been completed through the final vascular maturation stages, exposure to hyperoxia can prevent normal vascularization. New vessels fail to form, and the newly formed, immature vessels regress during this exposure to hyperoxia. Upon return of the infant to normal oxygen levels, uncontrolled neovascularization is initiated in the undervascularized retina, leading to the formation of disorganized and leaky vessels. These vessels, due to inappropriate development and patterning, cannot sufficiently oxygenate the retina and can hemorrhage eventually leading to the degeneration of retinal ganglion cells and photoreceptors (Ashton, 1966; Hellstrom *et al.*, 2001). Models of oxygen-induced retinopathy (OIR), described in Section 5, very closely recapitulate the events that occur during retinopathy of prematurity. These models have also been extended to the use of retinopathies in general whereby retinal ischemia drives excessive and abnormal neovascularization leading to several pathological events common to many ocular diseases. Models of normal retinal vascular development are described in Section 4.

The leading cause of vision loss in individuals under the age of 55 is diabetic retinopathy, a potentially blinding complication of diabetes. In early stages, retinal capillaries are damaged as a result of the microvasculopathy characteristic of diabetes, leading to retinal hypoxia. As the hypoxia and microvascular changes progress, they can induce the proliferative stage of diabetic retinopathy, where abnormal, fragile new blood vessels grow along the retina and into the vitreous. Similar to the abnormal vessels in ROP, these immature vessels can eventually hemorrhage causing vision to become blurred or cloudy. The hemorrhages may eventually clear, but more often, bleeding continues in association with fibrosis. These fibrovascular scars can then lead to retinal detachment and permanent visual impairment. Retinal edema in the area of the macula, the region that enables detailed vision, can also occur as a consequence of the microvascular abnormalities, also contributing to significant visual morbidity (Sarraf, 2001). There are currently no perfect models for diabetic retinopathy; however, the OIR models are used to study specific pathological events that commonly occur during diabetic retinopathy. Photocoagulation of retinal veins, described within Section 6, can be performed to initiate ocular angiogenesis with certain pathological events similar to diabetic retinopathy, or more closely associated with retinal vein occlusion. Models which specifically target vascular leak and can be used to test methods of reducing vascular leak associated with abnormal ocular neovascularization are described in Section 7. Other models such as streptozotocin-induced diabetic rats and mice are good at recapitulating certain events in diabetic retinopathy as well, but are not included in this chapter (Feit-Leichman et al., 2005).

The leading cause of vision loss in individuals over the age of 65 is age-related macular degeneration (ARMD), where visual loss occurs as a result of atrophic changes in the macula, and choroidal neovascularization. The choroid is the thin, highly vascular layer of the eye lying just posterior to the neurosensory retina and the retinal pigment epithelium (RPE). Its vessels provide oxygen to the outer third of the retina including the photoreceptors. Normally the choroidal vessels are restricted from directly contacting the adjacent retina by an anatomic barrier, Bruch's membrane. However, in ARMD patients, Bruch's membrane is weakened and endothelial cells from choroidal vessels are activated to proliferate. The resulting new vessels break through the weakened regions of Bruch's membrane, and fibrovascular tissues are deposited in the sub-RPE space (Husain et al., 2002). Like diabetic retinopathy, there are no perfect models for studying age-related macular degeneration. However, various models including the laser choroid model, described in Section 6, are used to recapitulate choroidal neovascularization.

Also described in this chapter is the corneal micropocket assay, Section 3, which is extensively used to study the effects of various stimulatory, or angiostatic, cytokines, growth factors, and small molecules on angiogenesis. We also describe methods for visualizing ocular vessels in live animals, Section 8, and finish with a discussion on techniques which can be used to isolate and study progenitor cells in various models of ocular angiogenesis.

2. TECHNIQUES COMMONLY USED IN OCULAR ANGIOGENESIS PROTOCOLS

Similar techniques are commonly used with many of the ocular angiogenesis protocols described in this chapter. The detailed protocols for each of these common techniques are described below and referred to in individual protocols presented in this chapter.

2.1. Intravitreal injections

Intravitreal injections are a very useful technique for the direct administration of various molecules including peptide, small molecule, or antibody angiostatics, various stimulatory cytokines, or even cells. For this technique, a dissection microscope should be used to monitor the injections and visualize the needle as it enters the vitreal cavity of the eye.

1. It is recommended that the animals be placed under sedation using either gas anesthesia or ketamine and xylazine injections.
2. For young animals in which the eyes are not yet open (up to ~2 weeks of age), open the lid with blunt forceps by gently creating a fissure between the eyelids along the slit from which the eyelids will eventually open.
3. Use a 5-μl Hamilton syringe with a 33-gauge sharp needle (Hamilton 33-gauge, 0.5″, point-style-4 needles) for the injections.
4. Under pupil dilation, penetrate the eye in the vitreous cavity by the limbus.
5. Direct the needle toward the center of the eye and the posterior pole. Extreme caution should be taken to avoid puncturing the lens as this will cause inflammation and experimental variability. The lens in rodents is quite large and fills a good portion of the vitreal cavity.
6. Slowly inject 0.5 μl into the vitreal cavity.
   - This should be done with minimal, constant force using a slow, constant flow rate.
   - Expel the solution into the vitreous over the course of ~5 to 10 s, and then pause 5 s with the needle inside the vitreal cavity.

7. Remove the needle from the eye very slowly, again over the course of ~ 5- 10 s, to minimize leakage of the material injected back through the injection site.

Intravitreal injections can be performed at any age as required for different protocols, although the younger, and therefore smaller, the mouse, the more technically challenging the injections may be. Due to the small volume vitreal cavities of mice and rats—approximately 5 μl and 50 μl, respectively, for adults—only small volumes should be injected intravitreally. Larger volumes cause a dramatic change in ocular pressure, and generally result in reflux, which causes a large percentage of the injected solution to leak back out. This can add substantial variables to the experiment. As a general rule, 0.5-μl volumes are injected for both mouse and rat injections, although if necessary, the volumes can be increased slightly for rats. Multiple injections are not advised, as this can cause major trauma to the eye. If multiple injections are necessary, maximal time should be allowed between injections to permit healing between injections and prevent phthisis of the eye.

2.2. Retina dissection

Many of the protocols described in this chapter utilize the retina as the model tissue. Thus, it is often imperative to dissect and isolate the retina free from other ocular tissues for analysis.

1. At the appropriate time of analysis the animals should be euthanized.
2. Immediately following euthanasia, enucleate the eye and place it in 4% PFA for 10 to 15 min. (The eye can be removed fairly easily by proptosing the eye and using blunt, curved tweezers to gently remove the eye.)

(The following steps should be done in sterile PBS using a dissection microscope.)

3. Grasp the center of the cornea using fine forceps with 0.12-mm teeth (Storz forceps suture Castroviejo 0.12 mm, E1796) and completely remove the cornea from the eye by cutting with fine dissection scissors for 360 degrees (Storz curved iris scissors, E3402).
4. Grasp the sclera and choroid using fine tweezers (jeweler-type forceps, E1974-7, E1947-5), and very gently remove the sclera and choroid from the retina by pulling the tissues until they are free from each other.
   • Try not to grasp the retina in this step, as this will damage the retina.
   • On opposite sides of the globe, grasp the sclera (free from the retina) and gently pull. This will eventually cause the entire retina to be released from the sclera.

5. Remove the lens from the retina.
6. If isolating retinas from younger mice, the hyaloidal vasculature should be removed as well.
7. Postfix the retina using methanol for 10 min, or 4% PFA for 1 h as per the specific staining protocol to be used.

2.3. Retinal whole mounts

Visualizing retina whole mounts can be a very useful technique for evaluating ocular angiogenesis in the entire retina. This is a modification of a technique first described to us by Marcus Fruttiger and William Richardson (Fruttiger *et al.*, 1996).

The following steps are to be performed after staining and final washes are complete. (Staining and washes can be easily performed in wells of a 24- or 48-well plate.)

1. Place each retina in a single well of a six-well Teflon slide with 8-mm diameter wells (Electron Microscopy Sciences, cat# 63423-08).
2. The center of the retina should be in the center of the well.
3. "Flower" the retina by making four to six evenly spaced radial cuts from the retinal periphery two-thirds of the way into the retina center.
4. Use these cuts to flatten the retina by unfolding each petal of the retinal flower (each section within the cuts) using fine forceps or a paint brush and gently flattening from the center outward.
   - The vitreous side of the retina should be facing up.
   - Keep the retinas moist, but flattening the retinas is easier if most of the liquid has been removed from the slide surface.
5. After all the retinas for a slide have been flattened, add a drop of slow-fade reagent (Molecular Probes Inc., Eugene OR) to each corresponding spot on the coverslip.
6. Apply the coverslip with the slow fade media drops onto the slide with retinas.
7. Remove excess slow fade media from the edges of the coverslip.
8. After the coverslip is satisfactorily on the slides with the retinas, and all retinas are flat and open vitreal side up, add a small amount of clear nail polish to the corners to prevent sliding of the retina(s).

3. Corneal Micropocket Assay

A key assay that is widely used to identify and characterize angiogenic and angiostatic agents is the rabbit cornea micropocket assay. The cornea is avascular, but is surrounded by a network of perilimbal vessels (Ruben, 1981), and any new blood vessels arising in the cornea after stimulation by angiogenesis-inducing tissues or factors can easily be identified. While the method was originally developed for rabbit eyes (Gimbrone *et al.*, 1974), which continues to be the most commonly used test animal, the method has now been adapted to mice (Muthukkaruppan and Auerbach, 1979; Muthukkaruppan *et al.*, 1982) and rats (Fournier *et al.*, 1981). Briefly, test materials (i.e., tumor cells, tissue, or purified factors) are introduced into a pocket made into the cornea eliciting growth of new vessels from the peripheral limbal vasculature. This vascular response can then be quantified in a highly reproducible, non-invasive manner (Shaffer, 1996). Because of these attributes, the corneal micropocket assay has been used extensively to assess the activity of various proangiogenic molecules, as well as test the ability of various angiostatic molecules to prevent corneal angiogenesis upon stimulation by cytokines (Friedlander *et al.*, 1995).

There are considerable advantages to the rabbit corneal micropocket assay, specifically the absence of an existing background vasculature in the cornea and the ability to monitor angiogenesis progression. However, the surgical procedure is demanding so that relatively few ($\sim$6 to 12 rabbits) can receive pellet implants in a single setting and inflammation can cause difficulty. Additionally, the fact that the normal cornea is an avascular structure makes this assay somewhat atypical, since normal tissues, with few exceptions, are vascularized.

3.1. Rabbit corneal micropocket assay protocol

3.1.1. Making six pellets

1. In a 1- to 15-ml Eppendorf vial, warm 6 ml of 95% ethanol at 37 °C for 15 min.
2. Add 0.72 g of hydron polymer (polyhydroxyethyl methacrylate, from Interferon Sciences, New Brunswick, NJ) and shake for 10 min.
3. Mix in 22.5 mg of sucralfate (Carafate, from Marion-Merrell Dow Pharmaceuticals, Cincinnati, OH) with PBS including cytokine/molecule of interest in 61.92 μl total solution.
4. Add 45 μl of hydron suspension and mix.
5. Pour mixture into six 17-μl casting molds and let dry under UV light overnight.
6. Take the pellets from the molds and leave them for 1 h before the surgery in UV light.

3.1.2. Surgical insertion of pellet into corneal micropocket
Instruments

Beaver mini-blade
Colibri forceps

0.12 suturing forceps
Lid speculum
Castroviejo measuring calipers
Iris spatula
Pellets

Surgical technique

1. Anesthetize white New Zealand rabbits (2.5–3.5 kg) with a 1-ml intramuscular injection of 50/50 mixture of ketamine HCL (100 mg/ml, Fort Dodge Laboratories Inc., Fort Dodge, IA) and xylazine HCL (20 mg/ml, Rugby Labs Inc., Rockville Centre, NY).
2. Anesthetize eyes with Opticaine tetrachloride, and irrigate topically with 0.9% saline throughout surgical procedure.
3. Insert lid speculum.
4. Image each cornea at 16× magnification under consistent lighting conditions with Kodak Ektachrome 64T slide film prior to surgery using a camera mounted on a Wild operating microscope.
5. Use the Castroviejo measuring caliper to measure 2 mm from the limbus.
6. Using the beaver mini-blade, create a vertical incision ~4 mm from the limbus up to the stroma avoiding penetration to the anterior chamber.
7. With an iris spatula dissect the cornea pocket in the direction of the limbus.
8. With the colibri and the 0.12 forceps insert the pellet. The pellet should be located 2 mm from the limbus.
9. Irrigate the cornea with saline and add one drop of Tobramycin (Tobrex 0.3%); Falcon Ophthalmics Inc., Fort Worth, TX).
10. Close corneal slit, and monitor rabbits until awake.

Imaging

1. Sedate rabbits with a 1-ml intramuscular injection of a 50/50 mixture of ketamine HCL (100 mg/ml, Fort Dodge Laboratories Inc., Fort Dodge, IA) and xylazine HCL (20 mg/ml, Rugby Labs Inc., Rockville Centre, NY).
2. Apply a drop of topical Opticaine tetrachloride.
3. Using a Wild M490 operating microscope, examine corneas on days 0 (postoperative), 2, 3, 5, 7, 10, and 14.
4. To yield a maximum field of view of neovascularization for the camera, rabbits should be positioned for photography such that the pellet and the limbus are oriented planar to the camera lens.
5. Euthanize rabbits at day 14 by lethal injection of sodium pentobarbital euthanasia solution into the marginal ear vein.

3.1.3. Rabbit corneal micropocket assay quantification

Using simple slide photography and a scanning imaging densitometer, one can obtain multiple data points from a single rabbit in a consistently reproducible, noninvasive manner that is sensitive to small changes in vessel area. Additionally, multiple observers may obtain consistent results (Shaffer, 1996).

Data acquisition

1. Using BIO-RAD's GS-670 imaging densitometer (or any other equivalent and available densitometer software), scan each slide by transmittance at a resolution of 64 μm and a gray scale of 256 shades.
2. Convert the scanned slide into a computer image using BIO-RAD's Molecular Analyst 1.1 software.
3. Using the software's magnification tool, expand image to 100%.
4. Adjust the histogram gray-scale plot to lighten the image until the greatest contrast of corneal surface to blood vessels is obtained. For density analysis, this procedure does not alter the underlying image data.
5. To obtain the resultant raw area, outline the area beginning at the limbus using the lasso tool. Do not include hemorrhage and noncorneal vessels in the analysis.
6. Measure the diameter of the cornea on a selected preoperative slide along the axis of the pellet. Calculate a conversion factor for raw area cm² to actual area mm² using the following equation:

$$\text{Conversion factor} = [(\text{actual diameter})^2 / (\text{slide diameter})^2] \times 100$$

7. Clear outline and retrace three times to determine the precision of the computer-assisted analysis.

Vessel quantity determination

1. Obtain measures of vessel quantity for rabbits at days 6, 8, 10, and 12. Days 2 and 4 may be difficult as massive corneal edema can be problematic and result in obscured vessel detail.
2. Utilizing the same slides from the area quantifications, following outlining, record both area and "volume" (expressed as cm² × optical density).
3. Divide the volume by the area to obtain the optical density of the region.
4. Using landmarks on the cornea, outline a similar region for each of the four postoperative slides, and record the same measurements.
5. To determine the background optical density for the avascular cornea, calculate the mean value of volume/area.

6. Multiply the difference in the background and neovascular optical density by the neovascular spot area to obtain a relative value for blood vessel quantity ($cm^2 \times$ optical density).
7. Multiply this value by the conversion factor to obtain actual relative vessel quantity.
8. Perform this procedure a total of four times for each slide.

4. NEONATAL MOUSE RETINAL DEVELOPMENTAL ANGIOGENESIS MODEL

The lack of directly observable models of developmental angiogenesis that can be manipulated and easily analyzed has proven to be a significant limitation for researchers investigating mechanisms of angiogenesis during development. Because they cannot fully account for the natural complex, context-dependent environment encountered by developing vessels the results observed from *in vitro* assays, while important, often fail to translate to similar results *in vivo*. Thus, there is a need for many different *in vivo* models of angiogenesis, many of which are described elsewhere in this volume.

The retina has several advantages that overcome many of the experimental limitations that can be associated with *in vivo* studies permitting the study of developmental neovascularization in its natural context. The biggest advantage is that mouse retinal vascular development occurs postnatally when well-defined vascular plexuses form in a highly reproducible temporal and spatial pattern (Dorrell *et al.*, 2002; Dorrell and Stone *et al.*, 1995; Friedlander, 2006). Postnatal vascular development facilitates direct treatment with exogenous materials, such as molecular agonists or antagonists (Otani *et al.*, 2002b), eliminating many of the difficulties associated with studying embryonic vascular development. Additionally, the retinal vascular plexuses are formed within well-defined regions in highly reproducible patterns (Dorrell and Friedlander, 2006); three distinct vascular plexuses are formed during the first 3 weeks after birth. Because these plexuses form in a characteristically uniform spatial and temporal fashion alterations in the location or pattern of these plexuses can easily be observed (Fig. 6.1). This is not necessarily true during vascular development of other tissues where, although endothelial cell guidance and vascular patterning is critical, alterations in the vascular patterns in response to exogenously administered factors are not as easily identified. Finally, a subpopulation of bone marrow–derived cells that can target the developing retinal vasculature has recently been identified (Dorrell *et al.*, 2004; Otani *et al.*, 2002a; Ritter *et al.*, 2006). This provides an opportunity to evaluate the role of specific factors on developmental neovascularization through targeted knock-ins or knock-downs. Thus, using the mouse retinal vascular development model, mechanisms of vascular development can be

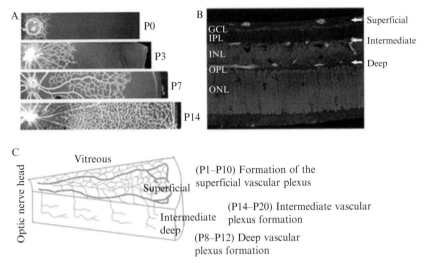

Figure 6.1   Postnatal mouse retinal vascular development occurs in a highly reproducible spatial and temporal fashion. (A) The early development of the superficial plexus using retinal whole mounts stained for collagen IV. (B) An adult-mouse retinal cross-section demonstrating the three distinct vascular plexuses that form during development. The superficial plexus forms within the ganglion cell layer (GCL), the intermediate plexus forms at the inner edge of the inner nuclear layer (INL), and the deep plexus forms near the outer plexiform layer (OPL) (blue, DAPI [nuclei]; red, CD31 [vessels]). (C) The sequence of retinal vascular development for BALB/cByJ mouse pups is shown in cartoon form. During week 1 after birth, the superficial vascular plexus forms as vessels migrate from the central retinal artery toward the retinal periphery. Just after week 1, the superficial vessels branch and endothelial cells migrate and form the secondary, deep vascular plexus at the outer edge of the inner nuclear layer (INL). During week 3 after birth a tertiary, intermediate vascular plexus is formed at the inner edge of the INL. (Adapted from Dorrell, M. I., and Friedlander, M. (2006). Mechanisms of endothelial cell guidance and vascular patterning in the developing mouse retina. *Prog. Retin. Eye Res.* **25**, 277–295.) (See Color Insert.)

analyzed *in vivo* by studying (1) the effects of exogenously administered compounds, (2) vascular development in various transgenic mice, and (3) targeted knock-ins or knock-downs using transfected vascular-targeting bone marrow cells. This model is particularly effective for the analysis of angiostatics on the formation of new vessels and the assessment of drug effects on established vasculature (Dorrell *et al.*, 2007).

4.1. Using the neonatal mouse model to assess the activity of angiostatic inhibitors

BALB/cByJ albino mice or C57BL/6J mice are used most frequently for the mouse retinal developmental angiogenesis model, although other strains can also be used. However, it should be noted that the developmental time course of different strains, including BALB/cByJ and C57BL/6J strains, can be slightly different. The superficial plexus of both BALB/cByJ and C57BL/6J strains forms during week 1 after birth. The superficial vessels sprout and begin the formation of the deep vascular plexuses around P7–P8. In the BALB/cByJ mouse pups, the deep plexus forms over the course of approximately 5 days (P8–P12) and the intermediate plexus forms subsequently during week 3 after birth (P14–P20) (Fig. 6.1). The formation of the deep and intermediate plexuses occurs more rapidly in the C57BL/6J strain. The deep plexus of C57BL/6J mouse pups becomes fully developed within 2 to 3 days (P8–P10) and formation of the intermediate plexus begins immediately thereafter. Formation of both plexuses is nearly complete by the end of week 2 after birth. Both strains are useful for studying vascular development. However, due to the distinct periods between formation of the deep and intermediate plexuses, as well as the relatively extended (but still short) developmental time period, the BALB/cByJ mouse model tends to be more useful for studying the effects of angiostatic molecules on neovascularization.

At postnatal day 7 (P7), the superficial vasculature has developed ~80% toward the retinal periphery, much of the central two-thirds of the superficial plexus has already matured (Ishida *et al.*, 2003b), and formation of the deep plexuses has yet to begin. By injecting various test molecules, such as angiostatics (Otani *et al.*, 2002b) or various combination of angiostatics (Dorrell *et al.*, 2007), at P7 the effects on formation of the deep vascular plexuses can be assessed (Fig. 6.2). In addition, one can assess the effects of the test molecules on established, but immature, as well as established mature vessels within the superficial plexus.

4.2. Neonatal mouse retinal developmental angiogenesis protocol

1. Mouse pups are injected intravitreally at postnatal day 7 (see Section 1 on intravitreal injections for details).
2. Following injections, the mouse pups are left with their mothers for 5 days.
3. At P12, euthanize the mice, enucleate the eyes, and isolate the retinas by dissection (see Section 1 on retina isolation for details).
4. The retinas are stained with any of a number of vascular markers such as collagen IV, CD31 (PECAM), Von Willibrand factor, or isolectin griffonia simplicifolia (any other valid marker of vessels can be applied here as well).
   - It should be noted that the staining methods (fixation, blocking, incubation times, and antibody concentrations) will vary depending on the vascular markers used. We commonly use the following collagen IV stain for analyzing angiostatic molecules in the neonatal mouse retinal model.

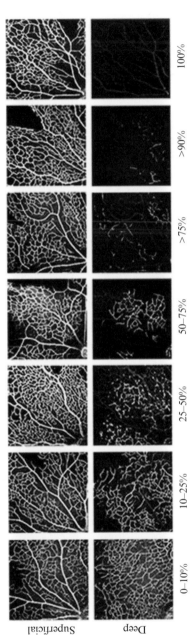

Figure 6.2 Levels of deep vascular inhibition. The superficial and deep vascular plexuses can be easily differentiated using confocal imaging of retinal whole mounts stained for the vasculature by collagen IV, CD31, Von Willebrand factor, isolectin *Griffonia simplicifolia*, or any other common vascular factor (shown images are stained with antibodies against collagen IV from Chemicon [AB756P]). Upon injection of angiostatic molecules at postnatal day 7 (P7), the effects on formation of the deep vascular plexus can be assessed 5 days later at P12. These effects can be semiquantified by a simple scoring method, or quantified more thoroughly by quantifying and comparing areas of pixilation using standard imaging and commonly available software that can analyze areas of pixilation. Adverse effects, or the lack thereof, on the superficial plexus, which is mainly established by P7, can also be assessed. (Adapted from Dorrell, M. I., and Friedlander, M. (2006). Mechanisms of endothelial cell guidance and vascular patterning in the developing mouse retina. *Prog. Retin. Eye Res. 25*, 277–295.)

- After retina isolation, postfix the retinas for 10 min in ice-cold methanol.
- Wash with PBS (2 × 5 min).
- Block with 20% fetal bovine serum and 20% normal goat serum in PBS containing 0.1% triton X-100 for ~2 h at room temperature.
- Incubate with primary antibodies against collagen IV (Chemicon, AB756P, 1:200 dilution) in 0.5X block solution (10% FBS + 10% NGS in PBS with 0.05% triton X-100) overnight at room temperature.
- Wash for 1.5 h in PBS (changing wash every 15–20 min).
- Incubate with appropriately labeled secondary antibodies in 0.5X block solution.
- Wash for 1.5 h in PBS (changing wash every 15–20 min).

5. Mount the retinas vitreal side up, one per well of a six-well slide, 8-mm diameter wells (Electron Microscopy Sciences, cat# 63423-08) with antifade media (see the previous section for details on mounting retinal whole mounts).

6. Imaging and analysis: Imaging can be done a number of different ways. In our experience consistently imaging the same quadrant of each retina (i.e., upper right quadrant) using 100× total magnification (10 × lens) works well (Fig. 6.2).
   - Using confocal microscopy focus on the superficial plexus. Image one quadrant of the superficial plexus.
   - Focus on the deep plexus within the same quadrant and obtain an image of the deep vasculature.
   - Analysis and quantification can be done by quantifying the total area of pixilation for each imaged area or by arbitrarily assigning different levels of deep vascular plexus formation across the entire retina (Dorrell *et al.*, 2007) (Fig. 6.2).

Notes: We have found that ~10 retinas per group are required to generate strong statistical differences, although this number is dependent on the strength and activity of the molecules being tested.

Contralateral eyes can be used to compare test molecules and controls. For example the test compound can be administered to the right eye of each animal while the left eye receives the vehicle. However, this should be avoided if the molecules are small and are able to easily cross into the blood and reach the contralateral eye through the circulation.

5. OXYGEN-INDUCED RETINOPATHY MODEL

Models of oxygen-induced retinopathy (OIR) have become integral to the study of pathological angiogenesis resulting from tissue ischemia. This model has been developed and reported in several animal species (Madan and Penn, 2003), including the kitten (Chan-Ling *et al.*, 1992), beagle puppy (McLeod *et al.*, 1998), rat (Penn *et al.*, 1993), and mouse (Smith *et al.*, 1994). Each of these models utilizing different animal species relies on a similar, basic premise by which exposure to hyperoxia during early retinal development stages results in the attenuation of normal retinal vascular development and the regression of immature, newly developed vessels. Upon removal from hyperoxia, a relative ischemic situation ensues as the retina is no longer immersed in the high oxygen environment and now lacks its normal vasculature which is required to adequately nourish the neural tissue in normoxic conditions. This ischemic situation initiates rapid, abnormal neovascularization which results in several vascular abnormalities including the formation of neovascular tufts which protrude out of the normal vascular plexuses into the vitreal cavity and vascular leak. The vascular changes are very consistent, reproducible and quantifiable, and thus it has become a well-accepted model for studying disease mechanisms and potential treatments for hypoxia-induced retinopathy. These models directly mirror the events that occur during retinopathy of prematurity (ROP), a condition involving pathological neovascularization that can affect premature infants (Ashton, 1966; Hellstrom *et al.*, 2001; Smith, 2002).

As a consequence of the underdeveloped respiratory and cardiovascular system, premature infants are often placed in hyperoxia for some time after birth. In certain cases, this can cause attenuation of the normal retinal vascular development. As the infant becomes healthier and is removed from the hyperoxic chambers ischemic retinopathy similar to that observed in the OIR models can ensue. In some cases this can lead to devastating vision loss and even blindness in certain individuals. This ischemia-mediated pathological neovascularization also makes the oxygen-induced retinopathy models useful for studying many different diseases with associated ischemic, pathological neovascularization including diabetic retinopathy, the leading cause of vision loss in younger adults. Over the last decade, the mouse model of OIR as described by Smith *et al.* has become the most common model for studying abnormal angiogenesis associated with oxygen-induced retinopathies (Smith *et al.*, 1994). In recent years the use of this model has been extended to the general study of ischemic vasculopathies and related anti-angiogenic interventions and it is now used extensively in both basic and applied research environments.

5.1. Mouse and rat OIR models

The most commonly used models of OIR are the mouse model, originally described by Lois Smith (Smith *et al.*, 1994), and the rat OIR model as described by John Penn (Penn *et al.*, 1992, 1993). There are specific advantages and disadvantages to the use of both. The main advantage of using the mouse OIR model is the powerful tool of mouse genetics;

transgenic mice can be used to directly study the roles of different genes on pathological neovascularization. Another advantage is the simplicity of the model, relative costs, and the size of the animals. However, a distinct disadvantage to the use of mice in the OIR model is the fact that in mice, the central retinal vessel are obliterated during exposure to hyperoxia. This is different from what happens in clinical ROP where the peripheral retinal vessels obliterate, rather than the central retinal vessels. However, despite this key difference, the mouse model nicely recapitulates many of the events that occur during ischemia-induced neovascularization. Unlike the mouse model, the rat model of OIR uses alternating hyperoxia–normoxia cycles and its main advantage is that similar to the events during ROP, it is the peripheral vasculature that becomes obliterated upon neonatal exposure to hyperoxia.

5.1.1. Mouse OIR protocol

1. C57BL/6J mouse pups and their mothers should be placed in 75% oxygen beginning at postnatal day 7 (P7).
   - C57BL/6J mice should be used since the pathology of other strains differs following this protocol. Some strains, such as BALB/cByJ, do not develop pathological NV at all despite hyperoxia-induced vascular obliteration (Ritter *et al.*, 2006).
   - Oxygen chamber: Automated oxygen monitors that control the influx of oxygen based on the oxygen levels within the chamber should be used.
   - Chambers should be used that allow circulation of the air throughout the chamber.
   - If possible, carbon dioxide levels should also be monitored and maintained at relatively normal levels. Although the chambers should be adequately air-tight so that the 75% oxygen levels can be maintained, the chamber should not be completely air-tight as this will cause lethal levels of carbon dioxide to be reached.
2. The mice remain in the oxygen chamber for 5 days with constant 75% oxygen levels.
3. At P12, remove the mice from oxygen and keep the mice in normoxia for an additional 5 days.

   Note: The incubation in hyperoxia can cause problems in the mothers. The mothers should be monitored closely during the last days in hyperoxia and the first few days upon return to normoxia. Replace the mothers with surrogate mothers as needed.

4. At P17, or 5 days after return to normoxia, euthanize the mouse pups and enucleate the eyes for analysis and quantification.

5.1.2. Classical method of quantifying pathological neovascularization

In the original publication by Lois Smith (Smith *et al.*, 1994), the area of obliteration was quantified using retinal whole mounts with the vessels visualized by perfusion with fluorescently labeled dextran. However, because the prelaminar neovascular tufts are not perfused sufficiently by dextran perfusion, quantification of the area of tuft formation using retinal whole-mounts was not possible. Thus, whole eyes were serially sectioned and the numbers of prelaminar nuclei were counted as an indication of the extent of pathological neovascularization.

Protocol for perfusion with fluorescein dextran

1. At P17 (or the appropriate time of analysis), anesthetize the mouse pups appropriately and perfuse PBS with high-molecular-weight (2×10^6 molecular weight) fluorescein dextran through the left ventricle.

   Note: The fluorescein dextran should be clarified by centrifugation prior to use.

2. Allow perfusion to occur for approximately 30 min and then euthanize the mice.
3. Enucleate the eyes and isolate the retina using the dissection protocol described earlier.
4. Postfix the retinas in 4% paraformaldehyde (PFA) for 2 h.
5. Flat mount the retinas as described earlier and image using fluorescent microscopy.
6. The areas of obliteration can be defined and quantified using standard software that allows quantification of areas, including Photoshop and ImageJ.

Quantifying the extent of pathological neovascular tuft formation by serial sections

1. Euthanize the mice at P17 or the desired time point for analysis.
2. Enucleate the eyes and immerse in 4% PFA for 24 h.
3. Embed the eyes in paraffin.
4. Section the eyes in 6- to 10-micron increments sagittally, through the cornea and parallel to the optic nerve.

   Note: No differences in quantification were found based on the nasal or temporal orientation of the eye during sectioning.

5. Stain the retina using standard hematoxylin- and eosin-staining techniques.
6. Identify the cross-section containing the optic nerve.
7. Count the number of nuclei that are located within the vitreal space (prelaminar nuclei) using four to eight slides on each side of the optic

nerve head approximately 30 microns apart. These nuclei are from the pathological neovascular tufts that have protruded into the vitreal space.

Note: Sections including the optic nerve head should be excluded since nuclei from the optic nerve head itself can inadvertently be included in the neovascular nuclei counts.

Note: Nuclei from new vessels penetrating the prelaminar space are easily identified by light microscopy. However, a small number of hyaloidal vessels remain in the vitreal cavity at P17 and it should be noted that nuclei from these vessels may also be found in the vitreal cavity. Only nuclei in the prelaminar space but relatively close to the retina should be included in the counts.

Protocol for quantifying areas of neovascular tuft formation using retinal whole mounts

Protocols have recently been developed that allow researchers to quantify the extent of neovascular tuft formation in retinal whole mounts using postfixation staining with isolectin griffonia simplicifolia (lectin). The advantages to using this method included ease of use, relatively rapid analysis, and the ability to assess the areas of remaining vascular obliteration and the extent of pathological neovascular tuft formation in the same eyes. Smaller areas of obliteration at P17 indicate faster physiological revascularization, assuming that the extent of original obliteration at P13 was equivalent between test and control animals. These methods have proven to be at least as reliable as the classical methods of quantification and are described thoroughly by Banin and colleagues (2006).

1. At P17, or the desired time point for analysis, euthanize the mice, enucleate the eyes, and isolate the retinas using the described dissection protocols.
2. Staining for isolectin griffonia simplicifolia:
   - Postfix the dissected retinas in 4% PFA for 1 h.
   - Briefly wash in PBS.
   - Stain using 1:200 dilution of 1 mg/ml fluorescently conjugated isolectin GS-IB4 (Invitrogen/Molecular Probes) in 1 × PBS for 4 h to overnight at room temperature.
   - Wash with 1 × PBS for approximately 30 min.
   - Whole mount according to the described protocol.
3. Image whole retinas using fluorescent microscopy.
   - The whole retina can be imaged using low-magnification lenses (4 × or 5 × lenses) and imaging each of the four quadrants.
   - Create a montage of the four images to obtain a single image of the whole retina.

- Focus just above the normal superficial plexus so that the prelaminar tufts are in focus and imaged at higher intensities. This will allow easier identification and quantification of the neovascular tufts versus normal retinal revascularization.

4. Quantify the extent of remaining vascular obliteration by tracing the central, avascular areas of the retina and determining the area using standard imaging and quantification software.

5. The neovascular tufts can be visualized as the intensely stained, abnormal collections of endothelial cells that overly the normal superficial vascular network (Fig. 6.3). Select these neovascular tufts and quantify the area of tuft formation using standard imaging and quantification software.

For details of the quantification, particularly using Photoshop analysis, see Banin *et al.* (2006).

5.1.3. Rat OIR model

While the response of the developing retinal vessels has varied with differing oxygen protocols, human-like patterns of vasoattenuation and subsequent neovascularization can be produced using the rat model of OIR. Similar to the mouse model, protocols using appropriately timed exposure to hyperoxia lead to retarded growth of both superficial and deep vessels. In the mouse model, this results in obliteration of the central retinal vessels while in the rat, this results in a peripheral avascular zone, more similar to the events observed in ROP (Madan and Penn, 2003). Also similar to the mouse model, removal from oxygen exposure to room air causes pathological neovascular growth that results in the formation of prelaminar neovascular tufts at the boundary between vascular and avascular areas. The original protocol described by Penn and colleagues (1993) utilized alternating exposure of 80% and 40% oxygen for 12-h cycles. The more refined, current protocol also uses alternating oxygen cycles, but these cycles now utilize 12-hour cycles of 50% and 10% oxygen (Madan and Penn, 2003). The timing of the oxygen exposure, both length and age at which the pups are initially exposed to hyperoxia, is also different from the mouse OIR protocol. The rat OIR model, has been used extensively to identify factors involved in hypoxia-induced neovascularization (Penn and Rajaratnam, 2003; Robbins *et al.*, 1998) as well as to test various ocular angiostatics (Penn *et al.*, 2001).

Protocol

1. Within hours after birth, litters of Sprague-Dawley albino rat pups should be randomized and placed with their mothers in chambers where oxygen levels can be controlled.

2. Alternate the oxygen levels in the chamber between 50% and 10% every 12 or 24 h for 14 days.

Figure 6.3 Mouse model of oxygen induced retinopathy. (A) Mouse pups are put into hyperoxia for 5 days from postnatal day 7 (P7) until P12. This results in attenuation of normal retinal vascular development (compare with Fig. 7.1) and obliteration of imma-ture central retinal vessels. Following return to normoxia at P12, the retina becomes hypoxic and pathological neovascularization is induced. (B) Characteristic retinas at various time points are shown using fluorescein dextran perfusion, and postfixative staining with isolectin *Griffonia simplicifolia* (lectin). Note the obliteration of the central vessels at P9 during hyperoxia incubation, the extensive pathologic neovascularization at P18 when stained with lectin and the natural healing that occurs in the mouse model. (C) The remaining area of obliteration and area of pathological NV (tufts) can be quan-tified by retinal whole mounts stained with lectin. This method is useful to compare dif-ferences between treated and control, or transgenic and WT retinas. (D) The classical method of quantifying the extent of neovascularization used serial sectioning and counted preretinal nuclei. (E) Note that perfusion should not be used to quantify the area of neovascular tufts since the majority of the tuft areas are not perfused. (Adapted from Banin, E., *et al.* (2006). T2-TrpRS inhibits preretinal neovascularization and enhances physiological vascular regrowth in OIR as assessed by a new method of quantification. *Invest. Ophthalmol. Vis. Sci.* **47**, 2125–2134; and Dorrell, M. I., *et al.* (2007). Combination angiostatic therapy completely inhibits ocular and tumor angiogenesis. *Proc. Natl. Acad. Sci. U S A* **104**, 967–972.)

3. Remove the rat pups and mothers from the oxygen chamber and maintain in room air (normoxia) for 4 to 7 days.
4. Euthanize rats at the desired endpoint.
5. Enucleate the eyes and dissect the retinas in 10% neutral buffered formalin.
6. Flatten the retinas and stain with adenosine diphosphatase (ADPase) or an equivalent method of visualizing the vasculature.
7. Visualize, image, and quantify the extent of remaining vascular obliteration and retinal neovascularization using quantification methods similar to those described for the mouse OIR model.

6. LASER-INDUCED RETINAL AND CHOROIDAL NEOVASCULARIZATION MODELS

Neovascularization (NV) secondary to vascular changes within the eye is responsible for several ocular pathologies, including age-related macular degeneration, retinal vein occlusion, premature retinopathy, and diabetic retinopathy. Clinically relevant animal models of NV are needed to help elucidate the mechanisms underlying such diseases. We have described various models which can be used to study developmental angiogenesis, and ischemia induced pathological angiogenesis in the retina (OIR). Other options include the use of growth factors to induce such vascular changes within animal eyes (Antonetti et al., 1998; Boyd et al., 2002; Caldwell et al., 2003; Fine et al., 2001; Qaum et al., 2001; Vinores et al., 1997). In this section, we will describe clinically relevant models that utilize laser treatment to mechanically induce retinal or choroidal NV, including central retinal vein laser photocoagulation and choroidal laser photocoagulation model. Most of these models are adapted from the seminal work of Stephen Ryan (1979).

6.1. Murine branch retinal vein occlusion model

Retinal vein occlusion in mice and rats causes microvascular changes that include venous dilation, vascular remodeling, and extravasation of fluid from retinal vessels, and serves as a useful model for the study of antiangiogenics in vivo. This model most closely mimics clinical branch retinal vein occlusion. This disease occurs when one or more major retinal vein(s) become blocked for various reasons, including blood clotting, thrombi, and particle blockage. This can lead to reduced blood flow, retinal ischemia, and neovascularization within the retina or iris vasculature. Similar to most other ocular vascular diseases, the newly formed vessels are generally abnormal, characterized by vascular leak, and can lead to hemorrhage, retinal traction, scarring, and subsequent visual complications (Central Vein Occlusion Study Group,

1997). In the central vein occlusion model, laser photocoagulation is used to obstruct major veins in the rodent retina, leading to similar vascular changes, including retinal neovascularization and vascular leak.

6.1.1. Protocol

1. Anesthetize the animal using an appropriate volume of intramuscular anesthesia (In C57BL/6J mice 6 to 8 weeks old, 50 mg/kg of ketamine HCl and 10 mg/kg of xylazine are used.)
2. Dilate pupils using two drops of cyclopentolate 1% and phenylephrine hydrochloride 10%.
3. Inject Rose Bengal (Sigma, cat. #R-3877) via tail vein to better visualization of the vasculature and to improve quality of the vascular closure by the laser.
4. Apply a glass coverslip coated with 2% methylcellulose solution over the cornea to flatten it and visualize the fundus using a slit lamp.
5. Once the retinal vasculature is visible by fundus imaging, the major veins can be identified (the arteries are less thick than the veins).
6. Apply laser spots aiming two to three major veins or one quadrant as they emerge from the optic nerve up to one disc diameter (try to avoid the arteries). Use optimized laser parameters: 180-mW diode laser, 50 microns in size, up to 0.10 s (or until venous occlusion is achieved). Parameters should be adjusted to the laser and the media clarity of the eye.

   Note: Obtaining the optimal laser spot is critical. If the laser photocoagulation is too weak, occlusion will not be achieved, but using laser settings that are too strong can extensively damage other areas of the retina or cause vitreous hemorrhage from the vein. Correct aim targeting solely the larger veins in regions within one disc diameter of the optic nerve is also critical for similar reasons.

   Note: Because of these and other potential complications, including accidental laser on the cornea, lens, retina, choroid, and or sclera, vitreous hemorrhage and possible perforation of the eye laser treatment should be performed by a fully trained ophthalmic surgeon.

9. After laser photocoagulation is complete, place hydroxypropylmethylcellulose ointment (Novartis) over the eye at the time of recovery.

   Note: Animals should be closely monitored following laser treatment for any signs of distress or discomfort, such as poor grooming habits and pawing of the eye.

10. The retina should be visualized over the course of several days in order to follow venous occlusion and subsequent remodeling of the vasculature and measurable extravasation of fluid. (We typically observe animals at days 0, 3, 5, 7, 14, 21, and 28 days after the procedure by

visualization of the fundus through a coverslip coated with methylcellulose and a slit lamp.)

11. At appropriate time points following the laser photocoagulation, the animals can be euthanized, the retinas isolated, and the vasculature stained for analysis according to the described retina dissection, staining, and mounting protocols.

12. Measurements of extravasated fluid (neovascular leak) can be made using either Evans blue as described in the leak assessment model section, or *in vivo* laser protocols as adopted from Miller *et al.* (1994) and Xu *et al.* (2001); respectively.

6.2. Choroidal laser photocoagulation in murine models

The leading cause of vision loss in older adults is caused by age-related macular degeneration (AMD). Although there currently is no suitable model for AMD, the laser-induced choroidal model can be used to study choroidal neovascularization (CNV). In neovascular AMD, which accounts for approximately 80% of severe vision loss, subretinal vessels originating from the choroid develop beneath the retina. The mouse model of laser-induced CNV is well characterized and, in the absence of models which perfectly recapitulate the various aspects of AMD, has been used extensively to study the role of individual angiogenic factors as well as the effect of anti-angiogenic agents or other therapeutic agents/techniques on choroidal neovascularization (Elizabeth Rakoczy *et al.*, 2006). Similar to the techniques used in the vein occlusion model, this model generally utilizes laser photocoagulation to induce break(s) in Bruch's membrane, the physical barrier that normally mediates separation of the choroidal vasculature from the RPE and neural retina. Inflammation, growth factor up regulation and the physical disruption of this barrier results in choroidal neovascularization that can be followed and studied to identify important neovascular factors, or assess various angiostatics.

6.2.1. Protocol

1. Anesthetize the animal using an appropriate volume of intramuscular anesthesia (In C57BL/6J mice 6 to 8 weeks old, 50 mg/kg of ketamine HCl and 10 mg/kg of xylazine are used.)

2. Dilate pupils using two drops of cyclopentolate 1% and phenylephrine hydrochloride 10%.

3. Apply a glass coverslip coated with 2% methylcellulose solution over the cornea to flatten it and visualize the fundus using a slit lamp.

4. Generally, three separate laser photocoagulation burns are performed to induce ruptures in Bruch's membrane at three separate locations.

5. A green diode laser set for 150 mW, 0.05-s duration, and 50-micron spot size should be focused on the RPE and applied approximately one disc

diameter from the optic nerve. (Parameters and quality of laser burns should be adjusted to the laser and the media clarity of the eye at the time of the procedure.)

Note: It is recommended to look for an air bubble at the time of the laser application (indicating the rupture of Bruch's membrane) to ensure quality and consistency.

6. After laser photocoagulation is complete, place hydroxypropylmethyl-cellulose ointment (Novartis) onto the eye.

Note: Animals should be closely monitored following laser treatment for any signs of distress or discomfort, such as poor grooming habits and pawing of the eye.

7. The retina should be visualized over the course of several days in order to follow venous occlusion and subsequent remodeling of the vasculature and measurable extravasation of fluid.
8. At appropriate time points following the laser photocoagulation, the animals can be euthanized, the retinas isolated, and the vasculature stained for analysis according to the described retina dissection, staining, and mounting protocols.
9. Various methods have been used to quantify the degree of neovascularization (Campa *et al.*, 2008).

7. RETINAL VASCULAR PERMEABILITY MODELS

Abnormal retinal vascular permeability characterizes a number of eye diseases such as diabetic retinopathy, retinal vascular occlusions, exudative macular degeneration, and inflammatory and neoplastic conditions (Chahal *et al.*, 1986; Cunha-Vaz, 1983). This increased vascular permeability leads to fluid and exudative deposits in the central retina, which is the leading cause of vision loss in retinal disease. The effects of these diseases highlight the need for representative models of vascular permeability. There are a variety of models demonstrating increased vascular permeability in the eye, as well as in other tissues, such as the lung, heart, and skin (Weis, 2007). Here we focus on models of ocular vascular permeability as a good way to assay the consequences of increased permeability in eye disease and to develop treatments to prevent leak and the resultant damage to surrounding tissues.

Several models of ocular permeability and blood–retinal barrier break-down exist, including *in vivo* laser-induced hyperpermeability (Ryan, 1979), streptozotocin-induced experimental diabetes, and intraocular injection of various growth factors, such as VEGF (Xu *et al.*, 2001). We will focus on intraocular VEGF injection–induced vascular permeability in mice. Vascular endothelial growth factor (VEGF) is known to be a potent regulator of vascular permeability (Dvorak *et al.*, 1995; Senger *et al.*, 1983) and increased levels of VEGF have been found to correlate with increases in ischemic disease and possibly exudative macular degeneration and uveitis as well (Antonetti *et al.*, 1998; Boyd *et al.*, 2002; Caldwell *et al.*, 2003; Fine *et al.*, 2001; Qaum *et al.*, 2001; Vinores *et al.*, 1997). This key role of VEGF in vascular leak has made it a popular cytokine to use in models of permeability. Quantification of retinal vascular leak in these rodent models can be achieved by utilizing a number of tracers such as Evans blue, FITC-dextran, and [^{3}H] mannitol (Dvorak *et al.*, 1995; Ishida *et al.*, 2003a; Xu *et al.*, 2001). The protocol described here is based on an already established protocol to assess vascular leak using quantification of Evans blue in rats (Xu *et al.*, 2001) with some modifications.

7.1. Retinal vascular permeability quantification using Evans blue

1. Under light anesthesia, typically 50 mg/kg ketamine and 5 mg/kg xylazine, inject recombinant human VEGF (R&D Systems) (we use 100 ng in a volume of 0.5 μl) into the vitreous cavity of one eye (BALB/cByJ mouse). Inject an equal volume of vehicle (PBS with 0.1 % BSA) into the contralateral eye (Hamilton syringe, 33-gauge needle, 0.5", and point style 4).

2. See the above protocol for intravitreal injections for specific details of this technique.

3. Prepare the Evans blue solution (Sigma) by preparing a solution of 20 mg/ml in PBS, sonicating for 10 min, and filtering to remove any undissolved Evans blue).

4. Inject 150 μl of Evans blue solution (20 mg/ml) into the tail vein 1 h after VEGF injection (BD, 30-gauge needle). To better visualize the tail vein, we use a heating pad on the tail for a few seconds before injection. Immediately following Evans blue (EB) injection, the mice turn visibly blue, confirming the uptake and circulation of the dye.

5. Reanesthetize mice 4 h after Evans blue injection and draw 50 to 100 μl of blood via intracardiac puncture (you only need 20 μl, but some will clot) (BD, 30-gauge needle). Save this for plasma isolation, continued at Step 12.

6. Sacrifice the mice and enucleate the eyes by proptosing the eye and gently removing with forceps. Place each eye in 4% PFA (we use a 16-well plate, one eye per well).

Note: Other protocols perfuse the mice prior to enucleation to remove Evans blue from the retinal vessels; however, we determined that similar results were found both with and without perfusion, and therefore this step was eliminated.

7. Dissect retina from whole eye by excising the cornea, lens, sclera, and vitreous, and put each retina into its own preweighed 1.5-ml micro-centrifuge tube (this starting tube weight will be needed later to calculate dry retina weight). Spin tubes briefly in desktop centrifuge and remove excess liquid from tube using a pipette. Rinse retina with PBS and remove PBS from tube. Place open tubes in speed-vac to dry with heat only and no vacuum.

8. Weigh tubes after retinas have dried and subtract the starting tube weight from this number to find the dry retina weight.

9. Add 200 μl formamide (Sigma) to each tube and place at 78 °C overnight. This will solubilize the EB dye conjugated to serum albumin in the retina.

10. Centrifuge each tube (including a blank of formamide alone) at 4 °C, 60,000 rpm in an ultracentrifuge for 45 min.

11. Remove 120 μl of supernatant immediately following spin and transfer to another tube to prevent the pellet from redissolving.

12. Measure absorbance of supernatant in microcuvette at 620 nm (the absorbance maximum for EB in formamide) and 740 nm (the absorbance minimum). The background-subtracted absorbance was determined by subtracting the reading at 740 nm from the reading at 620 nm.

13. Following collection of blood samples (Step 3), centrifuge 20 μl of blood from each mouse in ultracentrifuge for 15 min at 10,000 rpm at 25 °C.

14. Dilute the plasma (supernatant) 1:1000 in formamide through serial dilutions. We typically dilute 2 μl plasma immediately in 198 μl formamide (1:100), and leave it at room temperature overnight. The next day we add 50 μl of the diluted plasma to 450 μl formamide (1:10) in disposable cuvettes.

15. Measure absorbance of plasma in spectrophotometer at 620 nm and 740 nm.

We determined Evans Blue (EB) leakage by measuring intraretinal dye accumulation using the following equation:

$$[\text{retinal EB concentration}(\text{mg/ml})/\text{retinal weight}(\text{mg})]/$$

$$[\text{blood EB concentration}(\text{mg/ml}) \times \text{circulation time}(\text{h})]$$

The retinal EB concentration was determined using a standard curve for EB in formamide.

7.2. Vascular permeability in streptozotocin-injected diabetic rats

Vascular permeability and blood–retinal barrier breakdown are also measured in rats using a similar technique as described above. However, rather than using an intraocular injection of VEGF to induce leak, experimental diabetes is induced using a streptozotocin injection. In this model, VEGF expression in the retina is increased as a result of the induced diabetes rather than a direct VEGF injection. The resulting vascular permeability is then quantified by measuring either Evans Blue or FITC-dextran levels in the retina and blood plasma (Ishida *et al.*, 2003a).

7.3. *In vivo* laser model of retinal hyperpermeability

Retinal permeability was achieved by the occlusion of retinal vessels in one eye using a green wavelength diode laser (Iris) (we use a power of 180 mW, spot size of 50 μm, and duration of 1 s or until vein appears occluded). Parameters should be adjusted to the laser being used and media clarity of the eye. The laser should be centered on the major veins where they branch off of the vascular trunk adjacent to the optic nerve head (approximately one optic disc diameter). The endpoint of treatment is cessation of blood flow in the veins. Evans Blue is administered 3 days after laser treatment and permeability determined using the methods described above.

8. THREE-DIMENSIONAL *IN VIVO* IMAGING OF THE MOUSE OCULAR VASCULATURE

Much of what has been learned about the biology of the ocular vasculature has relied on fixed specimens collected from animal models. The retina is well suited for this type of analysis because of the relative ease with which it can be dissected free from surrounding tissues and the regular arrangement and reproducible developmental course of retinal blood vessel growth (Dorrell *et al.*, 2002). The retinal vasculature of the mouse eye has been visualized *in vivo* in numerous studies using both fundus photography and fluorescence angiography, which have been valuable in identifying gross changes in retinal vessels but lack the resolution necessary to visualize the smallest capillaries and can only produce two-dimensional images. Investigations into other vascular structures in the eye, such as the hyaloid, have largely depended on dissected tissue and/or histological sectioning to visualize these vessels. Analysis of fixed tissue, however, gives only a snapshot of the complex events that occur during the development and remodeling of these vessels. In addition, analysis of dissected tissue may not provide an accurate depiction of the conditions in the living animal, and artifacts introduced by death, dissection and/or fixation are possible.

We have recently published (Ritter *et al.*, 2005) and detail here a minimally invasive method of visualizing the vasculature of the mouse eye in the living animal that permits the study of vessels *in vivo*, without the need for sacrifice of the subject. This approach eliminates artifacts associated with the use of fixed tissue and facilitates our studies into several aspects of ocular vascular biology. We describe the details of the technique, and how it was applied to examine the mouse ocular vasculature during development, in a number of disease models and in the study of a possible therapeutic modality involving cell transplantation into the eye. Three-dimensional (3D) images produced using this technique are available for viewing online at http://www.iovs.org/cgi/content/full/46/9/3021/DC1.

8.1. Hyaloid regression

In the mature eye, the visual axis is clear due to the transparency of the cornea, lens, and vitreous, all of which are normally avascular. During development, however, the growing lens is supplied by a network of vessels collectively known as the hyaloidal vasculature. Originating from the optic disc, the branched hyaloidal vessels extend toward the posterior surface of the lens. The tunica vasculosa lentis is a major component of the hyaloidal vasculature and surrounds the lens (Mitchell *et al.*, 1998). The anterior portion of the tunica vasculosa lentis, also known as the pupillary membrane, covers the front of the lens and regresses by the time the mouse opens its eyes at 2 weeks. The posterior portion of the tunica vasculosa lentis lies anterior to the vasa hyaloidia propria, which branches from the hyaloid artery. The vasa hyaloidia propria regresses within the first 2 weeks while the hyaloidal vasculature completely disappears over about 5 to 6 weeks (Ito and Yoshioka, 1999). Complete regression of these developmentally associated vessels is roughly synchronized with maturation of the retinal vasculature and the onset of vision. While the hyaloid system is of great interest due to the natural process of regression it undergoes, its study has been somewhat limited by the fact that it is difficult to visualize in dissected specimens. These transient structures can be readily imaged *in vivo* (Fig. 6.4) and rendered into three dimensions (online Supplementary Videos 1 to 4). Different vascular structures can be digitally "dissected" from the 3D renderings and viewed separately as shown in Fig. 6.4C through F. Using color-depth coding, vascular systems can be distinguished more readily, even within a single image, by assigning color depending on depth (Fig. 6.5A and online Supplementary Video 5).

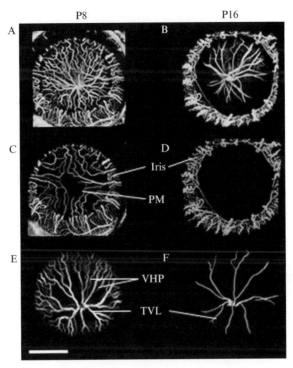

Figure 6.4 Developmental regression of ocular vessels imaged *in vivo* in three dimensions. (A–F) Mice at postnatal days 8 (P8) and 16 (P16) were imaged showing the transient vascular structures in the developing mouse eye. The complete ocular vasculature viewed on the visual axis is shown at P8 (A) and P16 (B). (C–F) To facilitate interpretation, structures in panels A and B were isolated in panels C–F. (C, E) The P8 mouse shows the presence of the pupilary membrane (PM) covering the anterior surface of the lens (C) and the *vasa hyaloidia propria* (VHP) (E), whereas both of these structures have completely regressed by P16 (D, F). The *tunica vasculosa lentis* persists at P16 (F), but will continue to regress as the eye reaches maturity at around 30 days.

8.2. Adult retinal vasculature

This method was used to generate images of the central retina in the adult mouse, a tissue that lies up to 2 mm deep in the eye. This depth of imaging is well beyond the usual limits of light microscopy, but the transparency of the optical axis allows effective imaging at this depth (Fig. 6.5B). Although the visible field size is limited, high-quality images of the optic disc region are obtained, while more peripheral regions of the retina can be viewed upon repositioning the subject (online Supplementary Video 6).

8.3. Retinal degeneration

We have applied this imaging technology to study a model of retinal degeneration which shows, as a major aspect of the model, significant loss of the retinal blood vessels in addition to the neurosensory degeneration.

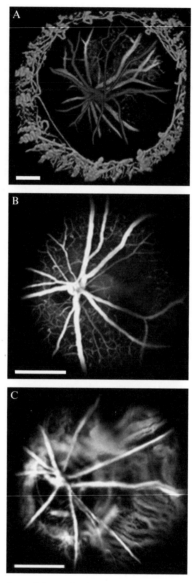

Figure 6.5 *In vivo* imaging of the retinal and choroidal vasculature. (A) Color depth-coded image of the ocular vasculature in a P16 mouse: red, iris vessels; blue, hyaloidal vessels; green, retinal vasculature. (B) Image of the central retinal vasculature in an adult mouse showing the major vessels as well as the retinal capillaries. (C) The choroidal vasculature imaged in a living mouse. Enhanced visualization of the choroid vessels was possible as a result of retinal degeneration that occurs in the strain of mouse used. Images comprised of 45 to 50 images captured in the z dimension and compiled into single images. (See Color Insert.)

When we imaged a 4-month-old mouse from this strain, we observed that only the major retinal vessels remained (Fig. 6.5C). With the loss of retinal tissue, we were able to clearly observe the vessels of the underlying choroid vasculature. The ability to visualize the choroid *in vivo* may be valuable in models of choroidal neovascularization.

8.4. Central retinal vein occlusion

In a model of central retinal vein occlusion (CRVO), a laser-induced occlusion was created near the optic disc in a mouse eye. Fluorescein-dextran was injected after the start of image capture and images were collected as the occluded vein was slowly filled by venules (Fig. 6.6 and online Supplementary Video 7). This approach identifies occluded veins *in vivo* which are often associated with edema and neovascularization in the retina.

8.5. Oxygen-induced retinopathy

Using an established model of oxygen-induced retinopathy in the mouse (Smith *et al.*, 1994) we have demonstrated the utility of this method by generating images of retinal pathology in the central retina. At postnatal day 14, or 2 days after removal from hyperoxia, vaso-obliteration is seen in the central retina along with tortuosity of retinal vessels (Fig. 6.7 and online Supplementary Video 8), which are alterations characteristic of this model. When the oxygen-induced retinopathy model is followed out to postnatal day 18 (6 days after return to normoxia), we observed evidence of pathological neovascularization and vascular leak (Fig. 6.7 and online Supplementary Video 8), which are typical features of this model. The hyaloid vessels remain tortuous and appear dilated.

8.6. Localization of injected cells *in vivo*

Bone marrow–derived myeloid progenitor cells expressing GFP have been shown in our lab to target the retina and differentiate into microglia, promoting vascular repair in the mouse OIR model (Ritter *et al.*, 2006). Using this *in vivo* imaging method, we can visualize the injected cells and ocular vessels simultaneously. This allows us to verify proper injection of cells prior to long term experiments and localize these cells in the living animal to determine their interaction with other structures in the eye, such as the hyaloidal vascular system (Fig. 6.8).

8.7. Serial imaging of a single eye during development

Due to the minimally invasive nature of this approach, we are able to observe single eyes *in vivo* over time through serial imaging. To accomplish this, we used a transgenic strain of mice that express green fluorescent

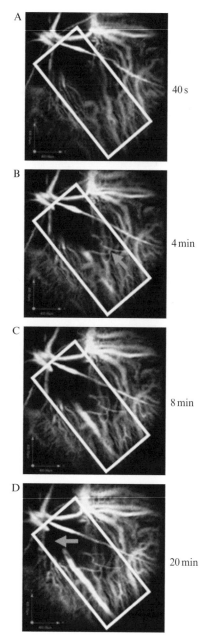

Figure 6.6 Using *in vivo* imaging to study a model of central retinal vein occlusion. After laser-induced occlusion of a retinal vein, fluorescent dye was injected into the mouse circulation during image capture. (A) Image collected 40 s after dye injection but prior to filling of occluded vein. (B) At 4 min after injection, filling of the occluded vein begins via a connecting venule (arrow). (C, D) Filling of the occluded vein continues but flow is stopped at the occlusion near the optic nerve head (arrow).

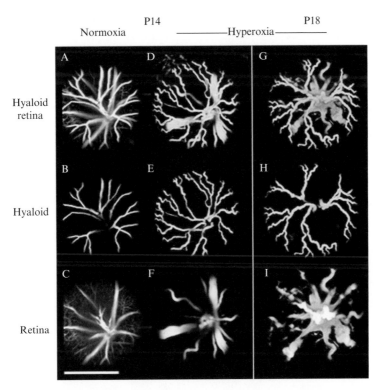

Figure 6.7 Studying oxygen-induced ocular pathology *in vivo*. Mice were imaged 14 days postnatal (P14) after exposure to normal oxygen conditions (A–C) or hyperoxia (D–F). Hyperoxia induces retinal vaso-obliteration and vascular tortuosity (F) as previously described (Smith *et al.*, 1994), but also seen *in vivo* is an alteration in retinal vein: artery size ratio, or venous dilation, and marked tortuosity of the hyaloid vessels (E). At postnatal day 18 (P18), evidence of the retinal neovascularization that characterizes this model is observed (I) along with persisting hyaloid tortuosity (H). Neovascularization is indicated by the enlarged regions of the retinal vessels. These enlarged regions are likely sites of dye extravasation from the neovessels which have been reported previously to be prone to leak fluid.

protein (GFP) exclusively in vascular endothelial cells (see Section 8.8), which eliminated the need for repeated intravenous injections of dye. Mice were followed from 15 days after birth for 6 days to image the regression of the hyaloidal vessels (Fig. 6.9).

8.8. Method details

8.8.1. Animal preparation and care

Developmental studies were performed on BALB/cByJ mice while the oxygen-induced retinopathy model was carried out using C57BL/6J mice. Animals were anesthetized with intraperitoneal administration of 30

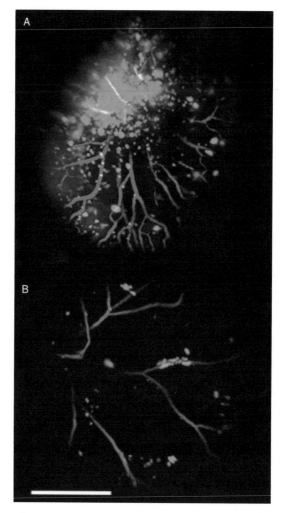

Figure 6.8 Localization of intravitreally injected cells *in vivo*. GFP-expressing bone marrow cells were injected into the vitreous of mice that received intravenous infusion of red dye. (A) Injected bone marrow cells (green) are shown to concentrate near the optic disc area from which the hyaloid vessels (red) emerge. (B) Two weeks after injection, GPF+ cells are shown to persist in the vitreous and, in some cases, target and adhere to segments of the regressing hyaloid vasculature. (See Color Insert.)

to 50 mg/kg ketamine and 2 to 5 mg/kg xylazine prior to the procedure. In most cases, FITC-labeled dextran (2000 kDa, 50 mg/ml in sterile PBS) was injected via the tail vein, or other accessible vessels, to visualize vasculature. The volume of dye was adjusted according to the size of the mouse and ranged from 10 μl for newborns to 200 μl for adults. For serial

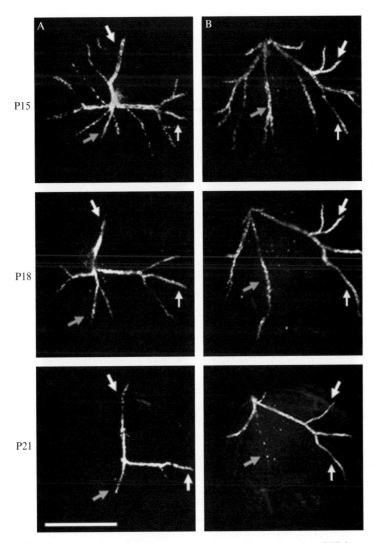

Figure 6.9   Serial imaging of single eyes over time. Mice expressing GFP in vascular endothelial cells were used to follow the regression of blood vessels in the eye. (A, B) Serial images of the hyaloid vasculature obtained from two mice are shown. Imaging spanned postnatal day 15 (P15) to P21, and relied on the intrinsically fluorescent vasculature eliminating the need for injected dyes. Arrows indicate vessels from each eye that persist over the course of the experiment.

imaging, animals expressing GFP under the promoter for the endothelial-specific protein, Tie2, were used. Topical anesthetic (tetracaine hydrochloride 0.5%, Alcon Laboratories, Fort Worth, TX) was applied to the eyes. Pupils were dilated with tropicamide (Mydriacyl, Alcon Laboratories) to improve visual access to the interior of eye. Eyes were kept moist with artificial tears and lubricants (Viscotears, Ciba Vision, Duluth, GA) throughout the procedure to prevent corneal drying or cataract formation. Mice were mounted and secured to the inverted microscope using a custom stage modification and were monitored visually to ensure sufficient degree of anesthetization during the procedure. The cornea was flattened with the use of a glass coverslip.

8.8.2. Retinal disease models

For the retinal degeneration model, rd/rd-Tie2GFP mice were used, which have a spontaneous degeneration reaching completion by 2 months. For the central vein occlusion model, large retinal veins of BALB/cByJ mice were occluded 3 days prior to imaging using a green diode laser at 180 mW with a 50-μm spot size and a duration of 1 s. The oxygen-induced retinopathy model used in our studies was described by Smith *et al.* (1994). C57BL/6J mice were place in 75% oxygen from postnatal day 7 through 12, at which time they were returned to room air.

8.8.3. Imaging equipment and procedures

Imaging was carried out using an inverted microscope and scanning laser confocal microscopy on a Nikon TE2000-U microscope equipped with a Bio-Rad Radiance 2100MP system. A stage modification was needed to allow positioning of the mice such that the visual axis could be placed in line with the optics of the microscope. This can be accomplished in a variety of ways with the critical feature being an opening in the stage of approximately 2.5 cm and the ability to position the stage slightly lower (toward the objective lens) than the normal range of the stage. Both fluorescein and GFP were excited using 488-nm light. The lowest effective light intensities were used to reduce any potential photo-damage to the eye. In most cases, a $4\times/0.2$ objective lens and a 5-μm step in the z dimension was used with the number of images collected totaling $\sim$150. In many instances, the laser power and/or detector gain were adjusted manually during collection to account for differences in fluorescence intensity as the focal plane moved from shallow (iris) to deep (retina) tissue. Three-dimensional images are generated by rendering xyz data sets into volumes using VOLOCITY software from Improvision. All images shown are either projections created from z-series data sets or snapshots taken from 3D renderings, except those in Fig. 6.3, which are single confocal images captured over time.

8.8.4. Z-dimension corrections

Our imaging setup creates an unusual optical path that likely introduces some spatial aberrations. The light used for excitation of the fluorophores in these experiments passes through several media with different properties, including glass, air, mouse cornea, aqueous, lens and vitreous. The light emitted from the fluorophores then passes back through these same media at a longer wavelength to the detectors. This convoluted optical path is the likely source of the altered spatial appearance of some of the ocular structures imaged in our studies. These alterations are generally manifested as reduced spatial separation in the z dimension, or the anterior to posterior dimension of the eye. We have introduced corrections in the data using VOLOCITY that involve increasing the um/pixel in the z dimension.

9. CELL-BASED MODELS AND OCULAR ANGIOGENESIS

Several recent studies have explored the use of cell-based approaches to maintain and repair abnormal retinal vasculature (Espinosa-Heidmann *et al.*, 2003; Otani, *et al.*, 2002a; Ritter *et al.*, 2006). The use of autologous adult bone marrow derived stem cell grafts for the treatment of retinal vascular and degenerative diseases represents a novel conceptual approach that may make it possible to "mature" otherwise immature neovasculature, stabilize existing vasculature to hypoxic damage, and/or rescue and protect retinal neurons from undergoing apoptosis. This may be possible through the use of autologous bone marrow, peripheral or cord blood–derived progenitor cells that selectively target sites of neovascularization and gliosis where they provide vasculo- and neuro-trophic effects facilitating vascularization of ischemic and otherwise damaged retinal tissue. Such treatments would have application to ischemic retinopathies such as ROP and diabetes as well as degenerative retinopathies such as ARMD and retinitis pigmentosa. Since these cells demonstrate specific targeting to sites of neovascularization and/or inflammation (or gliosis), they may also be used to deliver neurotrophic and/or angiostatic molecules in a form of cell-based therapy. One advantage of such an approach, in contrast to inhibiting or promoting angiogenesis with small molecules or recombinant proteins, is the ability of the cells to adapt and respond to a changing environment. The numerous factors produced by the cells can be appropriately modulated in response of changing conditions.

Several studies describe four basic populations of cells that may contain immature progenitor cells and mature differentiated cells which under appropriate conditions may have therapeutic application in the treatment of retinal diseases:

- Retinal stem cells that give rise to photoreceptors and other retinal neurons.
- Muller/glial stem cells that can differentiate into retinal neurons.
- Retinal pigment epithelial (RPE) stem cells that not only can serve to replace diseased RPE but may also be stimulated to differentiate into photoreceptors.

- Myeloid progenitor cells and monocytes that can differentiate into micro-glia and macrophages and promote vascular repair in a model of ischemic retinopathy.
- Endothelial progenitor cells (EPC) that can differentiate into endothelial cells and contribute to the retinal vasculature or exert a neurotrophic effect.

9.1. Cell description and purification

9.1.1. Retinal progenitor cells

Recently retinal progenitor cells have been identified in the ciliary margin of the adult retina; these cells can be expanded clonally in culture and give rise to a variety of retinal cell types including rod photoreceptors, bipolar neurons and Muller glia characterized by real stem cells properties (Ahmad *et al.*, 2004; Tropepe *et al.*, 2000).

9.1.2. Muller glial stem cells

The molecular profiling of developing mammalian retinas shows a high degree of similarity between the gene expression profiles of Muller glia and mitotic retinal progenitor cells in the mouse (Blackshaw *et al.*, 2004). Since the Muller glia are the cells that commonly proliferate in response to the retinal injury these cells could differentiate along a number of pathways and replace damaged retinal neurons.

9.1.3. Retinal pigment epithelial

RPE cell transplantation has been evaluated for its potential to serve as cell-based drug delivery vehicles, replace diseased RPE and provide a source of cells whose phenotypic differentiation may be manipulated by various cytokines and trophic substances. Cells with many molecular and functional characteristics of RPE cells have been isolated from spontaneously differentiating human embryonic cell lines and have been considered as a potential source of transplantable RPE cells for subretinal transplantation into human retinas.

9.1.4. Adult bone marrow/cord blood–derived progenitor cells

Ritter and colleagues suggested an alternative approach to the problem of ischemia-induced neovascularization in the eye (Ritter *et al.*, 2006); rather than prevent or eliminate the neovasculature, they propose to improve and stabilize the vascular response to hypoxia. This would be possible through the use of autologous bone marrow (BM)– or cord blood (CB)–derived hematopoietic stem cells (HSC) that selectively target sites of neovascularization and gliosis where they provide vasculo and neurotrophic effects.

They demonstrated that the adult BM contains different populations of endothelial and myeloid progenitor cells that can target activated astrocytes and participate in normal developmental or injury-induced, angiogenesis in the adult (Dorrell *et al.*, 2004; Otani *et al.*, 2002a, 2004). Intravitreal injection of these cells from mice and humans can prevent retinal vascular degeneration observed in mouse models of retinal degeneration; this vascular rescue correlates with functional neuronal rescue as well.

The HSC fraction used in the study by Otani and colleagues not only inhibits angiogenesis when engineered to express an anti-angiogenic but also rescues and stabilizes degenerating vessels. Whenever the vascular degeneration was prevented there was also a trophic effect on the photo-receptors (Otani *et al.*, 2004), suggesting that autologous BM grafts of HSC fractions containing EPCs may provide trophic effects on associated neural tissue.

Ritter and coworkers have recently identified high expressing CD44 positive cells from adult bone marrow as a population of positively selected cells with all of the targeting and functional rescue characteristics of the Lin-HSC. This population has clear myeloid progenitor characteristics and shows properties of Lin-cells in terms of vascular targeting and vasculo- and neuro-trophic effects. They also showed that in certain models (OIR) the injected bone marrow cells, in addition to targeting retinal vasculature will migrate through the retina to the RPE and acquire the characteristics of macrophage-like cells. They demonstrated that these cells differentiated in the retina into microglia and facilitated the enhanced recovery of vasculature after hypoxic injury.

Recent studies have identified a population of progenitor cells in the cord blood (Aoki *et al.*, 2004; Murohara *et al.*, 2000). This population is very heterogeneous and express the antigen CD34 and CD11b; when these cells differentiate they upregulate expression of Tie-2 and secretion of Ang-1 (Hildbrand *et al.*, 2004). When grown on an extracellular matrix containing fibronectin or collagen in the presence of serum and endothelial growth factors they form clusters and endothelial colonies demonstrating a differentiation into endothelial cells *in vitro*. This differentiation potentially suggests that the progenitor cells circulating in the cord blood and in the peripheral blood may not only give rise to cells of the hematopoietic lineage but may also play a role in repairing injured vascular endothelium following injury and surgery. Indeed, inflammatory processes occurring at sites of surgical injury and wound healing provide stimuli for cord blood–derived endothelial progenitors to differentiate into mature endothelial cells (Crisa *et al.*, 1999).

In order to isolate the appropriate populations of progenitor cells the blood and the marrow are separated into various fractions. The blood is generally separated by centrifugation into red blood cells, plasma, and nucleated cells (white blood cells). Methods of purifying specific sub-population of cells are extensively used. Two such methods are magnetic-assisted cell sorting (MACS) and fluorescence-activated cell sorting (FACS).

Both methods use antibodies able to recognize cell surface antigens and can be used for positive or negative selection of subpopulations. MACS is a very gentle way to purify cells but is limited in purity of cell populations. FACS separates the cells considering the size and granularity and the expression of several different surface antigens. FACS has the advantage to distinguish and capture different levels of antigen expression (high and low) while MACS is better for positive and negative selection.

Several recent studies have explored the use of cell-based delivery of therapeutic genes to treat retina diseases. Hematopoietic stem cells and progenitor cells have been transduced with viral vectors *ex vivo* and still maintain function. Molecules with potential trophic and/or static activities could be cloned into lenti and/or adenoviral expression vectors that could be used to transduce the appropriate cell population *ex vivo*. The lentiviruses are used for stable hematopoietic and endothelial cell delivery; using HIV vectors Torbett's group (Miyoshi *et al.*, 1999) has shown that human CD34+ stem cells can be transduced and maintain stem cell potency *in vivo*.

The expression of Ad-delivered transgenes is generally of short duration *in vivo*, but is characterized by a local administration rather than systemic. Adenoviral vectors have been previously shown to achieve delivery of antiangiogenic molecules in several different models of retinal and choroidal neovascularization (Lai *et al.*, 2001).

10. CONCLUDING REMARKS

The eye is an excellent model for studying angiogenesis and observing the effects of angiostatic and angiotrophic agents on this process. It is one of the few places in the body where blood vessels may be directly observed in a physiologically relevant context and, as such, provides an opportunity to evaluate proof of concept regarding many vascular systems. Ocular models of angiogenesis are well characterized, readily quantifiable, approachable experimentally, and have broad application to angiogenesis in general.

REFERENCES

Ahmad, I., Das, A. V., James, J., Bhattacharya, S., and Zhao, X. (2004). Neural stem cells in the mammalian eye: Types and regulation. *Semin. Cell Dev. Biol.* **15,** 53–62.

Antonetti, D. A., Barber, A. J., Khin, S., Lieth, E., Tarbell, J. M., and Gardner, T. W. (1998). Vascular permeability in experimental diabetes is associated with reduced endothelial occludin content: Vascular endothelial growth factor decreases occludin in retinal endothelial cells. Penn State Retina Research Group. *Diabetes* **47,** 1953–1959.

Aoki, M., Yasutake, M., and Murohara, T. (2004). Derivation of functional endothelial progenitor cells from human umbilical cord blood mononuclear cells isolated by a novel cell filtration device. *Stem Cells* **22**, 994–1002.

Ashton, N. (1966). Oxygen and the growth and development of retinal vessels. *In vivo* and *in vitro* studies. The XX Francis I. Proctor Lecture. *Am. J. Ophthalmol.* **62**, 412–435.

Banin, E., Dorrell, M. I., Aguilar, E., Ritter, M. R., Aderman, C. M., Smith, A. C., Friedlander, J., and Friedlander, M. (2006). T2-TrpRS inhibits preretinal neovascularization and enhances physiological vascular regrowth in OIR as assessed by a new method of quantification. *Invest. Ophthalmol. Vis. Sci.* **47**, 2125–2134.

Blackshaw, S., Harpavat, S., Trimarchi, J., Cai, L., Huang, H., Kuo, W. P., Weber, G., Lee, K., Fraioli, R. E., Cho, S. H., Yung, R., Asch, E., *et al.* (2004). Genomic analysis of mouse retinal development. *PLoS Biol.* **2**, E247.

Boyd, S. R., Zachary, I., Chakravarthy, U., Allen, G. J., Wisdom, G. B., Cree, I. A., Martin, J. F., and Hykin, P. G. (2002). Correlation of increased vascular endothelial growth factor with neovascularization and permeability in ischemic central vein occlusion. *Arch. Ophthalmol.* **120**, 1644–1650.

Caldwell, R. B., Bartoli, M., Behzadian, M. A., El-Remessy, A. E., Al-Shabrawey, M., Platt, D. H., and Caldwell, R. W. (2003). Vascular endothelial growth factor and diabetic retinopathy: Pathophysiological mechanisms and treatment perspectives. *Diabetes Metab. Res. Rev.* **19**, 442–455.

Campa, C., Kasman, I., Ye, W., Lee, W. P., Fuh, G., and Ferrara, N. (2008). Effects of an anti-VEGF-A monoclonal antibody on laser-induced choroidal neovascularization in mice: Optimizing methods to quantify vascular changes. *Invest. Ophthalmol. Vis. Sci.* **49**, 1178–1183.

Chahal, P. S., Fallon, T. J., and Kohner, E. M. (1986). Measurement of blood-retinal barrier function in central retinal vein occlusion. *Arch. Ophthalmol.* **104**, 554–557.

Chan-Ling, T., Tout, S., Holländer, H., and Stone, J. (1992). Vascular changes and their mechanisms in the feline model of retinopathy of prematurity. *Invest. Ophthalmol. Vis. Sci.* **33**, 2128–2147.

Crisa, L., Cirulli, V., Smith, K. A., Ellisman, M. H., Torbett, B. E., and Salomon, D. R. (1999). Human cord blood progenitors sustain thymic T-cell development and a novel form of angiogenesis. *Blood* **94**, 3928–3940.

Cunha-Vaz, J. G. (1983). Studies on the pathophysiology of diabetic retinopathy. The blood-retinal barrier in diabetes. *Diabetes* **32**(Suppl 2), 20–27.

Dorrell, M. I., Aguilar, E., and Friedlander, M. (2002). Retinal vascular development is mediated by endothelial filopodia, a preexisting astrocytic template and specific R-cadherin adhesion. *Invest. Ophthalmol. Vis. Sci.* **43**, 3500–3510.

Dorrell, M. I., Aguilar, E., Scheppke, L., Barnett, F. H., and Friedlander, M. (2007). Combination angiostatic therapy completely inhibits ocular and tumor angiogenesis. *Proc. Natl. Acad. Sci. USA* **104**, 967–972.

Dorrell, M. I., and Friedlander, M. (2006). Mechanisms of endothelial cell guidance and vascular patterning in the developing mouse retina. *Prog. Retin. Eye Res.* **25**, 277–295.

Dorrell, M. I., Otani, A., Aguilar, E., Moreno, S. K., and Friedlander, M. (2004). Adult bone marrow-derived stem cells use R-cadherin to target sites of neovascularization in the developing retina. *Blood* **103**, 3420–3427.

Dvorak, H. F., Brown, L. F., Detmar, M., and Dvorak, A. M. (1995). Vascular permeability factor/vascular endothelial growth factor, microvascular hyperpermeability, and angiogenesis. *Am. J. Pathol.* **146**, 1029–1039.

Elizabeth Rakoczy, P., Yu, M. J., Nusinowitz, S., Chang, B., and Heckenlively, J. R. (2006). Mouse models of age-related macular degeneration. *Exp. Eye Res.* **82**, 741–752.

Espinosa-Heidmann, D. G., Caicedo, A., Hernandez, E. P., Csaky, K. G., and Cousins, S. W. (2003). Bone marrow-derived progenitor cells contribute to experimental choroidal neovascularization. *Invest. Ophthalmol. Vis. Sci.* **44,** 4914–4919.

Feit-Leichman, R. A., Kinouchi, R., Takeda, M., Fan, Z., Mohr, S., Kern, T. S., and Chen, D. F. (2005). Vascular damage in a mouse model of diabetic retinopathy: Relation to neuronal and glial changes. *Invest. Ophthalmol. Vis. Sci.* **46,** 4281–4287.

Fine, H. F., Baffi, J., Reed, G. F., Csaky, K. G., and Nussenblatt, R. B. (2001). Aqueous humor and plasma vascular endothelial growth factor in uveitis-associated cystoid macular edema. *Am. J. Ophthalmol.* **132,** 794–796.

Fournier, G. A., Lutty, G. A., Watt, S., Fenselau, A., and Patz, A. (1981). A corneal micropocket assay for angiogenesis in the rat eye. *Invest. Ophthalmol. Vis. Sci.* **21,** 351–354.

Friedlander, M. (1999). New Pharmacological Approaches to the Treatment of AMD. *Eye Research Seminar, Research to Prevent Blindness* 40–42.

Friedlander, M., Brooks, P. C., Shaffer, R. W., Kincaid, C. M., Varner, J. A., and Cheresh, D. A. (1995). Definition of two angiogenic pathways by distinct alpha v integrins. *Science* **270,** 1500–1502.

Fruttiger, M., Calver, A. R., Krüger, W. H., Mudhar, H. S., Michalovich, D., Takakura, N., Nishikawa, S., and Richardson, W. D. (1996). PDGF mediates a neuron-astrocyte interaction in the developing retina. *Neuron.* **17,** 1117–1131.

Gimbrone, M. A., Jr., Cotran, R. S., Leapman, S. B., and Folkman, J. (1974). Tumor growth and neovascularization: An experimental model using the rabbit cornea. *J. Natl. Cancer Inst.* **52,** 413–427.

Hellstrom, A., Perruzzi, C., Ju, M., Engstrom, E., Hard, A. L., Liu, J. L., Albertsson-Wikland, K., Carlsson, B., Niklasson, A., Sjodell, L., LeRoith, D., Senger, D. R., *et al.* (2001). Low IGF-I suppresses VEGF-survival signaling in retinal endothelial cells: Direct correlation with clinical retinopathy of prematurity. *Proc. Natl. Acad. Sci. USA* **98,** 5804–5808.

Hildbrand, P., Cirulli, V., Prinsen, R. C., Smith, K. A., Torbett, B. E., Salomon, D. R., and Crisa, L. (2004). The role of angiopoietins in the development of endothelial cells from cord blood CD34+ progenitors. *Blood* **104,** 2010–2019.

Husain, D., Ambati, B., Adamis, A. P., and Miller, J. W. (2002). Mechanisms of age-related macular degeneration. *Ophthalmol. Clin. North Am.* **15,** 87–91.

Ishida, S., Usui, T., Yamashiro, K., Kaji, Y., Ahmed, E., Carrasquillo, K. G., Amano, S., Hida, T., Oguchi, Y., and Adamis, A. P. (2003a). VEGF164 is proinflammatory in the diabetic retina. *Invest. Ophthalmol. Vis. Sci.* **44,** 2155–2162.

Ishida, S., Yamashiro, K., Usui, T., Kaji, Y., Ogura, Y., Hida, T., Honda, Y., Oguchi, Y., and Adamis, A. P. (2003b). Leukocytes mediate retinal vascular remodeling during development and vaso-obliteration in disease. *Nat. Med.* **9,** 781–788.

Ito, M., and Yoshioka, M. (1999). Regression of the hyaloid vessels and pupillary membrane of the mouse. *Anat. Embryol. (Berl.)* **200,** 403–411.

Lai, C. M., Brankov, M., Zaknich, T., Lai, Y. K., Shen, W. Y., Constable, I. J., Kovesdi, I., and Rakoczy, P. E. (2001). Inhibition of angiogenesis by adenovirus-mediated sFlt-1 expression in a rat model of corneal neovascularization. *Hum. Gene Ther.* **12,** 1299–1310.

Madan, A., and Penn, J. S. (2003). Animal models of oxygen-induced retinopathy. *Front Biosci.* **8,** d1030–d1043.

McLeod, D. S., D'Anna, S. A., and Lutty, G. A. (1998). Clinical and histopathologic features of canine oxygen-induced proliferative retinopathy. *Invest. Ophthalmol. Vis. Sci.* **39,** 1918–1932.

Miller, J. W., Adamis, A. P., Shima, D. T., D'Amore, P. A., Moulton, R. S., O'Reilly, M. S., Folkman, J., Dvorak, H. F., Brown, L. F., and Berse, B. (1994). Vascular endothelial growth factor/vascular permeability factor is temporally and spatially corre-lated with ocular angiogenesis in a primate model. *Am. J. Pathol.* **145**, 574–584.

Mitchell, C. A., Risau, W., and Drexler, H. C. (1998). Regression of vessels in the tunica vasculosa lentis is initiated by coordinated endothelial apoptosis: A role for vascular endothelial growth factor as a survival factor for endothelium. *Dev. Dyn.* **213**, 322–333.

Miyoshi, H., Smith, K. A., Mosier, D. E., Verma, I. M., and Torbett, B. E. (1999). Transduction of human CD34+ cells that mediate long-term engraftment of NOD/SCID mice by HIV vectors. *Science* **283**, 682–686.

Murohara, T., Ikeda, H., Duan, J., Shintani, S., Sasaki, K., Eguchi, H., Onitsuka, I., Matsui, K., and Imaizumi, T. (2000). Transplanted cord blood-derived endothelial precursor cells augment postnatal neovascularization. *J. Clin. Invest.* **105**, 1527–1536.

Muthukkaruppan, V., and Auerbach, R. (1979). Angiogenesis in the mouse cornea. *Science* **205**, 1416–1418.

Muthukkaruppan, V. R., Kubai, L., and Auerbach, R. (1982). Tumor-induced neovascu-larization in the mouse eye. *J. Natl. Cancer Inst.* **69**, 699–708.

Otani, A., Kinder, K., Moreno, S. K., Nusinowitz, S., Banin, E., Heckenlively, J., and Friedlander, M. (2004). Rescue of retinal degeneration by intravitreally injected adult bone marrow-derived lineage-negative hematopoietic stem cells. *J. Clin. Invest.* **114**, 765–774.

Otani, A., Kinder, K., Ewalt, K., Otero, F. J., Schimmel, P., and Friedlander, M. (2002a). Bone marrow-derived stem cells target retinal astrocytes and can promote or inhibit retinal angiogenesis. *Nat. Med.* **8**, 1004–1010.

Otani, A., Slike, B. M., Dorrell, M. I., Hood, J., Kinder, K., Ewalt, K. L., Cheresh, D., Schimmel, P., and Friedlander, M. (2002b). A fragment of human TrpRS as a potent antagonist of ocular angiogenesis. *Proc. Natl. Acad. Sci. USA* **99**, 178–183.

Penn, J. S., and Rajaratnam, V. S. (2003). Inhibition of retinal neovascularization by intravitreal injection of human rPAI-1 in a rat model of retinopathy of prematurity. *Invest. Ophthalmol. Vis. Sci.* **44**, 5423–5429.

Penn, J. S., Rajaratnam, V. S., Collier, R. J., and Clark, A. F. (2001). The effect of an angiostatic steroid on neovascularization in a rat model of retinopathy of prematurity. *Invest. Ophthalmol. Vis. Sci.* **42**, 283–290.

Penn, J. S., Tolman, B. L., and Lowery, L. A. (1993). Variable oxygen exposure causes preretinal neovascularization in the newborn rat. *Invest. Ophthalmol. Vis. Sci.* **34**, 576–585.

Penn, J. S., Tolman, B. L., Lowery, L. A., and Koutz, C. A. (1992). Oxygen-induced retinopathy in the rat: Hemorrhages and dysplasias may lead to retinal detachment. *Curr. Eye Res.* **11**, 939–953.

Qaum, T., Joussen, A. M., Clemens, M. W., Qin, W., Miyamoto, K., Hassessian, H., Wiegand, S. J., Rudge, J., Yancopoulos, G. D., and Adamis, A. P. (2001). VEGF-initiated blood-retinal barrier breakdown in early diabetes. *Invest. Ophthalmol. Vis. Sci.* **42**, 2408–2413.

Ritter, M. R., Aguilar, E., Banin, E., Scheppke, L., Uusitalo-Jarvinen, H., and Friedlander, M. (2005). Three-dimensional *in vivo* imaging of the mouse intraocular vasculature during development and disease. *Invest. Ophthalmol. Vis. Sci.* **46**, 3021–3026.

Ritter, M. R., Banin, E., Moreno, S. K., Aguilar, E., Dorrell, M. I., and Friedlander, M. (2006). Myeloid progenitors differentiate into microglia and promote vascular repair in a model of ischemic retinopathy. *J. Clin. Invest.* **116**, 3266–3276.

Robbins, S. G., Rajaratnam, V. S., and Penn, J. S. (1998). Evidence for upregulation and redistribution of vascular endothelial growth factor (VEGF) receptors flt-1 and flk-1 in the oxygen-injured rat retina. *Growth Factors* **16**, 1–9.

Ruben, M. (1981). Corneal vascularization. *Int. Ophthalmol. Clin.* **21**, 27–38.

Ryan, S. J. (1979). The development of an experimental model of subretinal neovascularization in disciform macular degeneration. *Trans. Am. Ophthalmol. Soc.* **77,** 707–745.

Sarraf, D. (2001). *In:* Diabetes and Eye Disease,   pp. 3–5, 8–20. American Academy of Ophthalmology, San Francisco, CA.

Senger, D. R., Galli, S. J., Dvorak, A. M., Perruzzi, C. A., Harvey, V. S., and Dvorak, H. F. (1983). Tumor cells secrete a vascular permeability factor that promotes accumulation of ascites fluid. *Science* **219,** 983–985.

Shaffer, R., and Friedlander, M. (1996). A method for the *in vivo* quantitation of angiogenesis in the rabbit corneal model. *In:* Molecular, Cellular and Clinical Aspects of Angiogenesis, (M. E. Maragoudakis, ed.), Plenum Press, New York.

Smith, L. E. (2002). Pathogenesis of retinopathy of prematurity. *Acta Paediatr. Suppl.* **91,** 26–28.

Smith, L. E., Wesolowski, E., McLellan, A., Kostyk, S. K., D'Amato, R., Sullivan, R., and D'Amore, P. A. (1994). Oxygen-induced retinopathy in the mouse. *Invest. Ophthalmol. Vis. Sci.* **35,** 101–111.

Stone, J., Itin, A., Alon, T., Pe'er, J., Gnessin, H., Chan-Ling, T., and Keshet, E. (1995). Development of retinal vasculature is mediated by hypoxia–induced vascular endothelial growth factor (VEGF) expression by neuroglia. *J. Neurosci.* **15,** 4738–4747.

Tropepe, V., Coles, B. L., Chiasson, B. J., Horsford, D. J., Elia, A. J., McInnes, R. R., and van der Kooy, D. (2000). Retinal stem cells in the adult mammalian eye. *Science* **287,** 2032–2036.

Vinores, S. A., Youssri, A. I., Luna, J. D., Chen, Y. S., Bhargave, S., Vinores, M. A., Schoenfeld, C. L., Peng, B., Chan, C. C., LaRochelle, W., Green, W. R., and Campochiaro, P. A. (1997). Upregulation of vascular endothelial growth factor in ischemic and non-ischemic human and experimental retinal disease. *Histol. Histopathol.* **12,** 99–109.

Weis, S. M. (2007). Evaluating integrin function in models of angiogenesis and vascular permeability. *Methods Enzymol.* **426,** 505–528.

Xu, Q., Qaum, T., and Adamis, A. P. (2001). Sensitive blood-retinal barrier breakdown quantitation using Evans blue. *Invest. Ophthalmol. Vis. Sci.* **42,** 789–794.

CHAPTER SEVEN

Mouse Models of Ischemic Angiogenesis and Ischemia-Reperfusion Injury

Joshua I. Greenberg, Ahmed Suliman, Samuel Barillas, *and* Niren Angle

Contents

Abstract

Ischemia and ischemia-reperfusion (I/R) events are distinct but interrelated processes etiologic to the most prevalent human diseases. A delicate balance exists whereby ischemic injury can result in beneficial angiogenesis or in

Section of Vascular and Endovascular Surgery, Department of Surgery, School of Medicine, University of California, San Diego, San Diego, California

Methods in Enzymology, Volume 444

ISSN 0076-6879, DOI: 10.1016/S0076-6879(08)02807-3

detrimental reperfusion injury overwhelming the organism. Here, we describe *in vivo* models of ischemia and ischemia-reperfusion injury with emphasis on murine hindlimb ischemia models. We also provide a brief introduction to murine myocardial ischemia experiments. Each model is described in the context of human disease. Emphasis is made on the strengths and weaknesses of the available techniques, particularly as it relates to data analysis, interpretation, and translational relevance.

1. INTRODUCTION

The design and execution of *in vivo* animal experiments requires substantial effort and resources such that selection of the most appropriate model is critical. In the design of murine ischemia experiments it is thus imperative to clearly distinguish between ischemia and ischemia–reperfusion (I/R) events. Both events share in common the compromise of tissue perfusion, but subsequent reperfusion invokes a distinct set of pathophysiological elements.

Ischemia results from an inadequate blood supply to a tissue bed. Mechanical factors such as intentional occlusion (e.g., a tourniquet or operative clamp) or intrinsic factors such as clot or atheroma can induce ischemia. The result is tissue hypoxia and, if not reversed, tissue necrosis. Tissue hypoxia is not the only detrimental component of ischemia, also included are metabolic by-products such as organic acids and intracellular ions such as potassium which worsen the insult. In the setting of pure ischemia most of the injury is limited to the target tissue; but if reperfusion is allowed, loco-regional edema and stasis which marks the so-called "no re-flow phenomenon" can result (Blaisdell, 2002). Pronounced interstitial edema causes injury by exerting physical pressure on vital structures (e.g., nerves) and the leakage of cytokine mediators into the circulation results in systemic inflammation (Maseri, 1997). It must also be kept in mind that any compromise to the circulatory system of an animal such as from septic shock or hemorrhage can result in tissue ischemia. The subsequent fluid resuscitation of the animal is a defined I/R event but is clearly more related to the biology of inflammation than to that of angiogenesis. It is worth noting, however, that many of the mediators such as vascular endothelial growth factor (VEGF) that influence ischemic tissue remodeling are also involved in the exacerbation of the reperfusion insult (Weis and Cheresh, 2005).

Each organ-based model that we will describe is introduced in the context of the relevant human disease. The required equipment and supplies and the expertise of personnel can be found at the beginning of each protocol. Extensive discussion of data analysis and interpretation can be found in the text after each protocol, particularly with reference to hindlimb experiments. Because most transgenic mice are of the C57/BL6 genetic background, this work describes experiments with this bias. Not surprisingly, the influence of mouse strain vastly impacts these models results and must be considered. Because the experience of the authors is based mostly in murine hindlimb models, these are emphasized in the text. Finally, the procedures described herein are not painless and as such must be performed under institutional protocol in an Association for Assessment and Accreditation of Laboratory Animal Care (AAALAC)–accredited facility with properly trained and supervised individuals.

2. OVERVIEW OF METHODS IN HINDLIMB I/R INJURY

The armamentarium of techniques in which temporary hindlimb ischemia can be induced in an animal is vast and requires careful consideration. The specific disease to be modeled must be considered when preparing for a hindlimb I/R experiment. Ischemia-reperfusion of one or both hindlimbs is most easily instituted through mechanical and circumferential compression of the thigh using a pneumatic tourniquet or simply an orthodontic rubberband under tension. The former provides reproducible tension, but is costly, and has a finite lifespan (approximately 10 experiments); the latter is very difficult to standardize and induces greater focal tissue injury. The largest shortcomings of mechanical compression are tissue injury (even with a pneumatic cuff), complete vascular exclusion, and limited translational relevance. We believe that there is minimal translational relevance to mechanical compression because few diseases or injuries other than perhaps a severe crush injury are well modeled with combined arterial, venous, and lymphatic disruption. The advantages of mechanical compression are speed, minimal set-up, simplicity, and lack of surgical intervention. A severe systemic shock can be induced by reperfusion after 1 h of bilateral lower extremity mechanical compression (Wakai et al., 2001).

It is important to understand hindlimb vascular anatomy, routes of collateralization, and inter-strain variations, when planning hindlimb experiments. Inline flow to the lower extremity derives from the external iliac, femoral, and finally saphenous arteries in the mouse. The arborization of numerous branches from the internal iliac artery (pelvic collaterals) and the femoral artery (including epigastric arteries) makes simple arterial occlusion in the groin an often undesirable experiment. In our laboratory, prolonged interruption of the common iliac artery with a microvascular clip has produced minimal ischemic injury due to pelvic and abdominal wall collaterals in C57BL6 mice. This is discordant from our experience in humans where the same insult in even a young person would invariably result in critical limb ischemia Fig. 7.1 demonstrates strain-dependent

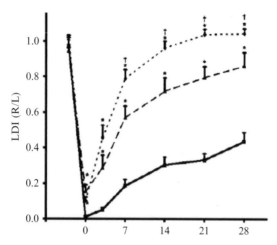

Figure 7.1 Serial measurements of hindlimb perfusion as measured by laser Doppler. Thin broken line represents C57/Bl6 mice, thick broken line represents 129S strain, and solid line represents Balb/c mice. $\star p < 0.05$ versus BALB/c. $p < 0.05$ versus 129S2 mice. (From Helisch, A., Wagner, S., Khan, N., Drinane, M., Wolfram, S., Heil, M., Ziegel-hoeffer, T., Brandt, U., Pearlman, J. D., Swartz, H. M., Schaper, W. (2006). Impact of mouse strain differences in innate hindlimb collateral vasculature. *Arterioscler. Thromb. Vasc. Biol.* **26**, 520–526, with permission by Lippincott Williams and Wilkins.)

ischemia as measured by laser Doppler flowmetry during murine hindlimb ischemia experiments. It is likely that the inherent resilience to ischemia is derived from strain-dependent differences in mouse collaterals has impacted the methods that we use to model human disease (Helisch *et al.*, 2006). Indeed, the most common transgenic and knock-out mouse background is C57/BL6, which is also quite a resilient strain as it relates to arterial ischemia. We have found that true isolated arterial insufficiency of the lower limbs can be reproducibly induced with transient infrarenal aortic occlusion. This models both acute arterial occlusive disease and could also serve to model operative aortic occlusion such as that in major aortic surgery.

2.1. Procedure: Hindlimb tourniquet-induced ischemia

2.1.1. Materials

Sphygmomanometer
Small-diameter inflatable cuff (cat #DC1.6, Hokanson Cuffs, Bellevue, WA)

Mice of uniform age and gender should be used in order to minimize the inherent variability to which tourniquet experiments are already prone. Interprocedural and interanimal variability stem from differences in the extent and location of tension and the amount tissue trauma that results.

The animal should be anesthetized for the duration of the procedure which can be either via inhalational (nose cone) or intraperitoneal ketamine/ xylazine. The tourniquet (source) is applied to 200 mmHg and fixed in position with a sphygmomanometer. The animal is then placed on a recirculating warmer and strictly monitored to keep its surface between 37 and 42 °C. Ischemia proceeds for 60 to 360 min and can be monitored qualitatively by observing skin color and quantitatively by laser Doppler flowmetry as described below. Reperfusion should proceed for 24 to 48 h after which most signs of reperfusion injury have been reversed. The exact length of ischemia and reperfusion should be titrated empirically according to end-points but we achieve a severe I/R injury with 60 min of ischemia and 24 h of reperfusion. Mortality from unilateral tourniquet-induced ischemia should be less than 10%.

2.2. Procedure: Transient occlusion of the murine infrarenal aorta

2.2.1. Materials

Sterile microvascular forceps, needle drivers, and scissors
Clean microvascular clamp suitable for 0.2- to 0.9-mm vessels (AROSurgical, cat# TKF-1-30, Newport Beach, CA)
1-{3/4}" sterilized paper clip
{1/4}" tape
Sterile normal saline
5-cc syringe
6-0 to 7-0 silk suture for division of arterial branches
5-0 nylon suture for skin closure
Clean cotton-tip swabs for operative sponges
Dissecting microscope (6–12×) with adequate illumination
Water-recirculating heating pad
Recommended: laser Doppler flowmeter

Mice 6 to 8 weeks of age are preferentially used with an age limit of 12 weeks. In our experience, mice must also be fit and well-nourished to tolerate prolonged aortic occlusion. A recovery cage placed atop a recirculating heating unit kept at a constant temperature of 37 to 40 °C should be configured before the procedure. Anesthesia consists of isoflurane mixed in an automated vaporizer with oxygen administered by way of a nose cone. A surgical stereomicroscope with variable magnification (at least 6×) is necessary. The animal is taped in position at all four extremities and the abdomen is prepped with 10% providine-iodine solution.

The peritoneal cavity is entered through the midline with one scissor-cut through the skin and underlying peritoneum, tenting up the abdominal wall to avoid underlying visceral injury. The small intestines are preferably teased into the lower abdomen with swabs or eviscerated if the procedure is short. The skin is retracted wide and laterally and secured to the table by one end of a paper clip bent at a right angle (the other end is taped to the table). Pulsation of the aorta should be obvious in the midline before it is cleared of fibro–adipose tissue or the magnification should be increased. The greatest risk of irreversible injury occurs at this stage when the vena cava is not exposed and one hastily attempts the midline aortic dissection. Dissection begins just below a point where the left renal vessels meet the midline which marks the proximal infrarenal aorta. Carefully incising the peritoneum over the aorta allows entry into the retroperitoneum. Forceps can be used to expose the aorta inferiorly but it is not necessary and would endanger the left renal vein if further exposure cephalad is undertaken. At this point, there should be some mobility to the aorta which can be enhanced by freeing any attachments on its left lateral wall. By elevating the vessel anteriorly and avoiding right lateral and posterior dissection, a posterior pocket is indirectly created. The posterior jaw of the microvascular clamp can be used to gently complete the posterior aortic pocket so long as the aorta is displaced anteriorly and no resistance is met. Aortic occlusion is achieved with the tip of the clamp angled onto but not occluding the vena cava. The absence of distal aortic pulsation can be subtle but should be demonstrated. Pallor of the foot skin should be evident and confirmation of hindlimb ischemia by laser Doppler flowmetry can also be used. The animal is closed with a running 5-0 suture incorporating peritoneum, fascia, and skin in a single layer and placed in the recovery cage.

One hour of infrarenal aortic occlusion is sufficient to produce extensive I/R injury. Careful monitoring of temperature to keep it within 37 to 40 °C is crucial. The animal should awaken during this period and demonstrate impaired hindlimb movement and overall weakness. The animal is reopened under judicious use of anesthesia (e.g., half the previous concentration of inhalational gas) and the clamp is carefully removed. Next, 0.5 ml of normal saline is left to absorb through the peritoneal membrane and the abdomen is reclosed in an identical fashion. The animal is then returned to the recovery cage until it has fully awoken. Mortality should not exceed 15%. Postoperative analgesia is provided by buprenorphine at 0.05 mg/kg intraperitoneally once.

3. EVALUATING HINDLIMB ISCHEMIA/ REPERFUSION INJURY

3.1. Quantification of edema

The simplest and most efficient method of characterizing hindlimb edema is the wet-to-dry ratio. Similar to the modified Miles assay discussed below, this assay is somewhat crude. Moreover, in our experience, the substantial interanimal variability in wet-to-dry ratios makes it only useful in models of severe hindlimb injury (i.e., the tourniquet model). We have found variability most minimized when an entire hindlimb is excised from the hip after the skin is removed. It is also reasonable to excise a single muscle such as the gastrocnemius so long as muscles of similar size are compared. The tissue is then weighed and placed in a standard laboratory oven at 78 °C for 24 h before it is reweighed to obtain the wet-to-dry ratio.

It is likely that the best representative of interstitial water content is the rodent T2-weighted magnetic resonance image but this requires extensive resources and local expertise (Ziv et al., 2004).

3.2. Measurement of capillary leak

The simplest method of visualizing capillary leak is with a modified Miles assay. We find, however, that this assay is rather crude and must be corroborated by other measurements of vascular permeability. In a solution, 100 μl of 1% Evans blue dye (1 mg/ml; Sigma, St. Louis) is injected into the tail vein of each mouse. The extravasation of Evans blue dye onto adductor muscle can then be recorded with a digital camera. Tissues can also be homogenized in trichloroacetic acid:ethanol (1:1 v/v) and read at OD 600 on a spectrophotometer. A positive control for capillary leak can consist of intravenous injection of VEGF-165.

For in vivo microscopy, the whole surface of the adductor muscle is excised and cut with a razor blade on a murine brain sectioning mold (Harvard Apparatus) and placed on a microscope slide. For contrast enhancement of the microvascular network a single bolus of 40-kDa, fluorescein-isothiocyanate (FITC)–labeled dextran (100 μg, Molecular Probes, San Diego) and rhodamine lectin (1.25 μg, Molecular Probes, San Diego) is injected intravenously 5 min prior to overdose with inhalational halothane. Slides are scanned on a confocal microscope and microscopic images at multiple sites and multiple depths of the muscle are chosen to provide a qualitative estimation of microvascular perfusion and vascular barrier dysfunction (Fig. 7.2). It is crucial to note that a properly performed image analysis including a z-stack (spaced at no greater than 2-mm intervals) is important to avoid interpreting tissue processing artifact as extravasation. We have recently begun using a dual-photon microscope which appears to hold promise in reducing background florescence. It is also possible to examine the entire limb on a multiphoton microscope because of its superior tissue penetration. Quantitation of extravasated FITC-dextran can also be performed by homogenization of hindlimb muscle in phosphate-buffered saline. Postcentrifuged supernatants can be aliquoted and measured at OD 488 on a spectrophotometer for a comparison of vascular permeability between samples.

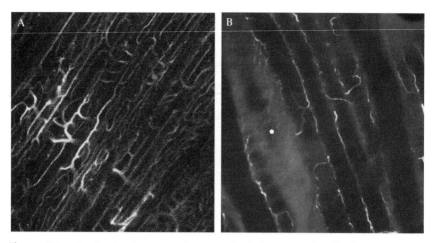

Figure 7.2 Confocal microscopy images of whole mount hindlimb tissues using a Confocal C1 Scanner at 20× objective. Animals were injected with FITC–dextan (40 kD) and Rhodamine lectin 5 min prior to tissue procurement. (A) Hindlimb tissue after sham I/R. (B) Tissue after 1 h of ischemia and 4 h of reperfusion. Asterisk denotes region of putative vascular extravasation.

3.3. Assessing tissue viability and survival

Tissue viability and apoptosis can be assessed with a modification of the *in vitro* MTT (3-(4,5-dimethylthiazol-2-yl)-2,5-diphenyl tetrazolium) assay in which small pieces of thigh musculature are incubated with 300 μl of the reagent at 37 °C for 3 h with gentle tumbling. The tissue is allowed to dry before 3 ml if 2-propanol is used to extract the formazan salt during gentle tumbling at 37 °C for 6 h. Aliquots can be read at OD 570 nm in a spectrophotometer and are best normalized to animal dry weight (see above). Apoptosis can be assessed histologically by commercially available apoptosis detection kits (e.g., ApopTag from Chemicon, Temecula, CA).

3.4. Neutrophil infiltration

Sequestration of polymorphonuclear cells is pathognomotic for an acute inflammatory state. The simplest method for identifying the extent of tissue neutrophil activity is the myeloperoxide assay (MPO) measurement of oxidative burst. Muscle is sonicated for 60 s and clarified by centrifugation (14,000×g, 10 min, room temperature), the resulting supernatants are tested immediately for MPO levels. The assay mixture consists of o-dianasidine (2.85 mg/ml; Sigma), hydrogen peroxidase (0.85%) and a 1:40 dilution of samples in PBS. Absorbances are read at OD 450 5 min after substrate addition, and results are expressed as absorbance units per gram of tissue. Inflammation can be further characterized by analyzing protein and mRNA levels of cytokines such as TNF-alpha and MCP-1 by enzyme-linked absorbant assay (ELISA) or real-time PCR.

3.5. Biophysical measurements of muscle function

A detailed description of biophysical adjuncts used to quantify contractility functionality after hindlimb I/R is beyond the scope of this discussion. It is important, however, to note that there are a number of biophysical assays available that range from tension/force devices to complex electrophysiologic analyses (Hayashi *et al.*, 1998).

4. Overview of Methods in Chronic Hindlimb Ischemia

Chronic hindlimb ischemia can be induced with materials and personnel similar to that required for Hindlimb ischemia-reperfusion experiments. Experimental designs can vary greatly, but it is typically convenient to include the contralateral, unoperated limb, as an "internal" control to minimize interanimal variation in the analysis. As previously emphasized, animal strain and genetic complement will produce vast differences in hindlimb ischemia outcomes from full recovery of function to necrosis and autoamputation (refer to Fig. 7.1 of flow differences). It is thus important carefully follow and document the appearance and functional status of all animals according to criteria that we discuss later.

4.1. Procedure: Ligation and excision of the murine femoral artery

4.1.1. Materials

Sterile microvascular forceps, needle drivers, and scissors
1-{3/4}" sterilized paper clips
{1/4}" tape
6-0 to 7-0 polypropylene suture for division of arterial branches
6-0 nylon suture for skin closure
Clean cotton-tip swabs for operative sponges
Dissecting microscope (6–12×) with adequate illumination
Water-recirculating heating pad
Laser Doppler flowmeter or transcutaneous oxygen meter

Mice 6 to 8 weeks of age are preferentially used with an age limit of 12 weeks. Anesthesia consists of either inhaled isoflurane or intraperitoneal xylazine/ketamine (dose). A surgical stereomicroscope (Zeiss, Germany), with variable magnification (20× to 50×) is used. We find it more convenient for a right-handed operator to perform the full ligation and excision on the left femoral artery. The distal most lower extremities and separated with tape for convenience after the induction of anesthesia. Next, the legs and lower abdomen are prepped with providine-iodine solution and the skin is pinched to ensure adequate depth of anesthesia. A vertical incision with narrow-blade scissors is performed over the mid-thigh with care to avoid the underlying structures. Once the femoral neurovascular bundle is identified, liberal separation of the skin from the vasculature is achieved by spreading the scissors towards the abdomen and then the knee. Paper clips are opened up at right angles and used at one end to retract the skin, while the other is taped to the table. Using sharp dissection, connective tissue is freed from the proximal most femoral artery down to the proximal popliteal artery at the knee. Care must be taken to circumferentially dissect the vessel as any hidden muscle perforator left undisturbed will jeopardize the extent of ischemic injury. We prefer to individually ligate (but not divide until the excision step) each branch with a single 6-0 polypropylene suture (Fig. 7.3A). Attempts should be made to leave the femoral vein intact, but this feat can extend the length of the procedure many times over and so often is abandoned. We do not believe that this confounds the experiment because of the robust venous collaterals that remain. Indeed, excision of the human superficial femoral vein when necessary is usually well tolerated. If laser Doppler analysis is available, it is useful to compare flow between the sham and ischemic limbs to ensure that all collaterals have been ligated. The procedure is completed when hemostasis is ensured and the skin is closed with interrupted monofilament suture (we use 6-0 nylon).

4.2. Evaluating the angiogenic response after hindlimb ischemia

Data analysis during and at the conclusion of hindlimb experiments involves at the minimum an assessment of function and vessel growth; attempts should also be made to document flow and physiologic recovery.

4.2.1. Active foot movement score

To assess functional recovery of the limbs, a scoring system based on active foot movement must be used. Movement of the occluded hindlimb can be scored in one of four different categories: 0 = full range of motion at the knee (flexion and extension) and paw level (flexion and extension); 1 = diminished flexion and extension of the calf, and paw extension and flexion are intact; 2 = diminished flexion and extension of the knee, paw flexion preserved, and no paw extension; and 3 = no flexion of the calf, and no flexion or extension of the paw (paw drop) dragging (Crawford *et al.*, 2007). Attempts should be made to carefully identify and document the hindlimb

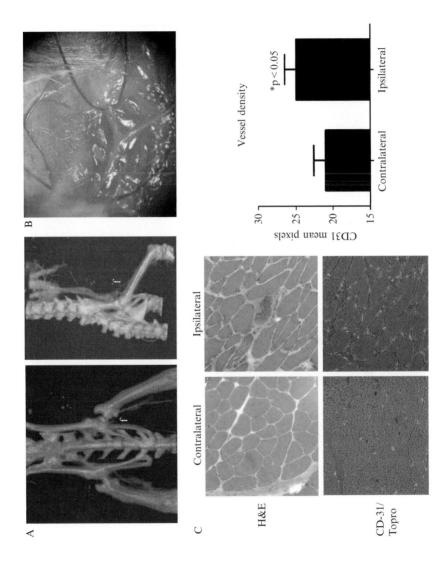

A

B

C   Contralateral   Ipsilateral

H&E

CD-31/
Topro

Vessel density

*p < 0.05

Contralateral   Ipsilateral

CD31 mean pixels

skin for the presence and extent of any of ulceration and necrosis that define the condition of critical limb ischemia. Lack of skin changes does not imply lack of ischemia, but it does suggest that the limb will remain viable. Muscle wasting can also be assessed postmortem by weighing gastrocnemius muscle is experimental versus sham limbs to assess ischemic atrophy.

4.2.2. Morphometry of collateral arteries

Hindlimbs are perfusion fixed using 2% paraformaldehyde in PBS. Promptly upon euthanasia the adductor muscles are harvested, and cryopreserved in OCT compound (Tissue-Tek). Serial transverse sections (5 μm thick) are subjected to, at the minimum, CD-31 and nuclear antigen identification for capillary enumeration (Fig. 7.3B). Immunofluorescence or immunohisto-chemistry is a suitable technique, but we prefer capillary counts for at least five high-power fields per sample. When possible, it is preferable to use more sophisticated data analysis software for more efficient and unbiased analysis. Finally, in an effort to quantify the proliferative response to hindlimb ischemia, it is useful to supplement 5-bromo-2′-desoxyuridine (BrdU, Sigma) to the drinking water (80 mg/100 ml) of mice 1 day before surgery and changed every 3 days if possible.

4.2.3. Microangiography (rhodamine lectin confocal microscopy)

The verification perfusion via angiogenic vessels is important and can be accomplished with "microangiography," simply performed according to the protocol used in hindlimb I/R with the omission of FITC-dextran.

4.2.4. Magnetic resonance (or computed tomography) angiography

A number of commercially available small-animal computed tomography (CT) and magnetic resonance imaging (MRI) devices are very useful to assess ischemic angiogenesis. The advantages of these devices over conventional angiography are superior soft tissue resolution in multiple planes which can be digitally reconstructed and rendered in three dimensions. We have utilized the GE eXplore RS rodent CT scanner, which can complete image acquisition within 15 min for a resolution of 95 microns (Fig. 7.3C). Three-dimensional time-of-flight MRI scans can also be obtained with intravenous

---

Figure 7.3 (A) Dissection of femoral artery and its branches with ligatures encircling the artery under 8× magnification. (B) Coronal and saggital murine CT angiograms obtained with GE eXplore RS rodent CT scanner at 95-μm resolution opacified with intravenous contrast (Fenestra, ART, Montreal), demonstrating normal vascular anatomy and the location of the iliac and femoral vessels (f). (Images courtesy of UCSD Small Animal Imaging Research Program.) (C) Transverse frozen sections of murine gastrocnemius muscle stained with hematoxylin and eosin (H&E) and for vascular endothelium (CD-31) and nuclei (Topro, Molecular Probes, San Diego) 28 days after femoral artery ligation and excision. (See color insert.)

gadolinium where available according to more complex protocols. The score differentiated "small," "medium," and "large" collaterals on a relative basis according to the caliber of vessels observed by an unbiased observer. Veins are differentiated from arteries by anatomic location and course.

4.3. Analysis of blood flow

The authors do not recommend any single tool for quantification of flow, but do recommend the use of such a tool. Perimed (Sweden) seems to be a common source of laser Dopplers for small-animal experiments and users can refer to the manufacturer for specifications and operating instructions. Blood flow is displayed as changes in laser frequency which is normalized in either a digital readout or represented by different color pixels if a camera is interfaced. Results of laser Doppler blood flow measurement should be expressed as the ratio of ischemic to nonischemic tissue. An important adjunct to the measurement of blood flow is transcutaneous oxygen pressure (TcO_2) measurements. Measurements can be repeatedly performed similar to laser Doppler flowmetry and also provide a reliable estimation of tissue perfusion.

5. OVERVIEW OF MURINE MODELS OF ACUTE MYOCARDIAL INFARCTION AND MYOCARDIAL I/R INJURY

The animal model of myocardial infarction (MI) plays an important role in study the leading cause of death in Western culture: coronary artery disease. The model for acute MI in the mouse typically involves ligation of the left anterior descending coronary artery (LAD) (Salto-Tellez et al., 2004). The operator is typically a trained surgeon or highly skilled and motivated technologist (or research associate). The most difficult problems facing the researcher performing these procedures is the expense and expertise required for small animal endotracheal intubation and mechanical ventilation.

5.1. Procedure: Interruption of the murine left coronary artery

5.1.1. Materials

Dissecting microscope (6–12×) with adequate illumination
Sterile microvascular forceps, needle drivers (including Castroviejo type) and scissors
Self-retaining retractor
10-0 polypropylene suture
Clean cotton-tip swabs for operative sponges

Water-recirculating heating pad
Miniventilator
Laser Doppler flowmeter or transcutaneous oxygen meter
5-0 Vicryl for rib approximation and deep closure
6-0 nylon suture for skin closure
6-0 polypropylene suture on an atraumatic needle
Small-diameter tubing

The mouse is placed in a supine position with extremities taped to the table before intubation and ventilation using a miniventilator hooked up to an oxygen concentrator apparatus. The ventilator should be set to deliver positive end-expiratory pressure (PEEP) with a stroke volume of 200 μl at 200 strokes per min. The third intercostal space is exposed and delicately dissected 3 mm from the sterno-costal junction, avoiding injury to the left internal mammary artery. Thoracotomy proceeds laterally on the upper border of the fourth rib to avoid damaging the intercostal nerves and vessels on the lower border of the third rib. Self-retaining microretractors are used to separate the third and fourth rib enough to gain adequate exposure of the operating region yet preserving rib integrity. The heart is gently squeezed out by pressing the thorax lightly; then the heart is held with the thumb and forefinger of the left hand. A 10-0 polypropylene suture is then passed under the LAD at 1 mm distal to left atrial appendage and the LAD is doubly ligated. Care is taken to avoid entry into either pleural cavity. The chest wall is closed by approximating the third and fourth ribs with one or two interrupted stitch using a 5-0 Vicryl suture. Muscle is re-approximated with the same Vicryl suture. The skin is closed with 6-0 polypropylene continuous sutures. The mouse is disconnected from the ventilator and spontaneous breathing should resume almost immediately. Recovery is in a cage warmed to 37 to 40 °C on a water-recirculating heating pad.

The procedure for myocardial I/R injury proceeds in a similar fashion except for the technique of arterial interruption. In this case, a stitch is taken around the left coronary artery (4-0 suture with an atraumatic needle). A snare formed by the two ends of the suture is threaded through a length of tubing and tightened for 60 min before it is reopened.

5.2. Evaluating myocardial ischemia after left coronary ligation (or transient occlusion)

5.2.1. Myocardial staining

Both ventricles are cut into three or four pieces and stained with 0.5% nitroblue tetrazolium (NBT, Sigma) for 5 to 10 min at 37 °C. The tissues are washed with PBS and images are taken with a digital camera. The myocardium is then fixed in 4% paraformaldehyde at 4 °C for 24 h and embedded in paraffin. Sections (5 μm thick) are cut in cross-section and stained with hematoxylin and eosin for histologic evaluation. The following parameters are determined with assistance of computerized morphometry: infarct size, or total length of the scar as a percentage of the LV circumference averaged over the endocardial and epicardial tracing; thickness of the infarct and the septum averaged from three measurements (margins and center of infarct and septum, respectively); ratio of LV cavity area to total LV area to quantify LV dilation; and infarct expansion index, the ratio of septum to scar thickness multiplied by this ratio of LV dilation.

5.2.2. Imaging and biophysical analyses

Technology for the real-time evaluation of myocardial work, performance, and viability has become very sophisticated, and is beyond the scope of this review (Scherrer-Crosbie et al., 2007; Nahrendorf et al., 2007). Typically small animal echocardiography with or without radioactive tracers is an extremely valuable tool, but requires local expertise and substantial resources. MRI is also a particularly valuable to assess myocardial edema after I/R injury (Hiba et al., 2007). It is, however, important to consider these tools before embarking on myocardial ischemia experiments as they have become prevalent.

6. Conclusion

Animal models of human diseases are just that—controlled simulations of complex processes. It is an exciting time for the study of genetically engineered mice with increasingly sophisticated instrumentation. But no less fundamental is the need to understand the limitations of each animal model in terms of extent of ischemic injury and translational relevance. For instance, hindlimb tourniquet-induced ischemia might produce significant edema, but also demonstrate minimal translational relevance to the disease being modeled. In the setting of transgenic or knock-out mice with a suspected vascular phenotype, the choice of ischemia versus ischemia/reperfusion experiments is often best guided by less complex earlier experiments. For instance, a positive result from a tumor angiogenesis or aortic ring sprouting assay might prompt the consideration of a femoral artery ligation and excision model to evaluate the angiogenic response to ischemia. An animal strain curiously resilient to inflammation or vascular leak might resist infrarenal aortic I/R injury. Finally, we recommend titrating all time points and repeating experiments with a new litter of mice to optimize and protect the reproducibility of exciting new findings. We believe that the tools described here offer an important bridge between vascular biologists and vascular clinicians that will continue to improve our knowledge of human disease.

REFERENCES

Blaisdell, F. W. (2002). The pathophysiology of skeletal muscle ischemia and the reperfusion syndrome: A review. *Cardiovasc. Surg.* **10**, 620–630.

O'Crawford, R. S., Hashmi, F. F., Jones, J. E., Albadawi, H., McCormack, M., Eberlin, K., Entabi, F., Atkins, M. D., Conrad, M. F., Austen, W. G., Jr., and Watkins, M. T. (2007). A novel model of acute murine hindlimb ischemia. *Am. J. Physiol. Heart Circ. Physiol.* **292**, H830–H837.

Hayashi, M., Hirose, H., Sasaki, E., Senga, S., Murakawa, S., Mori, Y., Furusawa, Y., and Banodo, M. (1998). Evaluation of ischemic damage in the skeletal muscle with the use of electrical properties. *J. Surg. Res.* **80**, 266–271.

Helisch, A., Wagner, S., Khan, N., Drinane, M., Wolfram, S., Heil, M., Ziegelhoeffer, T., Brandt, U., Pearlman, J. D., Swartz, H. M., and Schaper, W. (2006). Impact of mouse strain differences in innate hindlimb collateral vasculature. *Arterioscler. Thromb. Vasc. Biol.* **26**, 520–526.

Hiba, B., Richard, N., Thibault, H., and Janier, M. (2007). Cardiac and respiratory self-gated cine MRI in the mouse: Comparison between radial and rectilinear techniques at 7T. *Magn. Reson. Med.* **58**, 745–753.

Maseri, A. (1997). Inflammation, atherosclerosis, and ischemic events—exploring the hidden side of the moon. *N. Engl. J. Med.* **336**, 1014–1016.

Nahrendorf, M., Badea, C., Hedlund, L. W., Figueiredo, J. L., Sosnovik, D. E., Johnson, G. A., and Weissleder, R. (2007). High-resolution imaging of murine myocardial infarction with delayed-enhancement cine micro-CT. *Am. J. Physiol. Heart Circ. Physiol.* **292**, H3172–H3178.

Salto-Tellez, M., Yung Lim, S., El-Oakley, R. M., Tang, T. P., Almsherqi, Z. A., and Lim, S. K. (2004). Myocardial infarction in the C57BL/6J mouse: A quantifiable and highly reproducible experimental model. *Cardiovasc. Pathol.* **13**, 91–97.

Scherrer-Crosbie, M., Rodrigues, A. C., Hataishi, R., and Picard, M. H. (2007). Infarct size assessment in mice. *Echocardiography* **24**, 90–96.

Wakai, A., Winter, D. C., Street, J. T., O'Sullivan, R. G., Wang, J. H., and Redmond, H. P. (2001). Inosine attenuates tourniquet-induced skeletal muscle reperfusion injury. *J. Surg. Res.* **99**, 311–315.

Weis, S. M., and Cheresh, D. A. (2005). Pathophysiological consequences of VEGF-induced vascular permeability. *Nature* **437**, 497–504.

Ziv, K., Nevo, N., Dafni, H., Israely, T., Granot, D., Brenner, O., and Neeman, M. (2004). Longitudinal MRI tracking of the angiogenic response to hind limb ischemic injury in the mouse. *Magn. Reson. Med.* **51**, 304–311.

CHAPTER EIGHT

NONINVASIVE IMAGING OF BLOOD VESSELS

Milan Makale

Contents

Abstract

Angiogenesis is a key component in several major clinical conditions including cancer, diabetic retinopathy, rheumatoid arthritis, endometriosis and psoriasis. All these diseases could be managed much more effectively if their angiogenic capacities were somehow curtailed. Hence there is great interest in developing a fuller understanding of angiogenesis and designing agents to suppress, guide, and normalize this process. Although much has been learned from *in vitro* methods, the perspective is limited because angiogenesis depends on active blood flow and a variety of circulating precursor cells provided by the intact host. Therefore, noninvasive *in vivo* methods that provide information over days and weeks are needed. Accordingly, the rodent dorsal skinfold tissue window chamber facilitates the imaging of new vessels around implanted cells, around an injury, or around a simple device impregnated with growth factors. Tissue oxygen levels can be measured during the course of angiogenesis using a window chamber that is also fitted with a miniature multiple electrode sensor. The present review describes window chamber methods and hardware, the measurement of oxygen, and the introduction into the chamber of tumors,

Moores Cancer Center, University of California, San Diego, La Jolla, California

Methods in Enzymology, Volume 444
ISSN 0076-6879, DOI: 10.1016/S0076-6879(08)02808-5

growth factors, and organs to induce angiogenesis. The application of multi-photon microscopy to intravital imaging is discussed, along with a description of how to modify a standard brightfield or fluorescence microscope for multiphoton imaging of window chamber microvessels.

1. INTRODUCTION

The development of new blood vessels, angiogenesis, is of vital importance to normal development, to the healing of damaged tissues, and to the progression of diseases with major clinical and societal impact, including cancer, diabetic retinopathy, rheumatoid arthritis, endometriosis, and psoriasis (Healy *et al.*, 1998; Laschke and Menger, 2007). Accordingly, an intensive effort is underway to fully characterize angiogenesis so that it may be curtailed, corrected, or effectively promoted. In cancer research, for example, the design of antiangiogenic agents is a major goal since it has been demonstrated that the establishment and spread of primary and metastatic tumors often critically depends on local vascularization (Coomber *et al.*, 2003). When cancerous growths exceed 0.5 mm in diameter, diffusion alone cannot satisfy their oxygen requirements, and the tumor cells secrete various molecules such as vascular endothelial growth factor (VEGF) to stimulate angiogenesis (Raghunand *et al.*, 2003; Subarsky and Hill, 2003; Vaupel, 2006). Following the ingrowth of new vessels, the size of the tumor accelerates and rapidly proliferating cancer cells are either shed into the relatively porous new vessels or actively intravasate. This results in dissemination of the tumor cells throughout the body and metastasis (Bockhorn *et al.*, 2007; Li *et al.*, 2000).

In diabetes, the excessive growth of new microvessels together with abnormal vessel architecture is a serious complication causing destruction of the retina and blindness, loss of renal function, and poor perfusion of the extremities leading to peripheral ulceration and loss of limbs (Algenstaedt *et al.*, 2003; Dogra *et al.*, 2004; Feletou, 2003; Ikeda, 2005; Lawson *et al.*, 2005; Lefrandt *et al.*, 2003; Reusch, 2003). Hence, there is a need to suppress angiogenesis in diabetic patients and to correct the vessel deformities that significantly degrade microvascular function (Algenstaedt *et al.*, 2003; Dogra *et al.*, 2004; Feletou, 2003; Lawson *et al.*, 2005; Lefrandt *et al.*, 2003; Reusch, 2003).

On the other hand, the controlled promotion of angiogenesis and the prevention of microvessel rarefaction are also important goals. Loss of vasculature is a major aspect of diabetes (Bates and Harper, 2003), and is a major reason for the failure of bioengineered implants such as glucose sensors (Colton, 1996; Friedl, 2004; Gilligan *et al.*, 1994; Stenken *et al.*, 2002). Therefore, there is considerable interest and activity in developing agents that promote the formation of properly organized microvascular networks to alleviate conditions caused by inadequate perfusion.

A major obstacle to the understanding and effective control of angiogenesis is that much of the work in this field has been confined to *in vitro* systems and to relatively cumbersome indirect methods based on histology. Certainly, much can be learned from culturing blood vessel segments and various progenitor cells in artificial extracellular matrix (ECM) environments, and by serial histological sections taken from animals implanted with tumors or locally treated with VEGF. However, both of these approaches for studying angiogenesis have limitations. For example, in an intact animal, the circulating immune cells and bone marrow derived progenitor cells contribute to normal and pathological angiogenesis and the dynamics and timing of this are not fully understood and are difficult to realistically duplicate *in vitro* (Ceradini and Gurtner, 2005; Kashiwagi *et al.*, 2005; Stoll *et al.*, 2003). Moreover, the tissue ECM influences angiogenesis by providing direction to endothelial cells, via binding proteins, and by exerting mechanical forces (Krishnan *et al.*, 2007). The composition of the ECM cannot be fully duplicated *in vitro*, and subtle mechanical forces have to be defined *in vivo* before they can be faithfully emulated *in vitro* (Pedersen and Swartz, 2005). In addition, fluid forces significantly affect growing vessel tips and developing microvascular networks (Ives *et al.*, 1986; Muller-Marschhausen *et al.*, 2008; Ukropec *et al.*, 2002). Shear forces and pressures generated by moving blood and by bulk interstitial fluid flow are detected *in vivo* by vascular endothelial cells (Ives *et al.*, 1986; Muller-Marschhausen *et al.*, 2008; Pedersen and Swartz, 2005; Ukropec *et al.*, 2002). Presently an intact animal is the most effective way by which to incorporate and analyze such factors in angiogenesis.

In view of the foregoing, it is apparent that while *in vitro* methods for angiogenesis are extremely useful, there is a clear need for well-defined, noninvasive, *in vivo* models as well. Ideally such models would allow the real-time observation of developing blood vessels in the natural tissue environment and nondestructively over extended time periods. Within the last few years, such models have been made more feasible and potentially useful with the introduction of various fluorescently labeled cell lines, specific strains of mice, and powerful new methods of optical microscopy including laser scanning confocal and the laser scanning multiphoton instruments. The multiphoton microscope in particular holds promise because it enables thick, live tissue sections to be imaged with relatively low phototoxicity and photobleaching (Tsai *et al.*, 2002). Hence, several *in vivo* methods have been introduced, and one of the most promising of these is the rodent dorsal skinfold tissue window chamber. This tissue-based intravital method allows high-resolution imaging of implanted labeled tumor cells, extracellular matrix components, and developing blood vessels in the living animal over hours, days, and even weeks. Moreover, the window chamber can be modified to include a flat array of oxygen sensing electrodes to allow real-time oxygen and profiles associated with angiogenesis and developing microvessel architecture to be determined (Makale *et al.*, 2003).

The present review will discuss the application of the dorsal skinfold window chamber for the noninvasive study of blood vessels with a particular focus on multiphoton imaging technology. This chapter builds upon a previous review on the intravital imaging of cancer invasion (see Makale, 2007). The present article begins with a short history of the window chamber and its applicability to the imaging of blood vessels. Then the hardware and methods involved with the window chamber and associated oxygen sensors are described. Various means of inducing angiogenesis in the chamber are outlined with representative images. Finally a description of how to modify a standard brightfield or fluorescence microscope for multiphoton use is presented.

1.1. Window chamber methodology for angiogenesis

The rodent dorsal skinfold window chamber has a relatively long history. For reviews, see Makale (2007) and Menger et al. (2002). In 1943, Algire suspended the dorsal skin of the mouse into a longitudinal fold which contained a small area of subcutaneous tissue and microvasculature visible beneath a glass coverslip (Fig. 8.1). The chamber was used for microcirculatory studies (Endrich et al., 1979, 1980; Kerger et al., 1995; Menger et al., 2002), studies of endometriosis (Laschke and Menger, 2007), and studies of tumor microvasculature and angiogenesis (Dewhirst et al., 2000; Jain et al., 2002; Leunig et al., 1992; Pahernick et al., 2002). Li and colleagues (2000) injected 20 to 30 fluorescently labeled tumor cells into the window chamber to chronicle early tumor growth and angiogenesis, as well as tumor cell migration toward and around existing host microvessels. Window chambers have been adapted for the rat, the hamster, and the mouse (Kerger et al., 1995; Menger et al., 2002). The window chamber is comprised of two titanium plates sandwiching a lengthwise fold of loose dorsal skin, and a 1-cm diameter circular area of skin is removed and covered by a round cover glass (Fig. 8.1). Titanium is rigid yet light, and it is biocompatible. Titanium has a low coefficient of thermal conductivity so the chamber tissues do not lose heat excessively. Makale and Gough (Makale et al., 2003, 2006) developed the two-sided chamber, in which a transparent window is on one side of the fold and a planar oxygen sensor with up to 18 electrodes is on the other (Fig. 8.1). This arrangement allows the microvessels to be imaged and the oxygen levels in the sandwiched subcutaneous tissue to be measured. Such an arrangement potentially could be used with an implanted tumor to chronicle the change in tumor and tissue oxygen as the tumor grows and angiogenesis proceeds. The 18 detector electrodes can be arranged in a variety of configurations.

The window chamber has been used with confocal and multiphoton microscopy (Abdul-Karim et al., 2003; Friedl et al., 2003; Kashiwagi et al., 2005). Both modalities allow fluorescent cells to be imaged as well as unlabeled ECM, collagen for example, via autofluorescence and second harmonic imaging. The multiphoton microscope, while providing

A

B

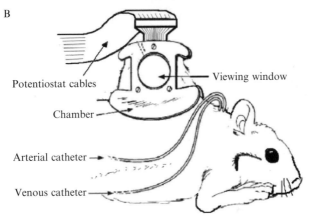

Potentiostat cables

Chamber

Viewing window

Arterial catheter

Venous catheter

Figure 8.1 (A) View of the rodent dorsal skinfold window chamber. Note the clarity of the viewing window. This apparatus is on a mouse, and for the hamster and rat is approximately 50% larger. (B) A rodent with a two-sided chamber containing a viewing window and sensor array with connector pin. The sensor is connected to the potentiostat and computer via a ribbon cable. This subject has lines in the jugular and carotid for blood sampling and injection of fluids/cells.

somewhat less color separation capability, is much less phototoxic than the confocal microscope. In the author's experience, properly designed multiphoton systems also offer a comparatively high signal-to-noise acquisition.

1.2. Oxygen in angiogenesis and oxygen measurement in the window chamber

Although much remains to be learned about angiogenesis, some of the triggers have been identified. For example, hypoxia often develops within and around growing tumors, wounds, and in diabetic tissues (Ikeda, 2005).

Fig. 8.2 shows extensive angiogenesis adjacent to a human tumor growing in the nude mouse (nu/nu) window chamber, and Fig. 8.3 is a multiphoton image of developing vessels in a mouse tumor implanted into a nu/nu mouse window chamber. Poorly oxygenated tumors are known to be angiogenic (Krishnamachary *et al.*, 2003; Subarsky and Hill, 2003), and the reduced microvessel function that occurs in diabetes causes hypoxia and elicits angiogenesis (Algenstaedt *et al.*, 2003; Dogra *et al.*, 2004; Feletou, 2003; Ikeda, 2005; Lawson *et al.*, 2005; Lefrandt *et al.*, 2003; Reusch, 2003). Low tissue oxygen tension (<7 mmHg) results in vascular endothelial cells activating major intracellular signaling molecules HIF-1α and Src (Hockel and Vaupel, 2001; Krishnamachary *et al.*, 2003; Raghunand *et al.*, 2003; Subarsky and Hill, 2003; Gray *et al.*, 2005). The rodent dorsal skinfold window chamber is well-suited to the study of the relationship between oxygen and the microvasculature. The chamber tissue oxygen levels may be measured via the phosphorescence decay method (Torres-Filho *et al.*, 1993; Kerger *et al.*, 1995), or by using a two-sided chamber in which one side is a glass viewing window and the other comprises a planar array of many oxygen electrodes (Makale *et al.*, 2003). The phosphorescent decay method only requires a typical one-sided chamber, while the oxygen sensor method necessitates chambers that have a window on one side and a flat oxygen sensor array on the other side, that is, a "two-sided" chamber. The phosphorescent method provides discontinuous measurements, while the oxygen sensor approach can provide continuous, real-time data.

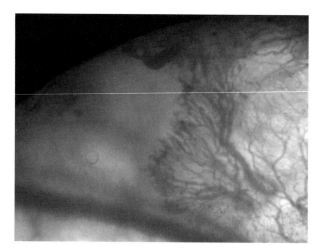

Figure 8.2 Human pancreatic cancer cells (FG tumor) growing in the mouse (nu/nu) window chamber. The normal tissue is on the right, the angiogenic vessels are clearly visible in the middle, and the tumor is the dusky region occupying the left half of the image.

Figure 8.3 A multiphoton image of a Lewis lung carcinoma (LLC) growing in the window chamber. This is a 40× image. The dark gray spindle shaped structures are individual cells, some are quite elongate, and interspersed among these is cellular debris. The developing tumor vessels are labeled light gray. The raw z-stack data were rendered using Imaris-Bitplane®.

2. METHODS

2.1. Design and fabrication of chambers

2.1.1. One-sided window chamber apparatus

The one-sided window chamber apparatus, shown in Fig. 8.4, consists of two complementary titanium alloy frames, each with a 1.2-cm diameter opening, and on one of the frames the opening is fitted with a titanium ring for attachment of the glass coverslip (Makale *et al.*, 2003). The ring extends 0.025" (0.6 mm) to the tissue to provide a seal, and it is machined so that the coverslip is held as close to the exposed tissue as possible. The glass coverslip is secured in place by a copper-beryllium retaining ring. The internal lip, on the tissue side, of the ring is only 0.007" (0.3 mm) thick. The contralateral fame has a circular opening only, without any ring or coverslip. The two frames are held together by three 10-mm screws that are separated by 4-mm hexagonal nuts. The frames are secured together with another set of hexagonal nuts. The frames themselves are 0.015" (0.4 mm) thick, and the entire assembly, including two frames, sealing ring, three screws, six nuts, glass coverslip, and retaining ring, weighs approximately 4 g (2.7 g without bolts,

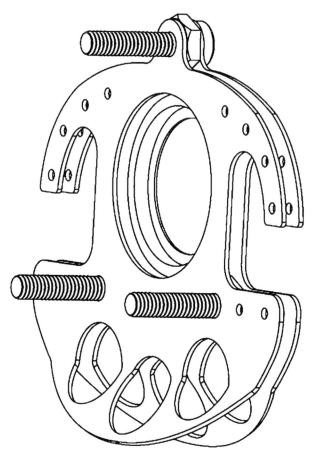

Figure 8.4 Schematic drawing of a one-sided window chamber apparatus. The left plate contains the viewing window over exposed skinfold tissue, while the companion side has is open to the intact skin of the contralateral side of the fold.

nuts, and clips). Each frame has a lateral, angled flange to permit the chamber apparatus to rest comfortably on the subject's back, and each flange has three circular areas of metal removed to reduce weight and allow air circulation. The top margin of each chamber plate has small holes drilled into it to allow placement of supporting suture material.

2.1.2. Two-sided window chamber apparatus

The apparatus, shown in Fig. 8.5, consists of two titanium-alloy frames, each having a 12.0-mm–diameter circular opening fitted with a ring for attachment of the window or sensor array (Makale *et al.*, 2003). Each ring projects 0.95 mm toward the tissue and, in conjunction with appropriate

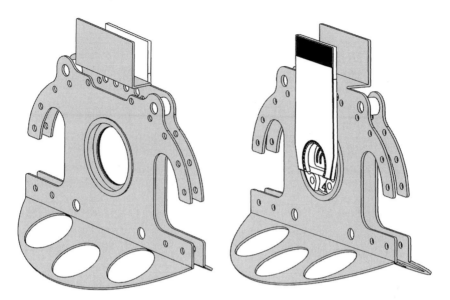

Figure 8.5 Schematic drawing of a two-sided chamber. The right plate contains the viewing window and the companion plate holds the planar sensor array. The connector stem of the array is shown.

spacers to separate the frames, prevents constriction of the skinfold, and allows perfusion of the microvasculature in the skin retractor muscle. A glass microscope coverslip is inserted into the window of one frame, and the sensor array disc is secured into the companion frame. The edges of the coverslip and sensor disc are coated with silicone adhesive before insertion into the ring to provide an airtight seal. An extremely important component of this system is the addition of a 100-μm-thick plastic disc, which is glued to the internal surface of the window to greatly reduce the thickness of the fluid layer between the tissue and sensor array surfaces. The frames are held together with four 10-mm screws and 4-mm hexagonal nuts.

2.2. Planar oxygen sensor

Implantation of a window chamber with a flat, round, multiple electrode oxygen sensor array may allow the blood vessels to be imaged around an angiogenic implant, and the change in oxygen delivery by the microvessels to be characterized in real time. The electrodes on the sensor can be arranged in almost any pattern, and since up to 18 electrodes may be installed, a fairly detailed profile of the tissue oxygen is feasible.

The configuration of the dorsal-tissue window chamber plates used in the UCSD Bioengineering Department Biosensors Lab allows the installation of a circular, planar sensor array 1.2 cm in diameter (Fig. 8.6A and B).

Any sensor type and surface membrane can be used. The UCSD Biosensors Lab fabricates sensor arrays by patterned thick film deposition of platinum paste on an alumina disc, which is then baked at 700 °C (Lin, 2000; Makale *et al.*, 2003). Eighteen disc-platinum working oxygen electrodes of 125-μm diameter are 1 to 2 mm apart, and there are six common platinum-disc counter-electrodes approximately 875 μm in diameter. A common potential silver reference electrode is electro-deposited on a platinum electrode base in the form of a ribbon, and then is chloridized to create the Ag/AgCl junction. A 25-μm layer of conductive electrolyte and a 25-μm layer of polydimethylsiloxane are deposited on the alumina disc. The various sensor

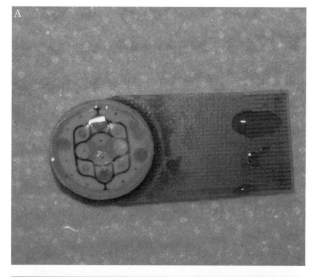

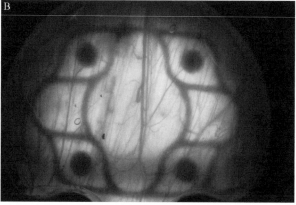

Figure 8.6 Continued

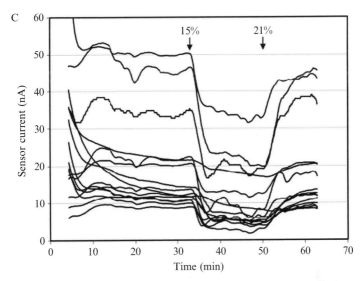

Figure 8.6 (A) Image of oxygen sensor. The sensor is 1.2 cm in diameter and has a green gold-tipped pin connector. The oxygen electrodes are the black dots. (B) A two-sided window chamber. The view is approximately 1.2× and through the glass coverslip, the two layers of retractor muscle, and the oxygen sensor array is visible against the opposite side of the retractor muscle. Tumor cells could be implanted in this muscle and the oxygen profiles associated with tumor growth and angiogenesis could be potentially be profiled. (C) Oxygen sensor output in real time taken from a hamster two-sided window chamber fitted with a planar 18 electrode oxygen sensor array. The animal was allowed to breathe room air at 21% oxygen concentration and then the oxygen level was reduced to 15% causing a corresponding fall in the tissue levels of oxygen as shown by the output of almost all the sensors. The oxygen level was then restored to 21% and the sensor output rose. Sensor output in nano amperes (nA) is directly proportional to and is an index of oxygen levels.

electrodes are physically linked to a pin connector that then communicates via a ribbon cable with a multichannel potentiostat. Data are displayed using a custom Labview® program on a PC computer. A sample oxygen recording from an oxygen sensor array in the dorsal skinfold window chamber is shown in Fig. 8.6C.

2.2.1. Sensor calibration

Sensors implanted in experimental subjects need to be carefully calibrated both before and after being tested *in vivo*. Calibration allows variability in sensor manufacture and inherent performance to be subtracted from the variability recorded to get *in vivo* an accurate measure of the variability due to tissue effects. In order to avoid the inconsistencies and ambiguity generated by liquid boundary layers, sensors should also be calibrated in the gas phase whenever feasible (Makale *et al.*, 2004). This will preclude the presence of a liquid boundary layer immediately adjacent to the sensor surface. This boundary layer is relatively static and does not effectively transfer changes in analyte concentration. Determination of sensitivity to analyte prior to sensor implantation and after sensor explantation allows separation of tissue mass transfer effects from sensor variance and drift (Makale *et al.*, 2004).

2.3. Preparation and installation of window chambers

2.3.1. Sterilization of chambers and surgical supplies

The window chamber and associated hardware are cleaned in an ultrasonic cleaner and carefully inspected for any adherent organic material. The chambers, coverslips, and all other hardware are sterilized by autoclaving. Surgical instruments, suture materials, and cotton swabs are autoclaved prior to surgery and kept sterile during the course of surgery using a bead sterilizer. The oxygen sensor must be sterilized by soaking for 24 h in a solution of 6% glutaraldehyde in phosphate-buffered saline (PBS). Aseptic technique is strictly followed, with the ultimate goal being the safety and comfort of the animal subjects.

2.3.2. Animal models

The one-sided window chamber can be installed on mice, rats, diabetic fat desert sand rats, and hamsters. The two-sided chamber has been installed on hamsters, and could potentially be modified to fit on mice and rats. For studies of human tumors, the athymic nude (nu/nu) mouse is used. Since these animals are somewhat more vulnerable to infection, they are provided the antibiotic SMZ in their drinking water upon arrival at the vivarium, and this is continued for the duration of the experiment. The Tie-2 GFP mouse expresses green fluorescent protein (GFP) in its vascular endothelial cells; this is particularly useful for angiogenesis studies. The animals are carefully monitored during all surgical procedures, and body temperature is maintained with a pad filled with circulating warm water. All animal protocols need to be approved by the appropriate institutional animal use review committee.

2.3.3. Surgical implantation of one-sided chambers

These chambers include a glass window on one side and intact skin on the other. The dorsal skin is elevated into a longitudinal fold and a disc of skin, fat, fascia and one skin–retractor muscle layer is removed from one side of the fold over the A0 vessels (Fig. 8.7A and B). The skin surrounding the exposed tissue field is coated with topical antibiotic ointment and a 0.5-ml mixture of sterile normal saline and Tobramycin broad spectrum antibiotic (4:1) is applied to the open tissue. A sterile 12-mm-diameter glass coverslip is placed over the exposed tissue, and the outer margins of the skin fold are

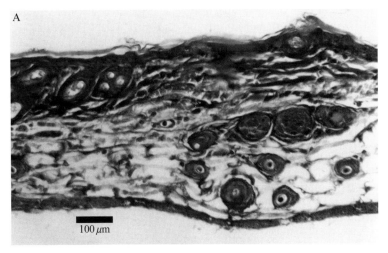

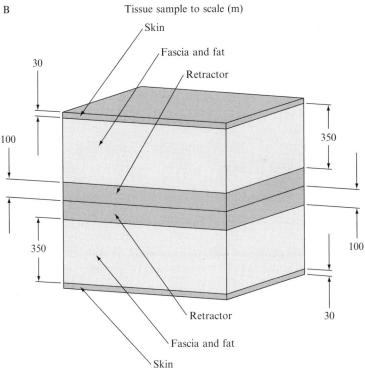

Figure 8.7 (A) Histological cross section through rodent (hamster) skin. The epidermal layer, the fat and fascia, and the retractor muscle are visible in this hematoxylin and eosin stained preparation. (B) Drawing of the folded dorsal skin showing the relative thickness of the major layers.

sutured to openings along the periphery of the chamber plates. The plates are also bolted together using the two lower chamber bolts through the skin.

2.3.4. Surgical implantation of two-sided chambers

The two-sided chambers, which have a combination of a glass coverslip and a sensor array, pose some technical challenges that do not exist with the one-sided preparations. Primarily, the two chamber plates must effectively seal the chamber volume against the entry of air without restricting blood flow. To accomplish this, a small piece of glass is glued to the interior surface of the coverslip to minimize the free volume within the chamber. The dorsal skin is drawn up into a longitudinal fold as with the one-sided approach, and two contralateral, identical, 1.2-mm-diameter discs of skin and subcutaneous fat and fascia are moved from both sides of the fold. Both retractor muscle layers are left intact, and a glass coverslip is secured on one side, and the planar sensor array is positioned on the other side.

2.3.5. Surgical recovery

After surgery is completed, the subject is subcutaneously injected with the medetomidine reversing agent, Antisedan® (atipimazole, 0.25 mg/kg). The animal is kept warm with a pad through which warm water is circulated. Bruprenex (buprenorphine, 0.2 mg/kg) is injected subcutaneously to alleviate discomfort.

2.4. Implantation of tumors and angiogenic factors in the window chamber

2.4.1. Tumor implantation

Human, hamster, and mouse cell lines can be grown in the window chamber (Figs. 8.2 and 8.3). Typically, our laboratory uses tumor cell suspensions that are concentrated from cell culture, although either cell culture suspensions or solid pieces of tumor removed from an animal may be used (Makale, 2007). The mouse Lewis lung carcinoma (LLC) grows very well in the athymic nude (nu/nu) or C57 black mouse, and the hamster melanoma grows in the hamster window chamber. Human lines that grow in the mouse (nu/nu) window chamber include pancreatic (FG and FGM), melanoma (M21 and M21L), lung, colon, brain, and prostate. At least 1 million cells of the M21 and M21L cell lines need to be implanted for growth to occur. Typically 300,000 LLC cells are implanted, which grow vigorously and elicit extensive angiogenesis. The FG and FGM cells grow avidly when 800,000 cells are implanted. Culture conditions for mouse Lewis Lung Carcinoma (LLC), the human pancreatic (FG and FGM) and the human melanoma (M21 and M21L) lines include Minimal Eagles Medium (MEM) to which is added the following:

10% FBS, L-glutamine 1:100 (200 mM stock), antibiotics 1:10 (10,000 IU penicillin, 10,000 μg/ml streptomycin, 25 μg/ml amphotericin). In some cases, the tumor cells are implanted without labeling, but for most work the tumor cell lines are stably transfected to express a variant of red fluorescent protein such as RFP or DsRed (Makale, 2007). The cells are grown under standard tissue culture conditions using 100- or 150-mm dishes.

To prepare a suspension of cells, tumor cells grown on the tissue culture dishes are rinsed with sterile PBS and then trypsinized for approximately 2 min at 37 °C. The FG and FGM cells may require 3 to 5 min of trypsinization because FG cells are particularly adherent. The cells are washed off the dishes with culture medium, counted using a hemocytometer, centrifuged into a pellet, and finally resuspended in PBS at a concentration of 200,000 cells/μl in a plastic tube that is placed on ice.

For implantation of the tumor cell suspension, the mice are subcutaneously injected with a light dose of Buprenex (buprenorphine 0.1 mg/kg) to alleviate any discomfort, and then are comfortably secured in a perforated clear Plexiglas holder inside a laminar flow sterile hood. The chamber coverslip is removed and an autoclaved 10-μl or 50-μl Hamilton syringe (Hamilton, Reno, NV) is used to inject the tumor cell suspension into the outer portion of the retractor muscle, usually between two major vessels and near the top of the window. A fresh, sterile coverslip is placed over the chamber after adding a few drops of sterile saline and Tobramycin antibiotic solution (4:1).

2.4.2. Preparation and installation of growth factor filter discs

Filter discs measuring 1 to 2 mm in diameter are punched from a Whatman #1 cellulose filter paper. A veterinary ear punch may be used for this purpose although it must first be cleaned with 70% ethanol. Hydrocortisone acetate (Sigma) is weighed to 100 mg and mixed into 95% ethanol to produce a final concentration of 12.5 mg/μl. The mixture may be heated at 60 °C for 2 h to effect complete entry of the cortisone into solution. The filter discs should then be placed in a 50-mm plastic Petri dish and taken to a sterile hood. The filter discs are then wetted with 4 μl of cortisone solution and allowed to completely dry in the hood. Then fibroblast growth factor (FGF) or vascular endothelial growth factor (VEGF) in PBS or water are added by micropipette to some or all of the filter discs to a concentration of 200 to 600 ng per disc. Control discs may be soaked with PBS. The filter discs may then be placed into the window chamber before drying. In any case they should be used reasonably promptly. The mouse should be lightly anesthetized with ketamine (35 mg/kg) and medetomidine (0.5 mg/kg) and the chamber coverslip removed inside a sterile hood. The filter discs are typically stacked, with two filters in each stack. Several stacks may be placed in each chamber. A fresh coverslip is placed over the stack and the dorsal skin is gently pressed against the coverslip to minimize air and allow the interstitial fluid to rehydrate the chamber. The addition of additional saline is avoided to preclude the possibility of diluting and washing away the growth factor(s).

2.4.3. Imaging of filter discs and vessels

Seven to 10 days after implantation the filter discs should be surrounded by new blood vessels as shown in Fig. 8.8. Our laboratory has found that FGF provides more consistent results than does VEGF. The mouse may be lightly anesthetized and comfortably placed within a clear perforated plastic holder and mounted on the stage of a brightfield microscope or a confocal microscope. The animals may be injected via the tail vein with FITC-lectin to illuminate the developing blood vessels.

2.4.4. Preparation of lymph nodes

The one-sided window preparation is performed on the host mouse in the usual fashion and just prior to the removal of nodes from the donor mouse. A mouse from any strain is deeply anesthetized with 75 mg/kg of ketamine and 1 mg/kg of medetomidine. The ventral fur is shaved with fine clippers and the remaining hair is removed using a commercially available depilatory (Nair®). The ventral aspect of the animal is cleaned with 70% ethanol and swabbed with Betadine. The axillary and inguinal nodes may be exposed by incising the skin with fine scissors and then using blunt dissection to free them from surrounding tissue. The nodes are then gently washed in warm sterile PBS, and immediately positioned in the center of the tissue window chamber field. The chamber is filled with sterile saline and a coverslip is positioned over the chamber. No air bubbles should be present and the

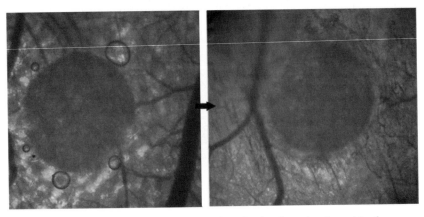

Figure 8.8 (Left panel) A VEGF-soaked filter disc has been implanted in the mouse (nu/nu) dorsal skinfold window chamber. (Right panel) The same filter disc as at left, after 1 week. Note the proliferation of small blood vessels.

chamber should be well-sealed with no leaks. The animal may then be allowed to recover with gentle warming.

2.4.5. Imaging of lymph nodes

After 14 days, the lymph node should begin to be vascularized with new vessels sprouting from the host, and should appear as shown in Fig. 8.9. The mouse tail vein is injected with 100 μl of rhodamine-lectin or FITC-lectin (Vector Labs FL1101). Approximately 15 to 30 min later, the mouse is anesthetized with ketamine (75 mg/kg) and medetomidine (1 mg/kg) and placed on a confocal or multiphoton microscope. The neovessels should be clearly visible (Fig. 8.9).

2.5. Imaging methodology

2.5.1. Microscopes and subject holders

The dorsal skinfold window chamber can be used with standard brightfield microscopes, fluorescence microscopes, and with confocal and multiphoton instruments. The planar-sensor array may be connected to the mouse during imaging to examine the effects of changing blood flow on oxygen profiles in the tissue. To use the chamber properly, each microscope will

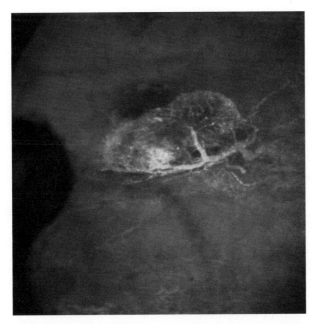

Figure 8.9   A 4× confocal microscope image of the dorsal skinfold chamber in a mouse that was injected in the tail vein with 0.1 ml of FITC-lectin. A C57 mouse inguinal lymph node was transplanted into the chamber 2 weeks previously, and it is apparent that it has become vascularized by the host.

necessitate the fabrication of a plastic device to hold the animal subject on the stage and to correctly position the chamber window in very close proximity to the microscope objective lenses. The animal must be comfortably and safely restrained, and the holder must allow free movement of the stage that is sufficient for the objective to be able to scan the entire chamber window. An example of a holder is shown in Fig. 8.10. This holder is machined from clear plastic, fits precisely into the stage of the multiphoton microscope, and all the surfaces and edges are smooth.

The window chamber works well with an upright microscope using both dry lenses and immersion lenses (water or oil), since the coverslip and surrounding frame make a convenient pocket for fluids. The lenses should be as narrow bore as possible, as wide lenses tend to collide with the frame opening and cannot be advanced sufficiently close to the chamber. Narrow bore, tapered lenses with a long working distance allow the entire window area to be traversed in an unimpeded manner.

2.5.2. Anesthesia

Animal movements must be minimized for useful imaging. The animal must be properly sedated so that it is relaxed and its voluntary movements are greatly diminished. Our laboratory has had good success with a

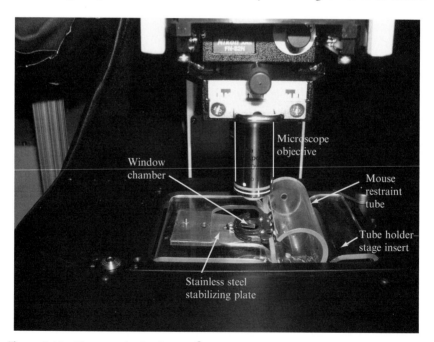

Figure 8.10 Photograph of a Plexiglas® mouse holder fitted to the stage of a multiphoton microscope. The stabilizing steel plate is bolted to the chamber and causes significant damping of breathing movements in the tissue chamber.

half-anesthetic dose of ketamine and medetomidine (35 mg/kg and 0.5 mg/kg, respectively). The mouse needs to be kept warm with this and any other anesthetic preparation. Another important issue is that breathing movements transmitted from the mouse, along the dorsal skin and into the window chamber, need to be minimized. This is done by bolting a heavy stainless plate to both the mouse holder and the window chamber. This securely positions the chamber and significantly dampens oscillation due to breathing (Fig. 8.10).

2.5.3. Maintaining normothermia during imaging
A small heating pad may be placed around the animal holder. The animal's core temperature should be monitored with a rectal probe and this should guide the heating cycles of the pad. The animal's respirations need to be closely attended to as well; vigorous or jerky respirations signal that the anesthetic level is insufficient, and extremely shallow, slow respirations indicate that the animal is hypothermic.

2.6. Multiphoton microscopy

The multiphoton microscope is increasingly used for all kinds of intravital imaging and is used with the window chamber (Abdul-Karim *et al.*, 2003; Imanishi *et al.*, 2007; Kashiwagi *et al.*, 2005). Multiphoton systems have superseded laser scanning confocal (pinhole) systems for some applications (Condeelis *et al.*, 2005; Imanishi *et al.*, 2007; Sahai *et al.*, 2005; Sipos *et al.*, 2007). This trend reflects the lower phototoxicity with multiphoton compared to confocal and standard fluorescence microscopy, and the fact that multiphoton images can be acquired relatively deeply (200 to 1000 μm) even in live tissues that have a relatively large degree of light scattering (Diaspro *et al.*, 2005; Imanishi *et al.*, 2007; Tsai *et al.*, 2002). In addition, via second harmonic generation, biological materials such as collagen can be effectively imaged without staining (Williams *et al.*, 2001). Submicrometer resolution is attainable with multiphoton instruments, allowing visualization of intracellular structures (Tsai *et al.*, 2002). For an excellent description on validation and testing of imaging and resolving ability of multiphoton systems refer to the review by Tsai *et al.* (2002).

2.6.1. Principle of operation and advantages
The two-photon principle was first described in 1931 by Maria Goeppert-Mayer (Schenke-Layland *et al.*, 2006; Williams *et al.*, 2001). In principle, a fluorophore can be excited by two photons with approximately twice the wavelength and half the energy of single photon excitation. However, the two photons must interact with the fluorophore almost simultaneously so that sufficient energy may be absorbed to elicit light emission. In a multiphoton microscope the flux of photons is only large enough within the focal

volume to generate two-photon excitation events. Specimen regions above and below the focal volume, which is sub-femtoliter in size, do not receive sufficient photon flux for two-photon excitation to occur (Diaspro et al., 2005). This fact, together with the reduced light scattering properties of infrared laser light, 700 nm and above, has far-reaching implications. First, the multiphoton microscope is by definition confocal, as only a highly defined volume within the focal plane emits light (Coombs et al., 2007; Williams et al., 2001). This means that there is no "flare," or interfering light emanating from tissue layers above and below the confocal volume entering the photodetectors. Since there is no excitation outside the focal volume, photobleaching does not occur outside the focal plane (Diaspro et al., 2005). Moreover, the multiphoton microscope does not have a pinhole as is the case with a laser scanning confocal microscope, so both ballistic (unscattered) and scattered photons are allocated to the excited focal volume (Diaspro et al., 2005; Martini et al., 2007). And, since focal volume boundary areas are not excited, and because the IR beam has relatively lower energy, there is much less photodamage to the specimen compared with confocal microscopy (Williams et al., 2001). This last attribute allows imaging of cells in live tissue to be performed over several hours, and is an important reason for the widespread use of multiphoton systems for intravital imaging (Imanishi et al., 2007). Finally, IR does not scatter as readily as shorter wavelength light, so a coherent beam may be directed deeper into the sample (Diaspro et al., 2005). With confocal systems, 99% of the emitted light is physically rejected by the pinhole system, yet some light emitted from specimen areas above and below the focal plane enters the pinhole and degrades signal-to-noise. The comparatively high-energy and shorter-wavelength laser light in confocal systems scatters and causes extensive photobleaching and photodamage (Imanishi et al., 2007). However, confocal systems use two or more lasers, so more fluorescent dyes may be more specifically activated at once, and emitted light may be very accurately separated using a spectral analysis system.

2.6.2. Design and construction of a multiphoton microscope

Multiphoton microscopes may either be purchased as a complete system or they may be custom built. The advantage of purchasing a complete system is that it may be configured specifically for a particular user and the provider ensures that all components work properly and reliably together. The system is designed to be "user friendly." The advantage of building a system is that it is much less expensive, and the user learns how the system works. However, building a multiphoton system requires time and a commitment to learning how the instrument works and how it should be properly designed. The operative principle should be simplicity. Complicated optics, optical telescopes, elaborate laser introduction lens arrangements, fiber optics, and descanning of sample emissions should all be avoided if possible.

The multiphoton microscope may be constructed entirely from purchased mechanical parts that are variously assembled by the user, or a high-quality microscope may be acquired and modified to have multiphoton capability. The latter option is simpler, although less versatile and any high-quality upright or inverted microscope will work. High-quality lenses with excellent optical qualities, high numerical aperture, and good IR and UV transmission properties that are compatible with the microscope are critical (Theer and Denk, 2006). The microscope must be isolated from floor vibrations, and therefore it is necessary to purchase an air isolation table. The air table should be placed in a room that can be entirely darkened, and the microscope, after it is bolted to the table, should be surrounded by either a darkroom curtain or a lightproof box. In order to maximize the signal-to-noise of the multiphoton system, light leaks must be minimized as much as possible.

Figure 8.11 shows the general layout of a multiphoton microscope. The primary component of the microscope is the laser light source.

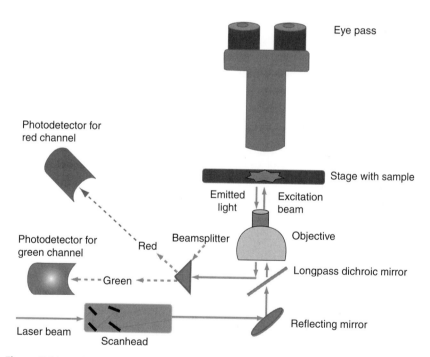

Figure 8.11 Schematic diagram depicting the general layout of a nondescanned multiphoton microscope system. The laser produces a beam that is attenuated, expanded, and then scanned across the specimen. The emitted light from the specimen is deflected by a long-pass dichroic mirror (beamsplitter) to the detector assembly, where the light is separated into red and green channels and detected by the photomultiplier tubes (PMTs).

For ease of operation the laser should require as little tuning and adjustment as possible, and should generate robust laser power levels between approximately 700 and 1020 nm. Pulsed lasers for multiphoton applications deliver a narrow Gaussian beam. A pulse duration of at least 80 femtoseconds (fs) is desirable so that optical compensation for beam broadening is not required. The beam can be attenuated using a neutral density wheel, a half-wave plate couple to a polarizing beam splitter, or with a Pockle cell. The principles behind these apparatuses are beyond the scope of the present article, but the half-wave plate and polarizing beam-splitter is a very effective method that is reliable and not overly costly. The system used to scan the excitation laser beam across the sample can be acquired from a variety of suppliers, but a good option is to purchase a microscope from a manufacturer that also offers confocal scanheads. Introduction of the beam into the scanhead should be as simple and direct as possible, and intervening optics in the scanhead designed to accept laser input via fiberoptic delivery should be removed. Otherwise the beam may be broadened, resulting in a less well defined confocal volume in the sample. A long focal-length lens ($\geq$500 mm) should be positioned between the laser and the scanhead to expand the beam so that it slightly overfills the back aperture of the microscope objective. This will ensure that the smallest, sharpest diffraction limited spot is created on the sample by the objective, ultimately resulting in a higher photon flux in a symmetrical focal volume.

After passing through the scanning system, the excitation laser beam goes through a long pass dichroic mirror and then enters a tube lens followed by the objective lens before it enters the sample. The dichroic mirror typically allows wavelengths longer than 700 nm to pass, while shorter wavelength light returned by the excited sample is reflected at some angle. The emitted light passes through an IR blocking filter, typically a short-pass 705-nm filter, to prevent stray laser light from entering the photodetection system. A collecting lens then focuses the light which enters a second mirror and is separated into two spectral ranges. A set of filters can be used to further restrict the two light ranges. The light then passes into the photodetectors. The collecting lens produces a slightly out of focus spot to project a comparatively uniform light field for detection. The illuminated spot is smaller than the sensitive area of the photodetectors, so that beam wander during raster scanning does not cause the light to deviate off the detector.

The multiphoton microscope needs to have some type of light detector so that emitted light may be acquired, digitized, and amplified for eventual display. Typically photomultiplier tubes (PMTs) are used, owing to their relatively large photoresponsive surface, and their sensitivity in the spectral ranges typically encountered in multiphoton microscopy (Diaspro *et al.*, 2005). These have to be attached to the microscope, and it is highly recommended that they be placed as close as possible to the microscope objective lens turret as possible. Placing the detectors behind the scanhead, that is descanning the emitted light and then detecting the light, will result in a very significant loss of signal and increase noise (Diaspro *et al.*, 2005; Majewska *et al.*, 2000; Masters and So, 1999; Tsai *et al.*, 2002).

The digitization of the acquired light signal, its integration into an image, and the display of the image, may be accomplished using commercial software, available through a microscope manufacturer that builds confocal instruments, or it may be programmed using an engineering language such as Labview. The software to operate the scanhead is complex and should be purchased from a confocal microscope manufacturer or some other recognized source.

REFERENCES

Abdul-Karim, M. A., Al-Kofahi, K., Brown, E. B., Jain, R. K., and Roysam, B. (2003). Automated tracing and change analysis of angiogenic vasculature from *in vivo* multiphoton confocal image time series. *Microvasc. Res.* **66,** 113–125.

Algenstaedt, P., Schaefer, C., Biermann, T., Hamann, A., Scwarzloh, B., Greten, H., Ruther, W., and Hansen-Algenstaedt, N. (2003). Microvascular alterations in diabetic mice correlate with level of hyperglycemia. *Diabetes* **52,** 542–549.

Algire, G. H. (1943). An adaptation of the transparent chamber technique to the mouse. *J. Natl. Cancer Inst.* **4,** 1–11.

Bates, D. O., and Harper, S. J. (2003). Regulation of vascular permeability by vascular endothelial growth factors. *Vasc. Pharmacol.* **39,** 225–237.

Bockhorn, M., Jain, R. K., and Munn, L. L. (2007). Active versus passive mechanisms in metastasis: Do cancer cells crawl into vessels, or are they pushed? *Lancet Oncol.* **8,** 444–448.

Ceradini, D. J., and Gurtner, G. C. (2005). Homing to hypoxia: HIF-1 as a mediator of progenitor cell recruitment to injured tissue. *Trends Cardiovasc. Med.* **15,** 57–63.

Colton, C. K. (1996). Engineering challenges in cell encapsulation technology. *Trends Biotechnol.* **14,** J58–J62.

Condeelis, J., Singer, R. H., and Seagall, J. E. (2005). The great escape: When cancer cells hijack the genes for chemotaxis and motility. *Annu. Rev. Cell Dev. Biol.* **21,** 695–718.

Coomber, B. L., Yu, J. L., Father, K. E., Plumb, C., and Rak, J. W. (2003). Angiogensis and the role of epigenetics in metastasis. *Clin. Exp. Metastasis* **20,** 215–227.

Diaspro, A., Chirico, G., and Collini, M. (2005). Two-photon fluorescence excitation and related techniques in biological microscopy. *Q. Rev. Biophys.* **38,** 97–166.

Dewhirst, M. W., Klitzman, B., Braun, R. D., Brizewl, D. M., Haroon, Z. A., and Secomb, T. W. (2000). Review of methods used to study oxygen transport at the microcirculatory level. *Int. J. Cancer* **90,** 237–255.

Endrich, B., Intaglietta, M., Reinhold, H. S., and Gross, J. F. (1979). Hemodynamic characteristics in microcirculatory blood channels during early tumor growth. *Cancer Res.* **39,** 17–23.

Endrich, B., Asaishi, K., Gotz, A., and Messmer, K. (1980). Technical report—a new chamber technique for microvascular studies in unanesthetized hamsters. *Res. Exp. Med. (Berl.)* **177,** 125–134.

Feletou, M., Boulanger, M., Staczek, J., Broux, O., and Duhalt, J. (2003). Fructose diet and VEGF-induced plasma extravasation in hamster cheek pouch. *Acta Pharmacol Sin.* **24,** 207–211.

Friedl, P., and Wolf, K. (2003). Proteolytic and non-proteolytic migration of tumour cells and leucocytes. *Biochem. Soc. Symp.* **70,** 277–285.

Friedl, K. (2004). Corticosteroid modulation of tissue response to implanted sensors. *Diabetes Technol. Ther.* **6,** 898–901.

Gilligan, B. J., Shults, M. C., Rhodes, R. K., and Updike, S. J. (1994). Evaluation of a subcutaneous glucose sensor out to 3 months in a dog model. *Diabetes Care* **17,** 882–887.

Gray, M. J., Zhang, J., Ellis, L. M., Semenza, G., Evans, D. B., Watowich, S. S., and Gallick, G. E. (2005). HIF-1α, STAT3, CBP/p300 and Ref-1/APE are components of a transcriptional complex that regulates Src-dependent hypoxia-induced expression of VEGF in pancreatic and prostate carcinomas. *Oncogene* **24,** 3110–3120.

Healy, D. L., Rogers, P. A., Hii, L., and Wingfield, M. (1998). Angiogenesis: A new theory for endometriosis. *Hum. Reprod. Update* **4,** 736–740.

Hockel, M., and Vaupel, P. (2001). Biological consequences of tumor hypoxia. *Semin. Oncol.* **28**(8), 36–41.

Ikeda, E. (2005). Cellular response to tissue hypoxia and its involvement in disease progression. *Pathol. Int.* **55,** 603–610.

Imanishi, Y., Lodowski, K. H., and Koutalos, Y. (2007). Two-photon microscopy: Shedding light on the chemistry of vision. *Biochemistry* **46,** 9674–9684.

Ives, C. L., Eskin, S. G., and McIntire, L. V. (1986). Mechanical effects on endothelial cell morphology: *In vitro* assessment. *In Vitro Cell Dev. Biol.* **22,** 500–507.

Jain, R. K., Munn, L. L., and Fukumara, D. D. (2002). Dissecting tumor pathophysiology using intravital microscopy. *Nat. Rev. Cancer* **2,** 266–276.

Kashiwagi, S., Izumi, Y., Gohongi, T., Demzou, Z. N., Xu, L., Huang, P. L., Buerk, D. G., Munn, L. L., Jain, R. K., and Fukumura, D. (2005). NO mediates mural cell recruitment and vessel morphogenesis in murine melanomas and tissue-engineered blood vessels. *J. Clin. Invest.* **115,** 1816–1827.

Kerger, H., Torres Filho, I. P., Rivas, M., Winslow, R. M., and Intaglietta, M. (1995). Systemic and subcutaneous microvascular oxygen tension in conscious Syrian golden hamsters. *Am. J. Physiol Heart Circ. Physiol.* **268,** H802–H810.

Krishnamachary, B., Berg-Dixon, S., Kelly, B., Agani, F., Feldser, D., Ferreira, G., Iyer, N., LaRusch, B., Pak, B., Taghavi, P., and Semenza, G. (2003). Regulation of colon carcinoma cell invasion by hypoxia-inducible factor 1. *Cancer Res.* **63,** 1138–1143.

Krishnan, L., Hoying, J. B., Nguyen, Q. T., Song, H., and Weiss, J. A. (2007). Interaction of angiogenic microvessels with the extracellular matrix. *Am. J. Physiol. Heart Circ. Physiol.* **293,** H3650–H3658.

Laschke, M., and Menger, M. (2007). *In vitro* and *in vivo* approaches to study angiogenesis in the pathophysiology and therapy of endometriosis. *Hum. Reprod. Update* **13,** 331–342.

Lawson, S. R., Gabra, B. H., Nantel, F., Battistini, B., and Sirois, P. (2005). Effects of a selective bradykinin B1 receptor antagonist on increased plasma extravasation in streptozotocin-induced diabetic rats: Distinct vasculopathic profile of major key organs. *Eur. J. Pharmacol.* **514,** 69–78.

Lefrandt, J. D., Bosma, E., Oomen, P. H. N., Hoeven, J. H., Roon, A. M., Smit, A. J., and Hoogenberg, K. (2003). Sympathetic mediated vasomotion and skin capillary permeability in diabetic patients with peripheral neuropathy. *Diabetologia* **46,** 40–47.

Leunig, M., Yuan, F., Menger, M. D., Boucher, Y., Goetz, A. E., Messmer, K., and Jain, R. K. (1992). Angiogenesis, microvascular architecture, microhemodynamics, and interstitial fluid pressure during early growth of human adenocarcinoma LS174T in SCID Mice. *Cancer Res.* **52,** 6553–65560.

Li, C. Y., Shan, S., Cao, Y., and Dewhirst, M. W. (2000). Role of incipient angiogenesis in cancer metastasis. *Cancer Metastasis Rev.* **19,** 7–11.

Martini, J., Andresen, V., and Anselmetti, D. (2007). Scattering suppression and confocal detection in multifocal multiphoton microscopy. *J. Biomed. Opt.* **12,** 034010-1–034010-6.

Menger, M. D., Laschke, M. W., and Vollmer, B. (2002). Viewing the microcirculation through the window: Some twenty years experience with the hamster dorsal skinfold chamber. *Eur. Surg. Res.* **34,** 83–91.

Majewska, A., Yiu, G., and Yuste, R. (2000). A custom-made two-photon microscope and deconvolution system. *Eur. J. Physiol.* **441,** 398–408.

Makale, M. T., Lin, J. T., Calou, R. E., Tsai, A. G., Chen, P. C., and Gough, D. A. (2003). Tissue window chamber system for validation of implanted oxygen sensors. *Am. J. Physiol Heart Circ. Physiol.* **284,** H2288–H2294.

Makale, M. T., Jablecki, M. C., and Gough, D. A. (2004). Mass transfer and gas-phase calibration of implanted oxygen sensors. *Anal. Chem.* **76,** 1773–1777.

Makale, M. (2007). Intravital imaging and cell invasion. *Methods Enzymol.* **426,** 375–401.

Masters, B. R., and So, P. T. C. (1999). Multi-photon excitation microscopy and confocal microscopy imaging of *in vivo* human skin: A comparison. *Microsc. Microanal.* **5,** 282–289.

Muller-Marschhausen, K., Waschke, J., and Drenckhahn, D. (2008). Physiological hydrostatic pressure protects endothelial monolayer integrity. *Am. J. Physiol. Cell Physiol.* **294,** C324.

Reusch, J. E. B. (2003). Diabetes, microvascular complications, and cardiovascular complications: what is it about glucose? *J. Clin. Invest.* **112,** 986–988.

Sahai, E., Wyckoff, J., Philippar, U., Seagall, J. E., Gertler, F., and Condeelis, J. (2005). Simultaneous imaging of GFP, CFP and collagen in tumors *in vivo* using multiphoton microscopy. *BMC Biotechnol.* 5–14 (published online).

Schenke-Layland, K., Riemann, I., Damour, O., Stock, U. A., and Konig, K. (2006). Two-photon micrscopes and *in vivo* multiphoton tomographs—powerful diagnostic tools for tissue engineering and drug delivery. *Adv. Drug Delivery Rev.* **58,** 878–896.

Sipos, A., Toma, I., Kang, J. J., Rosivall, L., and Peti-Peterdi, J. (2007). Advances in renal (patho)physiology using multiphoton microscopy. *Kidney Int.* **72,** 1188–1191.

Stenken, J. A., Reichert, W. M., and Klitzman, B. (2002). Magnetic resonance imaging of a tissue implanted device biointerface using *in vivo* microdialysis sampling. *Anal. Chem.* **74,** 4849–4854.

Stoll, B. R., Migliorini, C., Kadambi, A., Munn, L. L., and Jain, R. K. (2003). A mathematical model of the contribution of endothelial progenitor cells to angiogenesis in tumors: Implications for antiangiogenic therapy. *Blood* **102,** 2555–2561.

Subarsky, P., and Hill, R. P. (2003). The hypoxic tumor microenvironment and metastatic progression. *Clin. Exp. Metastasis* **20,** 237–250.

Theer, P., and Denk, W. (2006). On the fundamental imaging-depth limit in two photon microscopy. *J. Opt. Soc. Am.* **23,** 3139–3148.

Tsai, P. S., Nishimura, N., Yoder, E., Dolnick, E., White, G. A., and Kleinfeld, D. (2002). Principles, design, and construction of a two-photon laser-scanning microscope for *in vitro* and *in vivo* brain imaging. *In* ("*In Vivo* Optical Imaging of Brain Function.") (R. D. Frostig, ed.), CRC Press, Boca Raton, FL.

Ukropec, J. A., Hollinger, M. K., and Woolkalis, M. J. (2002). Regulation of VE-cadherin linkage to the cytoskeleton in endothelial cells exposed to fluid shear stress. *Cell Res. Exp. Cell Res.* **273,** 240–247.

Williams, R. M., Zipfel, W. R., and Webb, W. W. (2001). Multiphoton microscopy in biological research. *Curr. Opin. Chem. Biol.* **5,** 603–608.

CHAPTER NINE

INTRAVITAL VIDEOMICROSCOPY IN ANGIOGENESIS RESEARCH

Ian C. MacDonald* *and* Ann F. Chambers†

Contents

Abstract

Experimental studies on angiogenesis are clarifying many aspects of this important process and are leading to new approaches to use this information clinically. Histology of fixed tissues is a commonly used "gold standard" for assessing development of tumor vasculature during disease progression or changes in vasculature in response to genetic manipulation or therapy. However, histology provides only a static snapshot-in-time of vascular status, and

* Department of Medical Biophysics, University of Western Ontario, London, Ontario, Canada
† Department of Oncology, University of Western Ontario, London, Ontario, Canada

Methods in Enzymology, Volume 444
ISSN 0076-6879, DOI: 10.1016/S0076-6879(08)02809-7

can provide only limited information about vessel function or dynamics. Here we describe microscopy techniques and image processing approaches for using intravital video microscopy (IVVM) for the study of normal and tumor vascular morphology and function. IVVM provides powerful, high-resolution approaches for observing the vasculature in multiple organs or experimental animals. In addition to providing informative images, IVVM combined with video post-processing and image analysis approaches can be used to extract valuable quantitative information from video images. This information includes morphological parameters such as vascular diameter, density, branching, and three-dimensional vascular geometry, as well as functional and physiological information such as the identification of vessels that are perfused with red blood cells (RBCs) or plasma, rate of RBC flow, and oxygen status of RBCs. An added strength of IVVM is the ability to provide longitudinal information, looking at changes in vascular morphology and function over time in individual animals. In this chapter, we describe methods and analytical approaches for using IVVM to study vascular morphology and dynamics.

1. INTRODUCTION

Development of the microvasculature depends on a number of factors such as tissue type or disease processes such as cancer metastasis. This chapter describes methods we use to obtain digital video images of the microvasculature and mathematical procedures we use for displaying and analyzing images related to functional characteristics of the vessels. Application of these procedures to the microvasculature of different organs reveals characteristic features that distinguish them and which can be used to quantify the effects of disease or treatment. This work represents some of our "recipes" for analyzing the microvasculature, and the procedures described here are based on concepts that could easily be modified or processed using other imaging software.

1.1. Angiogenesis

Because of diffusion limitations, animals larger than a millimeter or so in radius must have a bulk transport system to service their cells. For mammals, the blood circulation carries molecular nutrients to within the required diffusion distance for the cells, removes waste products and distributes endocrine factors throughout the body. Although large vessels rapidly transport materials from one part of the body to another, it is at the level of the microcirculation, vessels less than 100 μm or so in diameter that diffusion is most effective. The metabolic rate limiting molecule in this process is oxygen, and hemoglobin carried in the circulating red blood cells (RBCs) maintains PO_2 levels in the blood, to drive diffusion at the tissue level. If the circulation is adequate for oxygen supply, it should be adequate for all other molecules. Thus, oxygen supply is the primary criterion for assessing the effectiveness of microvascular development and function.

In the fetus, vessels are initially formed *de novo* through a process known as vasculogenesis, but thereafter, new vessels develop through branching and extension of existing vessels, the process we know as angiogenesis. The typical sequence of events and factors influencing angiogenesis has been well-documented (Carmeliet and Jain, 2000; Folkman, 1997). Angiogenesis is a normal process during growth and development, but once the body reaches full size, the formation of new vessels is not normal in most tissues. Angiogenesis is, however, an integral part of wound healing or hyperplasia due to exercise or disease (Velazquez, 2007). Angiogenesis is of particular concern in cancer where it plays a major role in the growth and metastasis of solid tumors (Jain, 2005).

1.2. Tumor growth

During tumor growth an initiating cancer cell can divide to form a group of cells up to several hundred micrometers in diameter, while obtaining oxygen from surrounding tissues by diffusion. Beyond this size, cells in the center are too far from the surrounding supply, and will die, limiting the size of the tumor. If, however, there is a microvasculature within the tumor, tumor size is effectively unlimited (Chambers *et al.*, 2001; Folkman, 1992).

Vessels within a tumor may be coopted from the host tissues as cancer cells grow around them, displacing the original tissues. Alternatively, vessels may invade from surrounding tissues as the tumor expands. In either case, continued growth of cells within the tumor will require additional vessels. Just as the morphology of the microvasculature within different organs and tissues varies widely (Varghese *et al.*, 2005), vascular structure and function within tumors may vary, influenced either by the tumor type, the host tissues or a combination of both. Vessels may also arise through vasculo-genic mimicry (Hendrix *et al.*, 2003), where blood flows through channels lined with tumor tissue rather than an intact endothelium.

Newly formed vessels within tumors also play an important role in the spread of cancer, as the endothelium is poorly formed, allowing tumor cells to intravasate, enter the venous system and pass to other organs where they may form metastases (Chambers and Matrisian, 1997; Koop *et al.*, 1996; Weiss *et al.*, 1988; Wyckoff *et al.*, 2007).

1.3. Characterizing tumor microvasculature

Effective angiogenesis within a tumor will result in a microvascular network optimized for growth. Too poor or uneven a blood supply will result in necrosis, and while an oversupply may be wasteful, it will support additional growth, and maintain delivery of oxygen and nutrients even if some of the vessels fail or become occluded. The arrangement of the microvasculature may be tissue specific however and a number of parameters may be important in assessing the functional capabilities of the microcirculation (Groom et al., 1984, 1995; Minnich et al., 2001; Negrini et al., 2001; Poole et al., 1997; Pries et al., 1997; Secomb et al., 1995; Varghese et al., 2005). These include vessel spacing, diameter, volume density (V_V, fraction of tissue occupied by vessels), length density (J_V, length of vessel per unit volume of tissue), branch point density (BP_V, number of branch points per unit volume), vessel segment length (SL, length between branch points), tortuosity index (TI, segment length/distance between branch points), and, of course, blood flow (RBC velocity or flux). Quantification of such parameters is important for revealing differences between tumor types or host organs, and for assessing the appropriateness or effectiveness of antiangiogenic or vascular targeting agents.

For experimental modeling of angiogenesis in tumors, animals such as mice are used. Tumor cells are grown *in vitro* and injected to target a specific site. Tumors are allowed to grow for selected periods and they or any metastases that may form are then examined for vascular development.

1.4. Imaging

1.4.1. Histology

Histology is the traditional gold standard for assessing tumor microvasculature. All tissue within the tumor is accessible and staining can be used to identify specific cell types or metabolic processes and to differentiate between newly formed and established vessels. There are limitations to histology, however, in that endothelium often difficult to identify and the presence of RBCs is needed to confirm the presence of vessels. Functional differences in dynamic flow and RBC distribution within vessels cannot be determined, and any observations represent only a "snapshot" in time. In addition, postmortem processing artifacts can result in a loss of RBCs from the vessel lumen and tissue deformation. Also, traditional histological staining procedures cannot be used on living tissues.

1.4.2. Intravital video microscopy

Although intravital video microscopy (IVVM) is not normally used on large animals or for viewing within organs more than 100 μm from an exposed surface, murine models using IVVM provide good images in many tissues, such as muscle, liver, lung, gut, spleen, pancreas, and lymphatics (MacDonald and Chambers, 2006; MacDonald et al., 2002; Varghese et al., 2005). The main advantage of IVVM is that views are of tissues in a living organism that can be used to reveal the dynamic nature of blood flow, cell location, interactions, deformation, and even oxygen delivery. Images can be collected at video rates from selected depths within the tissue and can be digitally processed for quantitative functional assessment. IVVM is particularly useful for longitudinal observation, up to several hours during a single experiment, or on repeated occasions if sterile surgery is used for recovery between periods of observation or when window models are used.

1.4.3. Illumination/Contrast

Absorption contrast   Using traditional transillumination with white or filtered light, IVVM imaging contrast is provided by absorption of light. Natural pigments such as melanin or injected dyes such as Evans blue provide good contrast. Injected suspensions of colored microspheres or carbon particles may be taken up by endothelium or macrophages to reveal their location or behavior. Hemoglobin within the RBCs is particularly useful as it provides varying contrast at specific wavelengths. Because it is packaged within RBCs it can be seen moving within the vasculature, and since its absorption spectrum changes with oxygen saturation, can be used to measure oxygen delivery as the RBCs pass through the capillaries (Japee *et al.*, 2004, 2005).

Fluorescence contrast   Because most tissues do not absorb sufficient light to provide contrast, fluorescent probes are commonly used to provide contrast and to identify specific structures. Blood plasma can be labeled by intravenous injection of fluorescent dyes to reveal the lumens of vessels in high contrast to facilitate segmentation (MacDonald and Chambers, 2006; Varghese *et al.*, 2005). Cells can be labeled *in vitro* with fluorescent dyes (Morris *et al.*, 1993) or nanospheres (Naumov *et al.*, 2001), which provide high contrast when injected, or they can be transfected to express fluorescent proteins for permanent labeling (Vantyghem *et al.*, 2005). Fluorogenic substrates could also be injected, which only fluoresce when they take part in a specific metabolic process. Fluorescence contrast is often used together with absorption imaging to positively identify structures.

Refractive contrast   Slight local differences in the refractive index within all tissues can be used to provide contrast. This includes intravascular components such as RBCs, leukocytes, platelets, and cancer cells, endothelium, and extravascular tissues. At high numerical apertures, resolution and contrast are high and optical sectioning can reveal intracellular detail (Kachar, 1985; MacDonald and Chambers, 2006; Naumov *et al.*, 2002). Again, fluorescence contrast can be used together with refractive contrast to identify specific structures such as labeled cancer cells.

Real-time 3D   Beam splitters can be used to direct the image to multiple cameras. Because of the high refractive contrast within each optical section, by positioning the cameras at different image planes simultaneous video images can be obtained from different depths within the tissue, providing dynamic three-dimensional (3D) imaging.

1.4.4. Limitations to IVVM

Whereas histology and other imaging procedures such as ultrasound, magnetic resonance imaging, or x-ray computed tomography are used to image structures deep within the body (Graham *et al.*, 2005, 2008; Heyn *et al.*, 2006b), IVVM views are restricted to within 100 μm of the exposed organ surface. Except for surface structures such as the mouse ear circulation, anesthesia and surgery are required to expose the tissues for observation, either for acute experiments where the tissue is observed directly on the microscope, or to implant windows in the skin to allow visual access to internal structures. In addition, although some fluorescent probes or colored dyes have been used with IVVM, many of the common stains used with histology cannot be used *in vivo*.

1.4.5. Value of IVVM in angiogenesis research

Morphology   Because fluorescence IVVM can be used to positively identify and segment perfused vessels following injection of a fluorescent dye to label plasma in the microvasculature, it can be used to quantify the morphology and spatial distribution of newly formed vasculature in tumors. These measurements can include diffusion distances, vascular density, branching patterns, and segment lengths. Because refractive contrast can reveal microvascular endothelium, details of vascular structure such as vascular mimicry and aberrant wall structure can be seen.

Function   Because RBCs provide high color contrast against the fluorescent plasma, the dynamics of vascular flow networks can be quantified. Perfused vessels can be identified and by measuring the linear density and velocity of RBCs, their flux can be calculated (Japee *et al.*, 2004). Together with differential absorption of light by hemoglobin at different wavelengths and levels of oxygen saturation, these parameters can be used to model oxygen delivery to the tumor (Japee *et al.*, 2005).

2. METHODS

2.1. Scope of this chapter

There is a wide variety of approaches to obtaining microscopic video images of vasculature in live models. This chapter is not intended to be a review of all techniques. It will be limited to some of the data acquisition techniques presently used in our labs, and will focus on the analysis procedures for extracting quantitative information from video sequences of the microcirculation in acute mouse models. Many of these procedures can be translated directly to other chronic (window chamber) or nonmurine (chick chorioallantoic membrane) models (Koop *et al.*, 1995; MacDonald *et al.*, 1992). The advantage of the acute model is that it can reveal, in any tissue that can be accessed, the results of normal vessel development right up to the time of surgery.

2.2. Microscopes

The conventional upright microscope, especially when fitted with long working distance or immersion lenses, has the advantage that it can access most tissues that can be exposed, even within concave regions such as the abdominal cavity. The exposed tissue seldom presents a flat field of view however, and cardiac or respiratory motion makes it difficult to maintain focus and field of view, particularly at higher magnifications.

The inverted microscope avoids these problems by providing a stable viewing surface (MacDonald and Chambers, 2006; MacDonald *et al.*, 2002). The mouse is placed on a flat platform with a coverglass window (45 × 50 mm) in it. The tissue of interest is surgically exposed and positioned on the window. Skin or muscle flaps can be held in place with sutures taped to the platform. Liver can be stabilized if necessary by a cotton tipped applicator stick (Q-tip) taped to the platform. By placing a piece of paper (e.g., filter paper) between the tissue and the coverglass at a location adjacent to the region of interest, enough friction can be provided to eliminate lateral motion. The preparation can be covered by a clear film (saran) to help stabilize and prevent drying, and external heating and a saline drip can be used as necessary. When the platform is placed on the microscope stage with the tissue over the objective lens, the field of view is stable and remains in focus, despite motion in the organ in regions not in contact with the stationary coverglass. Lenses up to 100× oil immersion can be used.

To image lungs external ventilation is required, but by cutting away a portion of the rib cage and attaching a the tip of a small "finger" of acetate sheet to the upper side of the lung with a drop of cyanoacrylate glue (super glue), the lung can be held in position over the window. The other end of the acetate finger is taped to the platform, preventing lateral motion during respiration but allowing the upper surface to rise and fall. The lung will deform during respiration, but one region usually remains stationary. If apnea is maintained in positive pressure for brief periods, a wide field of view can be viewed and revisited during subsequent periods. A small piece of flexible film (saran) placed over the surgical site helps to stabilize the preparation and minimizes dehydration. If a saline drip is used, it is important it also use suction to remove excess fluid and prevent the tissue from floating on the coverglass.

2.3. Illumination/Contrast

2.3.1. Conventional transillumination

Conventional transillumination gives good color contrast with RBCs (Fig. 9.1A), particularly if a green filter is used. A traditional long working-distance condenser can be used to focus light on the tissue. This technique works well in thin tissues such as mesentery, skin, or cremaster

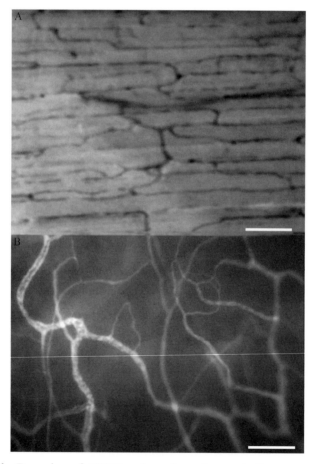

Figure 9.1 Examples of IVVM using absorption and fluorescence contrast. (A) Single frame from intravital videomicroscopy of red blood cells in rat muscle microcirculation. Absorption of green light by hemoglobin in cells provides color contrast against the light background. Little contrast from unpigmented white blood cells or muscle tissue. Scale bar = 50 μm. (B) Single frame from IVVM of subcutaneous microcirculation viewed by episcopic fluorescence illumination. View through plastic coverslip implanted in skin of mouse. Fluorescence from FITC-dextran tail vein injection. Dark spots in vessels due to presence of red blood cells. Scale bar = 100 μm.

muscle, where most of the light passes through tissues surrounding the blood vessels with little absorption of light. The high contrast makes these images good for tracking RBC location, motion, and density, and the video images are good for subsequent digital processing.

By using a beam splitter and two synchronized cameras, each fitted with narrow band interference filter, absorption at wavelengths that minimize or maximize the difference for high or low oxygen saturation can be measured (Japee *et al.*, 2004). This technique can be used for measuring the unloading of oxygen in the tissues.

In thicker tissues or those with a high blood content such as liver or spleen, green light will already have been absorbed by passage through RBCs in regions overlying the plane of focus, so color contrast is poor. Color contrast is also poor for nonpigmented tissues such as leukocytes, endothelium, or extravascular structures.

2.3.2. Fluorescence contrast

Fluorescence contrast is produced by exciting a fluorescent material at one region of the spectrum and observing the resulting fluorescence at another (longer wavelength) region. Because there are many combinations of excitation and barrier filters that can be used to identify specific fluorescent probes, multiple labels can be used in a single preparation. In addition to being useful for identifying specific structures, the use of episcopic illumination makes it possible to look into thick organs such as the kidney that cannot be transilluminated effectively.

To highlight the luminal geometry of the vasculature, high molecular weight (70 kD) fluorescein isothiocyanate-dextran (HMW FITC-dextran) can be injected intravenously to label the blood plasma (MacDonald and Chambers, 2006; Varghese *et al.*, 2005). This molecule, which is distributed rapidly within the vasculature, is too large to pass through the endothelium unless the vessel is damaged or very new and incompletely formed. Once distributed, the fluorescent probe produces a bright green image against a dark background (Fig. 9.1B) that is easy to segment for further analysis. On its own, the fluorescently labeled plasma gives no indication of flow, but the moving RBCs can be seen as they do not take up the dye and, in fact, absorb light produced by the green fluorescence. This makes FITC-dextran ideal for tracking the density and movement of RBCs in the microcirculation of normal and newly formed vessels.

Drawbacks of FITC-dextran include the fact that it obscures the presence of other green fluorescing structures such as GFP labeled cancer cells that may be traveling in the circulation during the metastatic process. There is also a problem of autofluorescence in some tissues, particularly the liver, which reduces contrast. To avoid these problems a red, rhodamine-based probe (e.g., tetramethylrhodamine [TMR]-dextran) can be used to label plasma.

For IVVM models of cancer metastasis, it is important to be able to positively identify tumor cells in the preparation. For experimental metastasis, where cells are injected directly in to the blood vessels to target the vasculature of a specific organ, the cells can be labeled with a suitable probe in culture before injection. Molecular probes that label the cytoplasm or nucleus of the cells will provide clear images, but may fade with time or quench with exposure to excitation. Fluorescent nanospheres (e.g., 50-nm diameter fluorescein latex or polycarbonate beads) are taken up in culture by many cell types. These can remain for a long time within the cell and are resistant to quenching, so can be used for long-term studies. By using this technique, dormant cancer cells have been observed in liver up to 11 weeks following mesenteric vein injection (Naumov et al., 2001). These labels are diluted by cell division, and after three or four generations, they can no longer be detected. This makes nanospheres labeling useful for identifying undivided dormant cells, but poor for observing tumor growth or the metastatic spread of cells from a primary tumor. Again, multiple colors can be used for specific contrast needs. Nanospheres can also contain contrast agents for magnetic resonance imaging such as superparamagnetic iron oxide (Heyn et al., 2006a). Although MRI has relatively low resolution compared to optical microscopy it can be used noninvasively to detect single cells at depth, and so complements observations made by IVVM. Cells can also be labeled in vivo, such as with rhodamine-G, which is taken up by both leukocytes and platelets in the circulation.

Permanent fluorescent labels for cells can be made by transfecting them to express fluorescent proteins (Naumov et al., 1999). Because the transfected genes are passed on to daughter cells the proteins can be expressed after many generations. These cells are useful for following the progress of metastasis in rapidly dividing cells, and for in vivo identification of tumor tissues where angiogenesis occurs.

Fluorescence contrast has the advantages that it is good for positively identifying cells and for segmenting vessels, and can be used on thick tissues. Fluorescence is also ideal for confocal microscopy, particularly for features where the surrounding tissue is not fluorescent. Recent advances in confocal techniques have made it possible to record multiple probes in real time at multiple focal planes. Disadvantages include tissue damage resulting from the high intensity needed for excitation, bleaching, and the lack of a background image from nonfluorescing tissues.

2.3.3. Refractive contrast

Refractive contrast makes use of variations in refractive index within tissue structure. The tissue is transilluminated predominantly from one side (oblique transillumination) (Kachar, 1985; MacDonald and Chambers, 2006; MacDonald et al., 2002). As it enters a cell with a high refractive index, light is refracted toward that side while passing through. On leaving the cell, the light returns to its original direction, adding to the light on the near side and reducing the light intensity on the far side. For objects at the focal plane, this gives a 3D shadowed appearance. Objects above or below the image plane are out of focus, and since they do not absorb light, they just contribute evenly to the background image. For this reason, refractive contrast works best at high magnifications and numerical apertures where there is a very shallow depth of focus and minimal noise from out-of-plane structures, so details of individual cells and their internal structures are revealed (Fig. 9.2A to E). Although the difference in light intensity is actually quite small and hard to see when viewed directly through the microscope eyepieces, cameras used for IVVM can be adjusted for brightness and contrast to enhance the captured image and subsequent image processing can reveal details not seen by conventional microscopy.

Typically, transillumination is provided by a fiber optic light source held at about 45 degrees. If the light is too far from the tissue, it acts more like a point source, reducing the numerical aperture of the system so diffraction bands form around contrasting structures. This is particularly noticeable in thin tissues but in thick organs, the light is scattered throughout the tissues, effectively increasing the numerical aperture. This can also be accomplished by placing a light scattering layer such as a thin piece of paper over the tissue. Structures such as cells show up particularly well when surrounded by a uniform medium of different refractive index, such as RBCs surrounded by plasma (Fig. 9.2F and G). Materials with large differences in refractive index from their surroundings—for instance, collagen fibers or lipid droplets (Fig. 9.2A)—also show up well.

Advantages of refractive contrast include the ability to image all structures without the need for absorption of light or fluorescence. It can be used in thick tissues such as liver or spleen where light passing through RBCs in overlying tissue has already been filtered by the hemoglobin so color contrast cannot be used. The shadowed image gives a 3D impression of the structures rather than a section through them so that 3D deformation of moving cells can be appreciated. Abnormal refractive index within cells, such as in damaged hepatocytes, may reverse the deflection of light rays, so the shadowing makes them appear concave rather than convex in the image. Where light is absorbed at the object plane, color contrast is seen as well. Fluorescence can be briefly added to the image to identify specific structures which can then be viewed by refractive contrast as they interact with other nonfluorescent structures.

2.4. Cameras

Continuing improvement in camera technology and digital recording makes detailed description currently available products of little value. Some comments on cameras and recording are useful however. For IVVM of the

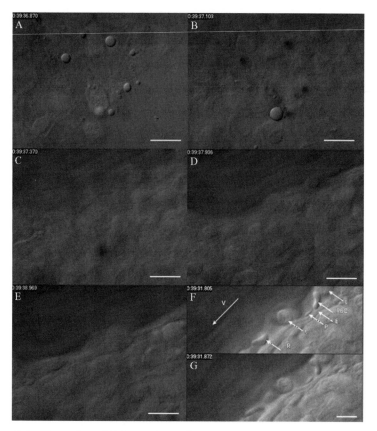

Figure 9.2 Examples of IVVM using refractive contrast. (A–E) Individual frames from Intravital videomicroscopy of murine inguinal lymph node viewed at different depths by oblique transillumination using 60× objective: (A) Focus on surface of lymph node capsule. Lipid droplets from dissection of overlying mammary fat pad show in high contrast due to high refractive index. (B) Focus at 5 μm inside capsule. Fibers of connective tissue and one lipid droplet seen along with shadows of droplets seen in (A). (C) Focus at 10 μm. Optical section entering fast flowing venule (top left) and reticular fibers and lymphocytes of subcapsular sinus. (D) Focus at 15 μm. Optical section deeper into lymph node showing fast flowing venule with leukocytes adhering to vessel wall and lymphocytes within lymphatic spaces in node. (E) Focus at 20 μm within lymph node at mid plane of venule. Oblique illumination provides refractive contrast showing granular appearance of cells within lymphatic spaces, leukocyte adhering to endothelial wall (center) of venule (top left) and dark granules of metastasizing melanoma cells at bottom right. (F–G) Different field of view at 40×. (F) Interaction of cells at wall of fast-flowing venule (V). Leukocytes (L) adhering to endothelium (E) create a vortex with a red blood cell (RBC) and platelet (P) suspended in the rotating plasma. Red blood cell (R) deformed as it is flows past obstructing leukocyte adhering to wall. (G) Same field of view two frames (67 ms) later showing movement of red blood cells and platelet. Bars 10 μm.

microvasculature, camera sensitivity should be adequate for the image intensity available; for instance, if recordings of faint fluorescence are required, and frame rates should be fast enough to record blood flow, e.g., as in a case where a cell 10 μm long moving at 0.5 mm/s will travel its own length in 1/50 s, requiring a recording rate of 50 frames per second for accurate tracking. For oblique illumination, black-and-white cameras with greater sensitivity work well, but for fluorescence multiple colors within each frame may be useful for following motion. High capacity storage now makes it possible to record digital video images for hours at a time.

2.5. Multispectral imaging

For measuring oxygen levels in the microcirculation, the absorption spectrum of hemoglobin can be used to determine saturation in RBCs (Japee *et al.*, 2005; Pittman and Duling, 1975). A beam splitter directs the image to two matched, registered cameras. One camera has an interference filter to measure the intensity of light transmitted through a RBC at 420 nm where the absorption is the same (isosbestic) for oxygenated and deoxygenated blood. The second camera simultaneously records at a wavelength of 436 nm where the absorption is maximally different. These readings, together with a measure of the background intensity when the RBC has moved from the region of measurement, can be used to determine oxygen saturation of the hemoglobin. By measuring RBC flux and the change in saturation as the cells pass through the tissues, oxygen exchange can be calculated.

2.6. Multifocal imaging

With confocal microscopy, relatively clear images can be obtained from object planes within solid tissues. Thus, it is common to move the lens or stage to step through sequential planes creating a Z stack of XY images that can be reconstructed to form a 3D image. For each image, deconvolution routines can make use of information in adjacent planes to remove noise and further improve image quality. For moving objects, this does not work because the images are not collected simultaneously. An alternate method for imaging different object planes is to change the lens-to-image distance by changing the camera position. A series of beam splitters is used to direct images to different cameras, each set at a different image plane, thus collecting simultaneous images from multiple object planes. Major differences in magnification can be corrected using adjustable transfer lenses, and the resulting images can be further registered using calibration images and affine transformations for translation, rotation, and magnification. By synchronizing the cameras, the spatial relationship of moving objects in different planes is not distorted. For given image plane separations, higher power lenses with higher numerical apertures produce thinner, more closely spaced object planes using any or all of the contrast methods described above. The synchronized video images are useful for dynamic 3D reconstruction for tracking rapidly moving cells and for obtaining concurrent flow measurement in vessels at different planes within the tissues. These images are also appropriate for deconvolution of moving objects.

2.7. Postprocessing of video images

When observing unprocessed video images obtained by IVVM, it is easy to subjectively identify vessels with flow in them and to see differences in RBC velocity and density. The size and arrangement of vessels is also apparent, with vascular geometry often characteristic of specific organs and optimized for their functions. Newly formed vessels within tumors, however, often appear abnormal (Peddinti et al., 2007), with large variations in diameter, density, tortuosity, and flow. Although tumors may appear to be oversupplied by vessels, they may or may not be adequately perfused. Low perfusion may interfere with drug delivery or the effectiveness of radiation therapy, whereas excess perfusion would make therapy with vascular targeting agents less effective if only a portion of the vessel function were disrupted and the remaining vasculature could maintain adequate perfusion. This makes it important, when studying models of tumor progression and metastasis, to be able to assess in quantitative terms the functional as well as the morphological characteristics of the microcirculation.

2.8. Measurements related to tumor angiogenesis

Visual inspection of the tumor microvasculature reveals a number of functional and morphological parameters related to the delivery of oxygen to tissues for which quantification is useful (Heyn et al., 2006b; Japee et al., 2004; Varghese et al., 2005).

2.8.1. Functional parameters

Identification of all perfused vessels   These are all vessels with flow through them, including intermittent flow and those with only plasma flow such as new vessels that are yet too small for RBCs to pass through. Low hematocrits are also found in small vessels due to the Fahareus effect, where RBCs preferentially move to the central axis leaving a layer of plasma next to the vessel wall or where side branches from these vessels are filled by plasma skimming. In capillary networks, uneven flow distribution at bifurcations can result in the slow flowing branch being filled mainly with plasma as RBCs preferentially take the faster branch.

Identification RBC perfused vessels   These are vessels with any RBC movement in them during the period of observation. Microvascular network architecture may affect the uniformity of RBC distribution throughout the tissue, as the number and velocity of RBCs may vary widely.

Measurement of RBC flux   The rate at which RBCs enter and pass through the microvasculature will determine the availability of O_2 to the tissues. RBC flux measurement is important for assessing both the development and function of the microvasculature in various tumor types and locations, and the effectiveness of antiangiogenic or vascular targeting agents. Because RBC flux through a vessel is the product of RBC velocity and RBC lineal density, ideally these parameters should be measured independently. In capillary networks, however, decreased lineal density generally occurs in vessels with low flows so their relative contribution is less and RBC flux is primarily a function of RBC velocity.

2.8.2. Morphological parameters

The functional parameters listed above will be determined primarily by the microvascular geometry as it has developed in different organs or tumor types, or has been affected by pathology or therapeutic intervention. These morphological parameters include the following:

- Volume density (V_V: volume of vessels per unit volume of tissue), which will depend on the total length and diameter of vessels in the tissues.
- Vessel diameter (D), which often appears much more variable in tumors, and is not related to flow rates as generally occurs in normal tissues.
- Lineal density (J_V, length of vessels per unit volume of tissue), which will decrease unless vessel growth keeps up with tissue growth. Lineal density will vary with both vessel segment length and the number of vessel segments per unit volume.
- Branch point density (BP_V, number of vessel bifurcations per unit tissue volume), which will decrease as tissues grow in volume unless capillary sprouts develop from existing vessels and connect to form new functional vessel segments.
- Segment length (SL, vessel length between branch points). Although lineal density can be maintained by increasing segment length, this will add to flow resistance, so the development of new parallel segments is more effective for maintaining tissue perfusion.
- Tortuosity index (TI, segment path length/shortest distance between branch points). Increased tortuosity is often associated with tumor growth, particularly noticeable in the host tissues adjacent to the tumor, possibly in response to an angiogenic stimulus from the tumor.
- 3D arrangement of the microvascular network geometry. At low magnifications and numerical apertures, a projected image of vessels within the

depth of field is obtained. At high numerical apertures, optical sectioning at discrete depth intervals provides a Z stack of images which can be integrated to provide 3D continuity of the vessels.

By determining the 3D geometry and RBC flux distribution in micro-vascular networks, mathematical modeling can be used for assessing the adequacy of oxygen delivery to the tissues. Although the information could be obtained by manual segmentation of the vasculature and frame-to-frame tracking of RBC motion, computerized digital image processing and motion analysis provide a practical means of quantifying microvascular function.

2.9. Image processing and analysis

2.9.1. Image registration

Successful segmentation, reconstruction and analysis of the microvascula-ture depend on having each frame of all concurrent video sequences in register. For multiple camera acquisition, the initial stage of the procedure is manual registration. This is accomplished by using a calibration grid to align the cameras as closely as possible, for translation, rotation and magnification. Beam splitters on the microscope allow all cameras to view the grid simultaneously. For a given objective lens, the calibration grid is first brought into focus on the eyepieces. This gives a primary image magnifica-tion equal to that specified on the objective at the standard microscope tube length. This image is then passed to the first camera by a relay lens. Typically the relay lens is set at twice its focal length, half way between the primary image and the camera, giving an image of the same magnification on the image sensor of the camera, and a field of view proportional to the image sensor size. If vignetting or loss of image quality at the periphery of the image is a problem, the relay lens to camera distance can be increased and the camera moved closer to provide a focused image with a smaller field of view (greater magnification). That image is used as a reference.

For simultaneous multispectral imaging at the same plane, a beam splitter is used and the second camera position is adjusted for a parfocal image. Translation, rotation, and magnification are adjusted to register the images.

For simultaneous multifocal image capture beam splitters direct the image to multiple cameras. The stage is adjusted to provide the desired Z stack separation for the reference and second cameras. Because the distance from the calibration grid to the objective has changed, the distance for the primary image plane of the grid has also moved with a corresponding change in magnification. The image is manually registered on the second camera by rotation and translation, and by adjusting the camera and relay lens position to compensate for the change in primary image magnification. Subsequent cameras to capture images at sequential object planes are registered to the reference camera in the same way.

Because manual registration is time consuming and can shift slightly between experiments, the calibration grid can be recorded briefly on each camera before use, and moderate variation in the rotation, translation, and magnification of the image can be corrected digitally after the video images have been obtained. Matlab (the Mathworks) includes image processing tools designed for these procedures. A calibration object with two identifiable points, (e.g., the ends of a line etched on a stage micrometer) is brought into focus on each camera where a single frame is recorded. Using the first image as a reference, Matlab routines identify the line and record the coordinates of the endpoints. Corresponding coordinates for the other cameras are also recorded, and the affine transformations required translate, rotate, and magnify them for registration with the reference image are calculated. The transformations are then applied to each frame of the video sequences taken on corresponding cameras, providing new set of video sequences with optimum camera registration.

2.9.2. Image stabilization

In addition to camera-to-camera registration, background motion in the tissue due to the respiratory or cardiac cycle means that frame-to-frame registration must also be performed. Again, manual stabilization of the image is performed first as described earlier. Residual periodic motion or gradual drifting of the image is corrected digitally following image capture using Matlab routines as described previously (Japee *et al.*, 2004; Varghese *et al.*, 2005).

For a stationary reference in the first frame of a video sequence, a region of interest is selected from the background tissue, which includes high contrast features and which ideally should not move during the entire video sequence. Pixels in the next frame are then shifted through a range of X and Y displacements and the pixel intensities in the region of interest are compared with those of the first frame using normalized cross correlation to find the best match. The optimal shift is applied to the whole image, and subsequent frames are stabilized in the same manner. Since images from the other cameras are synchronized, the same transformations can be applied to the corresponding frames for them as well.

2.9.3. Analysis of temporal variation in pixel intensity

Once the background image has been registered for all frames of video sequences from all cameras, digital image calculations can be performed on temporal variation in pixel intensities within the images. Sequences of 300 frames (10 s) are generally sufficient to characterize vascular parameters during that period. Standard Matlab routines are used to open the video sequence files, read the pixel intensities for the current frame into an array and use those values to update a set of output arrays as the routine loops through sequential frames of the sequence. For each pixel location (x, y) of image array, gray-scale values for all frames ($i = 1$ to N) of video clip are processed and recorded in new arrays as described in the following equations.

The minimum value for each pixel,

$$Min(x, y) = \left[\overset{N}{\underset{i=1}{Min}} I_i(x, y) \right]$$

The maximum value for each pixel,

$$Max(x, y) = \left[\overset{N}{\underset{i=1}{Max}} I_i(x, y) \right]$$

The sum of values for each pixel,

$$Sum(x, y) = \left[\sum_{i=1}^{N} I_i(x, y) \right]$$

The sum of squared values for each pixel,

$$SSQ(x, y) = \left[\sum_{i=1}^{N} I_i^2(x, y) \right]$$

Sums of frame-to-frame differences squared for each pixel,

$$DiffSQ(x, y) = \left[\sum_{i=1}^{N-1} [I_i(x, y) - I_{i+1}(x, y)]^2 \right]$$

All of the above arrays can be created on a single pass through all frames of the image file. Subsequent processing based on these arrays is used to produce the following.

Average intensity for each pixel location,

$$Avg(x, y) = \frac{Sum(x, y)}{N}$$

Range of intensities,

$$Range(x, y) = [Max(x, y) - Min(x, y)]$$

Variance for each pixel location,

$$\mathrm{Var}(x, y) = \frac{1}{N-1}\left\{ \mathrm{SSQ}(x, y) - \frac{[\mathrm{Sum}(x, y)]^2}{N} \right\}$$

Mean square frame-to-frame difference,

$$\mathrm{Diff}(x, y) = \frac{\mathrm{DiffSQ}(x, y)}{N}$$

2.9.4. Processed arrays displayed as images

Arrays of 8-bit video intensity values can be displayed directly as images for any individual frame from a video sequence (Fig. 9.3A), and for calculations of Min (x,y), Max (x,y) (Fig. 9.3B and C), and Avg (x,y).

The maximum-intensity image (Fig. 9.3C) shows the brightest intensity for each pixel location, including noise that could be present in any frame. The background is similar to that of any frame of the original video sequence but for vessels with fluorescently labeled plasma, the lumen is highlighted as plasma gaps between moving RBCs pass through.

The average-intensity image reduces noise in the image and provides a smooth outline of the luminal wall of the vessels. Although the minimum-intensity image (Fig. 9.3B) is not particularly useful on its own, when subtracted from the maximum-intensity image to produce the intensity range image (Fig. 9.3D), the background is minimized and regions where there is RBC motion are maximized, highlighting vessels with flow. Conversely, vessels with no flow or with only plasma in them do not show up. Because the values for range (x,y) are small relative to the other arrays, they are scaled to fill the dynamic range of the image and improve contrast (Fig. 9.3D). Range (x,y) can be influenced by a single value, so it may be affected by noise, increasing the background intensity.

The variance image (Fig. 9.3E) is particularly useful as it minimizes the background where variance is low, even if there is an occasional very high or low value, and highlights vessels where RBC motion creates a large variance in pixel intensities. Variance is good for highlighting where any amount of RBC flow occurs, but does not distinguish between a slow moving RBC that reduces the light intensity in a region of the vessel for half the observation period as opposed to a series of fast moving RBCs that do the same.

The frame-to-frame difference image (Fig. 9.3F) is more sensitive to RBC velocity. For slow-moving RBCs, a pixel will gradually become darker and lighter as the cell passes over it, so the frame-to-frame differences are small. For fast-moving cells, the change is rapid, so the differences are larger.

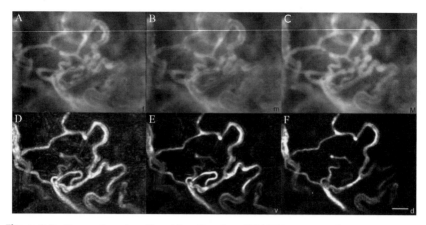

Figure 9.3   Images based on 10-s video clip of small B16F10 mouse melanoma metastasis viewed by IVVM using fluorescence contrast. (A) Single frame of microvasculature following intravenous injection of FITC dextran. Dark spots in vessels show regions perfused with red blood cells. (B) Minimum intensity image. Vessels are dark where red blood cells have been present at any time during the video clip. (C) Maximum intensity image. Vessels are bright where FITC-labeled plasma has been present during the video clip. (D) Intensity range image. Pixel intensity is proportional to the difference between (B) and (C). Contrast adjusted to fill dynamic range. (E) Variance image. Pixel intensity related to overall variance in light intensity as red blood cells pass through vessels. (F) Difference image. Pixel intensity related to frame-to-frame differences in light intensity as red blood cells pass through the vessels. Scale bar = 40 μm.

Because the differences are squared, the faster-moving cells are highlighted, and the more cells there are, the greater the intensity in the difference image. In this way, Diff (x,y) serves as a good index of RBC flux. In Fig. 9.3E, the slow-moving cells passing through the vessel loop on the right provide a high variance, while in Fig. 9.3F the low velocity results in reduced frame-to-frame intensity differences. Difference images are also more sensitive to image sharpness, so the movement of unfocused out-of-plane cells is minimized. Because changes in focus and RBC content within a vessel are less likely to change than flow velocity, variation within the difference image can be used to indicate variation in RBC velocity.

RBC velocity along each vessel segment can be measured directly using a procedure based on skeletonized images in conjunction with the original stabilized video data to create space–time images. Coordinates of each individual vessel segment centerline are used to record grayscale intensities along the vessel segment for each frame over the duration of the video sequence. These grayscale intensities are stored in an array of distance along the segment versus frame number and can be displayed as a space–time image. Cross-correlation is used to determine the distance that RBCs have traveled from one frame to the next. The mean value for the RBC shift (pixels per frame) multiplied by the frame rate (frames per second) provides the RBC speed in pixels/second.

Figure 9.4 illustrates the correlation between difference intensity and flow velocity. Figure 9.4A is a single frame from an IVVM clip of normal liver, showing RBCs in vessels labeled with FITC-Dextran. The maximum-intensity image (Fig. 9.4B) shows all vessels in the plane of focus, including some (upper left, lower center and mid right) with stationary RBCs. The high linear density of RBCs in some vessels show as a dark centerline where the RBCs are continuously present. The variance image (Fig. 9.4C) shows vessels with flow but not those with stationary RBCs. In the vessel on the extreme right, slight motion of a high-contrast RBC shows as a bright spot. Vessels with stationary or very slow RBCs do not show in the difference image (Fig. 9.4D), even some with a high variance. By measuring flow velocities in selected vessels (Fig. 9.4E) and correlating them with the corresponding pixel intensity in the difference image, a strong relationship is seen (Fig. 9.4F). This indicates that although variance images are best for mapping out the functional microvasculature, the difference image provides a good indication of RBC delivery.

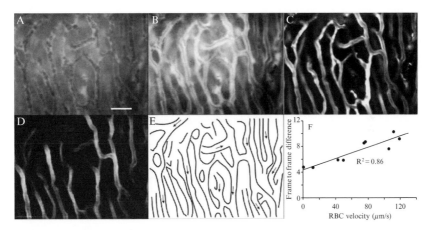

Figure 9.4 Series of images showing vascular geometry and quantification of blood flow in normal mouse liver circulation. (A) Single frame from video clip of liver viewed by FITC labeled plasma. Red blood cells (RBCs) show in dark contrast within vessels. Scale bar = 40 μm. (B) Maximum intensity image outlining all vessels. Moving RBCs seen as streaks along midline of vessels while stationary cells (lower center) have normal outline. (C) Variance image showing RBC flow in slow (top left) and fast (mid-right) but not stationary (lower center) vessels. (D) Frame to frame difference image showing vessel intensity related to RBC velocity. (E) Outline of vessels seen in (B) with arrows showing location of velocity measurement by space–time image analysis. (F) Correlation of RBC velocity measured from space–time images at locations indicated in (E), and pixel intensity at corresponding locations in difference image (D).

2.10. Morphological analysis of functional vasculature

For morphological analysis of the functional circulation (in this case all vessels with RBC flow) the variance image is the best choice. Although we have used the MATLAB Image Processing toolbox to analyze microvessel structure in three sequential steps: segmentation, skeletonization, and quantification, many of the procedures can be performed by other image processing software such as ImageJ (National Institutes of Health).

2.10.1. Segmentation

If residual variation in illumination, tissue motion or out-of-focal-plane blood flow results in uneven variance in nonvascular areas throughout the image, background subtraction can be performed using a Matlab procedure to normalize the variance intensity in regions between the vessels, allowing more accurate segmentation based on threshold selection. The variance image is divided into uniform regions with dimensions large enough to always include darker nonvessel areas. The minimum pixel intensity within each sample region is then used to generate an array representing the average background intensity of the relevant area within the variance image. A smoothed background image is then generated by filling the matrix using bicubic interpolation. Subtracting this background image from the variance image produces a "corrected" variance image with uniform background intensity, so the higher-intensity vessel regions can be uniformly segmented throughout the entire image.

Initial segmentation is done manually by setting the threshold intensity to produce binary images that include all distinct vessels. Slight variations in pixel intensity along the vessel boundaries and in regions of high background intensity result in artifacts showing as high spatial-frequency noise along the segmented boundaries and small, non-vessel areas above the threshold (Fig. 9.5A). The small areas can be eliminated by selecting only objects above a given size threshold, and the high-frequency noise can be eliminated by applying a rotationally symmetric Gaussian low-pass filter to smooth the binary image followed by an automatic threshold segmentation at a mid-gray intensity of 127. The resulting binary threshold image represents regions of the image where there were vessels with RBC flow in the original video sequence (Fig. 9.5B).

This semiautomatic segmentation procedure based on thresholding has been validated by testing against "gold standard" manual segmentation. Sensitivity and specificity both exceeded 90%. When the entire range of threshold values were tested against the gold standard, receiver-operator characteristic (ROC) curve analysis produced an area under the curve of about 98%.

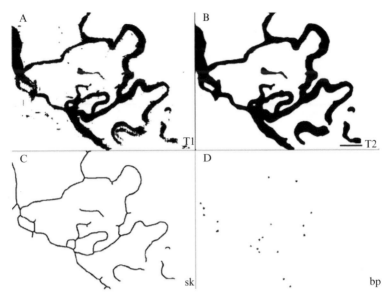

Figure 9.5 Series of images showing segmentation and skeletonization of variance image of tumor vasculature (3E). (A) Initial segmentation of variance image showing high spatial frequency noise in background and along vessel outlines. (B) Final segmentation following removal of small artifacts, application of Gaussian filter and threhold resegmentation. Scale bar = 40 μm. (C) Skeletonized variance image showing path of all RBC perfused vessels in image plane. (D) Map of branch points where three vessel segments meet within image plane.

2.10.2. Skeletonization

To generate skeletonized images, a morphological "thinning" function is repeatedly applied to segmented images, as described previously (Lam *et al.*, 1992) removing edge pixels from the image until only vessel centerlines are present (Fig. 9.5C). After skeletonization, all 3×3 pixel neighborhoods are classified, such that all skeleton pixels with a neighborhood sum equal to 4 are identified as branch points, while those with a neighborhood sum of 3 or less are identified as centerline pixels. This procedure is used to generate "vessel segment images" and "branch point images," shown as black pixels where the centerlines of vessel segments and branch points were located, respectively (Fig. 9.5D). These images can be superimposed on the original variance image, allowing visual assessment of the accuracy of the algorithm in localizing vessel centerlines and branch points. Note that vessel segmentation refers to selection of areas of the image where vessels occur whereas a vessel segment is the region of a vessel between branch points, and hence segmented image as opposed to vessel segment image.

2.10.3. Quantification

Routines for quantification are based on the concepts of quantitative stereology. Assumptions may have to be made about the effective thickness of optical sections, and orientation of vessels (primarily in the plane of focus, parallel to the tissue surface), but for comparative purposes, this approach reliably provides numbers for describing the vascular in different tissues.

Segmented and skeletonized images are used to measure vessel diameter (D), volume fraction (V_V), lineal density (J_V), branch point density (BP_V), segment length (SL), and tortuosity index (TI). Each parameter is calculated using units of pixels, and those requiring scaling are divided by the magnification factor (pixels/meter) to provide SI units.

The functional vessel diameter, D, defined as the vessel region swept out by passing RBCs (Vink and Duling, 1996), is assessed using a built-in MATLAB distance transform function. An initial estimate of vessel diameter is obtained at each centerline location as twice the distance from a centerline pixel to the nearest zero-valued background pixel in the segmented image. Since measurement of vessel diameter at branch points can be ambiguous, these regions were excluded by deleting centerline pixels from circular regions centered on branch points in the skeletonized image. The diameter of the circle is set to be slightly larger than an initial estimate of vessel diameter. Diameter measurements at each vessel centerline pixel are recorded, allowing calculation of average diameter or plotting of vessel diameters in a frequency histogram.

V_V, the volume of the vessels per unit volume of tissue, was equated to the proportion of object pixels within the image (A_A), using the Delesse principle, and corrected for the Holmes effect for vessels oriented in the focal plane using the following equation:

$$V_V = A_A/(1 + 4t/\pi D)$$

where V_V is vessel volume fraction, D is average diameter, and t is the thickness of the optical section (Weibel, 1979). Section thickness can be experimentally determined at the time of recording by focusing up and down on a small high-contrast object and measuring the range over which it remains in focus.

BP_V, branch points per unit volume, are simply calculated from the number of branch points per unit area divided by the section thickness. The section thickness depends on the numerical aperture and power of the objective lens as well as the threshold level, so absolute values may be difficult to assess without measuring them experimentally by focusing through the tissue at the time of data collection. If a multifocal microscope is used, a more accurate calculation can be made retrospectively.

J_V, the length of vessel per unit volume, is determined by summing the lengths of all vessel segment centerlines and dividing by the volume of the optical slice. Again, assumptions are made about the orientation of the vessels. This calculation underestimates J_V if the vessels are not in the plane of the optical section, but provided that vessel orientation is similar for all tissues, comparisons are valid.

SL is measured for individual vessel segments by classifying the connectivity of each pixel within the segment. The number of diagonally and orthogonally connected pixel pairs along the vessel centerline (N_d and N_o, respectively) between endpoints were counted, and SL was then calculated using the following equation to minimize the effect of vessel orientation on measured length (Kimura et al., 1999).

$$SL = \sqrt{N_d^2 + (N_d + N_o/2)^2} + N_o/2$$

Again, the true value of segment length is probably underestimated as segments passing out of the plane of focus are not measured to their terminating branch points (Fig. 9.5C). Since there are three segments per branch point and two branch points per segment, an alternative calculation would be to use $SL = 2J_V/3BP_V$, thereby avoiding the problem of partial segments.

The tortuosity index (Eze et al., 2000) is based on the ratio of SL to L, the straight-line distance between centerline endpoints, and was calculated for each vessel segment as the excess length of the vessel relative to the straight-line distance.

2.11. Applications

Intravital microscopy provides highly detailed views of the microvascular morphology and fluid dynamics characteristic of different organs. Descriptive interpretation of these views is subjective, however, and manual extraction of quantitative data is a very labor-intensive process. Because the image processing procedures described here provide a rapid and reliable quantitative assessment of vascular morphology and function, they can effectively be used to measure differences in the development of the microvascular in a variety of normal tissues and to quantify pathological changes due to injury or disease.

2.11.1. Quantitative assessment of normal tissues

Variation in microvascular morphology can occur both within and between organs. Figure 9.6 shows example variance images for five vascular beds recorded by IVVM in mouse along with frequency distributions of calculated values for vessel diameter, segment length and tortuosity index based on those images. From the images, vessel diameters appear larger in the liver and kidney than in the pancreas or muscle, while those seen in the testis

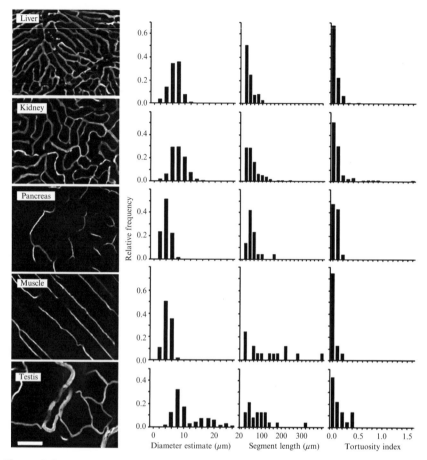

Figure 9.6 Frequency distributions of diameter, segment length, and tortuosity in a variety of anatomical locations. Variance images of liver, kidney, pancreas, hindleg muscle, and testis microvasculature are horizontally aligned with corresponding histograms. Scale bar = 100 μm.

appear to vary widely. Quantitative analysis confirms this impression with numerical values. Likewise, quantitative analysis supports the subjective impression that segment lengths are longer in muscle and testis. The subjective impression that the tortuosity of vessels in the liver and kidney is much greater than in the other organs is not supported by quantitative assessment, however. Although there are a few highly tortuous vessels in the kidney, the short segments in the highly branched liver are for the most part fairly straight, and based on the calculation, less tortuous than the longer vessels segment in the testis. These results show that within some organs such as the liver, vascular geometry is uniform while in others such muscle or testes, values for some parameters can vary widely.

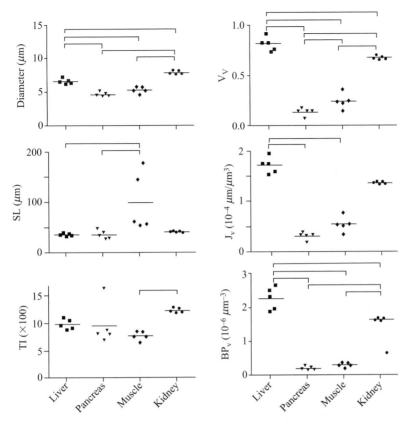

Figure 9.7 Comparison of stereological measurements of microvasculature parameters of various organs based on skeletonized variance images of functional vessels. Values represent results from five animals, brackets show significant differences (p < 0.05).

Figure 9.7 illustrates the similarity of mean values for vessel geometry within organs for five mice. For vessel diameter, vascular volume ratio (V_V) and branch point density (BP_V) values are consistent within organs but in most cases there are significant differences between organs. For segment length (SL), values are very consistent for liver, pancreas, and kidney, while those for muscle are greater and vary considerably among animals. The high vessel-length density (J_V) for liver and kidney cannot be attributed to segment length, but rather to the number of segments as indicated by the branch point density. These results show that quantitative analysis of variance images can serve as a sensitive measure of vascular geometry.

3. SUMMARY

In this chapter we have summarized IVVM microscopy approaches that can be used to study vascular morphology and function, including experimental design, microscopy and camera considerations, as well as quantitative video postprocessing approaches that can be used to extract information on vascular structure, function and changes over time, from video images obtained by IVVM of the vasculature of multiple tissues in experimental animals. These experimental tools provide a strong complement to information obtained from histology of single time points during an experiment, and are providing experimental approaches for obtaining new and valuable information about normal vascular function, as well as dynamic changes that occur during tumor growth and in response to genetic manipulation or antiangiogenic therapy.

ACKNOWLEDGMENTS

We thank Hemanth Varghese, Lisa MacKenzie, Joshua Fuller, and Jeff Hawel for their contributions to the work described here. Many of the concepts presented in this chapter have been described in Josh Fuller's master's of science thesis, "Quantification of Organ-Specific Structural and Functional Differences in Microvasculature of Tumor Metastases and Host Organs Using Variance Images" (Medical Biophysics, University of Western Ontario, 2006). Research described here from the authors' laboratories is supported by grant 016506 from the Canadian Breast Cancer Research Alliance "Special Competition in New Approaches to Metastatic Disease," with special funding support from the Canadian Breast Cancer Foundation and the Cancer Research Society, and by grant 42511 from the Canadian Institutes of Health Research. Infrastructure support is provided by awards from the Canadian Foundation for Innovation/Ontario Innovation Trust. AFC, the Canada Research Chair in Oncology, is supported by the Canada Research Chairs Program.

REFERENCES

Carmeliet, P., and Jain, R. K. (2000). Angiogenesis in cancer and other diseases. *Nature* **407,** 249–257.

Chambers, A. F., and Matrisian, L. M. (1997). Changing views of the role of matrix metalloproteinases in metastasis. *J. Natl. Cancer Inst.* **89,** 1260–1270.

Chambers, A. F., Naumov, G. N., Varghese, H. J., Nadkarni, K. V., MacDonald, I. C., and Groom, A. C. (2001). Critical steps in hematogenous metastasis: An overview. *Surg. Oncol. Clin. North Am.* **10,** 243–255, vii.

Eze, C. U., Gupta, R., and Newman, D. L. (2000). A comparison of quantitative measures of arterial tortuosity using sine wave simulations and 3D wire models. *Phys. Med. Biol.* **45,** 2593–2599.

Folkman, J. (1992). The role of angiogenesis in tumor growth. *Semin. Cancer Biol.* **3,** 65–71.

Folkman, J. (1997). Angiogenesis and angiogenesis inhibition: An overview. *EXS* **79,** 1–8.

Graham, K. C., Ford, N. L., MacKenzie, L. T., Postenka, C. O., Groom, A. C., MacDonald, I. C., Holdsworth, D. W., Drangova, M., and Chambers, A. F. (2008). Noninvasive quantification of tumor volume in preclinical liver metastasis models using contrast-enhanced x-ray computed tomography. *Invest. Radiol.* **43**, 92–99.

Graham, K. C., Wirtzfeld, L. A., MacKenzie, L. T., Postenka, C. O., Groom, A. C., MacDonald, I. C., Fenster, A., Lacefield, J. C., and Chambers, A. F. (2005). Three-dimensional high-frequency ultrasound imaging for longitudinal evaluation of liver metastases in preclinical models. *Cancer Res.* **65**, 5231–5237.

Groom, A. C., Ellis, C. G., and Potter, R. F. (1984). Microvascular geometry in relation to modeling oxygen transport in contracted skeletal muscle. *Am. Rev. Respir. Dis.* **129**, S6–S9.

Groom, A. C., Ellis, C. G., Wrigley, S. J., and Potter, R. F. (1995). Capillary network morphology and capillary flow. *Int. J. Microirc. Clin. Exp.* **15**, 223–230.

Hendrix, M. J., Seftor, E. A., Hess, A. R., and Seftor, R. E. (2003). Vasculogenic mimicry and tumour-cell plasticity: lessons from melanoma. *Nat. Rev. Cancer* **3**, 411–421.

Heyn, C., Ronald, J. A., Mackenzie, L. T., MacDonald, I. C., Chambers, A. F., Rutt, B. K., and Foster, P. J. (2006a). *In vivo* magnetic resonance imaging of single cells in mouse brain with optical validation. *Magn. Reson. Med.* **55**, 23–29.

Heyn, C., Ronald, J. A., Ramadan, S. S., Snir, J. A., Barry, A. M., MacKenzie, L. T., Mikulis, D. J., Palmieri, D., Bronder, J. L., Steeg, P. S., Yoneda, T., MacDonald, I. C., et al. (2006b). *In vivo* MRI of cancer cell fate at the single-cell level in a mouse model of breast cancer metastasis to the brain. *Magn. Reson. Med.* **56**, 1001–1010.

Jain, R. K. (2005). Normalization of tumor vasculature: An emerging concept in antiangiogenic therapy. *Science* **307**, 58–62.

Japee, S. A., Ellis, C. G., and Pittman, R. N. (2004). Flow visualization tools for image analysis of capillary networks. *Microcirculation* **11**, 39–54.

Japee, S. A., Pittman, R. N., and Ellis, C. G. (2005). A new video image analysis system to study red blood cell dynamics and oxygenation in capillary networks. *Microcirculation* **12**, 489–506.

Kachar, B. (1985). Asymmetric illumination contrast: A method of image formation for video light microscopy. *Science* **227**, 766–768.

Kimura, K., Kikuchi, S., and Yamasaki, S. (1999). Accurate root length measurement by image analysis. *Plant Soil* **216**, 117–127.

Koop, S., MacDonald, I. C., Luzzi, K., Schmidt, E. E., Morris, V. L., Grattan, M., Khokha, R., Chambers, A. F., and Groom, A. C. (1995). Fate of melanoma cells entering the microcirculation: Over 80% survive and extravasate. *Cancer Res.* **55**, 2520–2523.

Koop, S., Schmidt, E. E., MacDonald, I. C., Morris, V. L., Khokha, R., Grattan, M., Leone, J., Chambers, A. F., and Groom, A. C. (1996). Independence of metastatic ability and extravasation: Metastatic ras-transformed and control fibroblasts extravasate equally well. *Proc. Natl. Acad. Sci. USA* **93**, 11080–11084.

Lam, L., Lee, S. W., and Suen, C. Y. (1992). Thinning methodologies—a comprehensive survey. *IEEE Trans. Pattern Analysis Machine Intelligence* **14**, 869–885.

MacDonald, I. C., and Chambers, A. F. (2006). Breast cancer metastasis progression as revealed by intravital videomicroscopy. *Expert Rev. Anticancer Ther.* **6**, 1271–1279.

MacDonald, I. C., Groom, A. C., and Chambers, A. F. (2002). Cancer spread and micrometastasis development: Quantitative approaches for *in vivo* models. *Bioessays* **24**, 885–893.

MacDonald, I. C., Schmidt, E. E., Morris, V. L., Chambers, A. F., and Groom, A. C. (1992). Intravital videomicroscopy of the chorioallantoic microcirculation: A model system for studying metastasis. *Microvasc. Res.* **44**, 185–199.

Minnich, B., Bartel, H., and Lametschwandtner, A. (2001). Quantitative microvascular corrosion casting by 2D- and 3D-morphometry. *Ital. J. Anat. Embryol.* **106**, 213–220.

Morris, V. L., MacDonald, I. C., Koop, S., Schmidt, E. E., Chambers, A. F., and Groom, A. C. (1993). Early interactions of cancer cells with the microvasculature in mouse liver and muscle during hematogenous metastasis: Videomicroscopic analysis. *Clin. Exp. Metastasis* **11,** 377–390.

Naumov, G. N., MacDonald, I. C., Chambers, A. F., and Groom, A. C. (2001). Solitary cancer cells as a possible source of tumour dormancy? *Semin. Cancer Biol.* **11,** 271–276.

Naumov, G. N., MacDonald, I. C., Weinmeister, P. M., Kerkvliet, N., Nadkarni, K. V., Wilson, S. M., Morris, V. L., Groom, A. C., and Chambers, A. F. (2002). Persistence of solitary mammary carcinoma cells in a secondary site: A possible contributor to dormancy. *Cancer Res.* **62,** 2162–2168.

Naumov, G. N., Wilson, S. M., MacDonald, I. C., Schmidt, E. E., Morris, V. L., Groom, A. C., Hoffman, R. M., and Chambers, A. F. (1999). Cellular expression of green fluorescent protein, coupled with high-resolution *in vivo* videomicroscopy, to monitor steps in tumor metastasis. *J. Cell Sci.* **112,** 1835–1842.

Negrini, D., Candiani, A., Boschetti, F., Crisafulli, B., Del Fabbro, M., Bettinelli, D., and Miserocchi, G. (2001). Pulmonary microvascular and perivascular interstitial geometry during development of mild hydraulic edema. *Am. J. Physiol. Lung Cell Mol. Physiol.* **281,** L1464–L1471.

Peddinti, R., Zeine, R., Luca, D., Seshadri, R., Chlenski, A., Cole, K., Pawel, B., Salwen, H. R., Maris, J. M., and Cohn, S. L. (2007). Prominent microvascular proliferation in clinically aggressive neuroblastoma. *Clin Cancer Res.* **13,** 3499–3506.

Pittman, R. N., and Duling, B. R. (1975). A new method for the measurement of percent oxyhemoglobin. *J. Appl. Physiol.* **38,** 315–320.

Poole, D. C., Musch, T. I., and Kindig, C. A. (1997). *In vivo* microvascular structural and functional consequences of muscle length changes. *Am. J. Physiol.* **272,** H2107–H2114.

Pries, A. R., Schonfeld, D., Gaehtgens, P., Kiani, M. F., and Cokelet, G. R. (1997). Diameter variability and microvascular flow resistance. *Am. J. Physiol.* **272,** H2716–H2725.

Secomb, T. W., Hsu, R., Ong, E. T., Gross, J. F., and Dewhirst, M. W. (1995). Analysis of the effects of oxygen supply and demand on hypoxic fraction in tumors. *Acta Oncol.* **34,** 313–316.

Vantyghem, S. A., Allan, A. L., Postenka, C. O., Al-Katib, W., Keeney, M., Tuck, A. B., and Chambers, A. F. (2005). A new model for lymphatic metastasis: Development of a variant of the MDA-MB-468 human breast cancer cell line that aggressively metastasizes to lymph nodes. *Clin. Exp. Metastasis* **22,** 351–61.

Varghese, H. J., MacKenzie, L. T., Groom, A. C., Ellis, C. G., Chambers, A. F., and MacDonald, I. C. (2005). Mapping of the functional microcirculation in vital organs using contrast-enhanced *in vivo* video microscopy. *Am. J. Physiol. Heart Circ. Physiol.* **288,** H185–H193.

Velazquez, O. C. (2007). Angiogenesis and vasculogenesis: Inducing the growth of new blood vessels and wound healing by stimulation of bone marrow-derived progenitor cell mobilization and homing. *J. Vasc. Surg.* **45**(Suppl A), A39–A47.

Vink, H., and Duling, B. R. (1996). Identification of distinct luminal domains for macromolecules, erythrocytes, and leukocytes within mammalian capillaries. *Circ. Res.* **79,** 581–589.

Weibel, E. R. (1979). "Stereological Methods I." Academic Press, London, New York, Toronto.

Weiss, L., Orr, F. W., and Honn, K. V. (1988). Interactions of cancer cells with the microvasculature during metastasis. *FASEB J.* **2,** 12–21.

Wyckoff, J. B., Wang, Y., Lin, E. Y., Li, J. F., Goswami, S., Stanley, E. R., Segall, J. E., Pollard, J. W., and Condeelis, J. (2007). Direct visualization of macrophage-assisted tumor cell intravasation in mammary tumors. *Cancer Res.* **67,** 2649–2656.

CHAPTER TEN

In Vivo Measurements of Blood Flow and Glial Cell Function with Two-Photon Laser-Scanning Microscopy

Fritjof Helmchen* *and* David Kleinfeld[†]

Contents

Abstract

Two-photon laser scanning microscopy is an ideal tool for high-resolution fluorescence imaging in intact organs of living animals. With regard to *in vivo* brain research, this technique provides new opportunities to study hemodynamics in the microvascular system and morphological dynamics and calcium signaling in various glial cell types. These studies benefit from the ongoing developments for *in vivo* labeling, imaging, and photostimulation. Here, we review recent advances in the application of two-photon microscopy for the study of blood flow and glial cell function in the neocortex. We emphasize the dual role of two-photon imaging as a means to assess function in the normal state as well as a tool to investigate the vascular system and glia under pathological conditions,

* Department of Neurophysiology, Brain Research Institute, University of Zurich, Zurich, Switzerland
† Department of Physics, University of California, San Diego, La Jolla, California

Methods in Enzymology, Volume 444
ISSN 0076-6879, DOI: 10.1016/S0076-6879(08)02810-3

such as ischemia and microvascular disease. Further, we show how extensions of ultra-fast laser techniques lead to new models of stroke, where individual vessels may be targeted for occlusion with micrometer precision.

1. INTRODUCTION

All tissues are mixtures of various cellular components that perform specialized tasks. In the central nervous system (CNS), several cellular structures exist along with the electrically excitable neuronal cells that process information and fulfill the primary job of the CNS. All neuronal networks within the brain are tightly interwoven with networks of glial cells, and thus the underlying microvasculature. The three major subtypes of glial cells are astrocytes, oligodendrocytes, and microglial cells (Fig. 10.1).

Astrocytes have long been recognized for their predominant role in maintaining the homeostasis of neurons. They contribute to the regulation of local blood supply and provide nutrition to the neurons. More recently it has been recognized that the communication between astrocytes and neurons is much tighter than previously thought, for example via the release of gliotransmitters from astrocytes and through the ability of astrocytes to sense glutamate

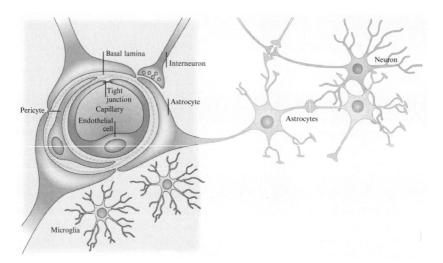

Figure 10.1 Vascular constituents of the rodent blood–brain barrier. The barrier is formed by capillary endothelial cells that form the capillary wall. These are surrounded by basal lamina, pericytes, and astrocytic perivascular endfeet. Astrocytes provide one cellular link to neurons; the other is supplied by direct connections from inhibitory interneurons. The figure also shows microglial cells that populate the brain. (Adapted from Abbott, N. J., Rönnbäck, L., and Hansson, E. (2006). Astrocyte–endothelial interactions at the blood–brain barrier. *Nat. Rev. Neurosci.* **7**, 41–53.)

(Volterra and Meldolesi, 2005). Astrocytes are also closely linked to the microvasculature system. Their "endfeet" enwrap the entire vasculature system so that they are in continuous communication with the endothelial cells. In the brain, endothelial cells from a tight blood–brain barrier, which leads to immunological isolation of the brain. Recent work has probed the *in vivo* dynamics of astroglial signaling in the control of blood flow (Takano *et al.*, 2006) and following sensory stimulation (Wang *et al.*, 2006). Lastly, recent evidence suggests that astrocytes have important physiological functions on the scale of individual synapses as well as on the level of neural circuits.

Oligodendrocytes are the myelin-producing cells of the CNS and play a role roughly analogous to Schwann cells in the peripheral nervous system. Oligodendroctytes form the myelin sheets around a neuronal axon, particularly in the long-range axonal tracts that form the white matter, and accelerate the propagation of action potentials. Finally, microglial cells are the immune-competent cells in the CNS. In contrast to astrocytes and oligodendrocytes, which are of ectodermal origin, microglial cells derive from the mesoderm and are part of the mononuclear phagocytic system. Microglial cells invade the brain during development and become permanent residents. They can be activated by a variety of stimuli, mostly in response to any kind of tissue injury, and quickly transform into phagocyting cells. Recent work, detailed below, has probed the *in vivo* dynamics of microglia during resting conditions and in response to brain vascular injury (Davalos *et al.*, 2005; Nimmerjahn *et al.*, 2005).

We now turn to the angioarchitecture of the vascular system *per se*. The topology of arteriole networks varies among different brain regions. In the neocortex, the topology is stereotyped and well understood (Fig. 10.2). The surface of cortex is covered by highly interconnected mesh-like networks of arterioles formed by anastomoses between branches of the great cerebral arteries (Brozici *et al.*, 2003). The most studied of these is the network formed by anastomoses of the middle cerebral artery, which supplies parietal cortex and parts of the striatum. Recent data support the notion that the surface network functions as an ideal grid, so that interruption to flow at a single point receives compensation from flow in neighboring regions (Schaffer *et al.*, 2006). The subsurface microvascular also forms a series of loops, in three rather than two dimensions. While these loops are quite tortuous, their detailed organization is a topic of ongoing research. Finally, the penetrating arterioles connect the surface arteriole network to the underlying microvessels that course throughout the parenchyma. The penetrating arterioles are bottlenecks, in that occlusion of a single penetrating vessel leads to a loss of supply to all microvessels in the territory fed by that arteriole (Nishimura *et al.*, 2007). Recent work, detailed below, considers the changes in blood flow dynamics that accompany a single, targeted occlusion (Nishimura *et al.*, 2006; Schaffer *et al.*, 2006) as a model of microstroke (Vermeer *et al.*, 2003).

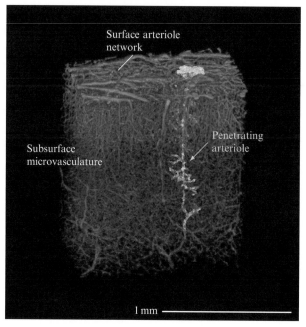

Figure 10.2   Major vascular topologies in the supply of blood to neocortex. The vasculature of a mouse was filled with fluorescently labeled agarose and a region imaged with
the all-optical histology method (see Tsai *et al.*, 2005). The reconstructed image was highlighted in the vicinity of a single penetrating arteriole (yellow hue). (From unpublished
data of P. S. Tsai and P. Blinder in the Kleinfeld laboratory.) (See color insert.)

2. Two-Photon Microscopy of Fluorescent Labels as a Tool for Brain Imaging

Historically, experimental methods to investigate glial function in the
intact brain have been limited. Different from neurons, glial cells are
electrically mostly silent, which makes them invisible to standard *in vivo*
electrophysiological methods, such as extracellular recordings. Similarly, a
fine-scale investigation of blood flow down to the level of individual
capillaries has been difficult, although coarse measurements of spatially
averaged blood flow in the brain are routinely performed, such as by laser
Doppler flowmetry. Advances in imaging technologies that have been
achieved during the past decade now provide a means to investigate blood
flow and glial function *in vivo*. The key technology is two-photon laser-
scanning fluorescence microscopy (TPLSM) (Denk *et al.*, 1990, 1994),
which can achieve penetration depths of 500 to 1000 μm into tissue
(Helmchen and Denk, 2005). Two-photon microscopy not only provides
micrometer spatial resolution deep within the tissue, but also permits

dynamic measurements over a wide range of time scales, ranging from milliseconds to years. Besides improvements in microscope technology (Helmchen and Denk, 2002), the development of various techniques for fluorescence labeling of specific tissue components *in vivo* has been of utmost importance. In particular, many variants of fluorescent proteins (FPs) have been created as anatomical markers or biosensors. These constructs can be specifically expressed in particular subsets of cells and are revolutionizing the field of neuroscience by opening new possibilities to study cellular dynamics in the living animal (Garaschuk *et al.*, 2007; Kleinfeld and Griesbeck, 2005; Miyawaki, 2005). In the following discussion, we restrict our focus to progress that has been made for *in vivo* labeling of the components of the vascular and the glial system in mice and rats. We highlight emerging applications for revealing their functions under both physiological and pathological conditions.

2.1. Animal preparation

Most measurements of blood flow dynamics and glial function involve acute experiments with anesthetized animals. Access to the cortex is through a craniotomy (Kleinfeld and Delaney, 1996), although with transgenic mice a thinned-skull preparation may be used with the advantage that the dura may be kept intact (Frostig *et al.*, 1993; Grutzendler *et al.*, 2002). A metal frame (Fig. 10.3A) is employed both as part of a chamber above the craniotomy, or thinned skull, and as a support for the animals head in the imaging apparatus (Fig. 10.3B). We typically mount the animal and all support gear, including air/gas supplies, on a single plate in our surgical suite and then move the entire plate to the microscope (Kleinfeld *et al.*, 2008) (Fig. 10. 3C). Anesthesia, airway condition, and body temperature are maintained throughout the experiment and blood gases can be collected, typically every 4 h with rats, to assess the health of the animals.

2.2. *In vivo* staining of blood vessels and glial cells

Labeling of the structures of interest with appropriate fluorescent markers is a prerequisite for dissecting cellular functions with the use of TPLSM. Such markers label cells or cellular structures as a means to identify cells and target them for experimental manipulations. Further, markers may also have a functional dependence, such that their fluorescence properties change according to changes in particular physiological parameters, such as intracellular calcium concentration. We focus on methods that are available for staining components of the vascular and glial system in the rodent brain; reviews of methods for *in vivo* staining of neurons and neuronal networks can be found elsewhere (Feng *et al.*, 2000; Garaschuk *et al.*, 2006b, 2007; Göbel and Helmchen, 2007).

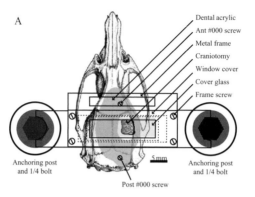

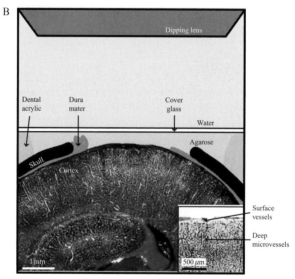

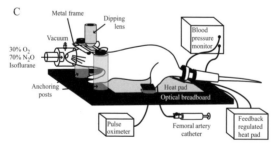

Figure 10.3 Method for *in vivo* imaging through a cranial window. (A) A metal frame with a glass window over the craniotomy immobilizes the head of the animal during imaging. (B) A cross-sectional view of the cranial window. The inset diagram shows an inverted coronal view of surface vessels and deep microvessels that are targeted for occlusion. (C) The metal frame attached to the skull is immobilized between two anchoring posts inserted into an optical breadboard. Anesthesia is maintained

2.2.1. Blood plasma

The blood vessel system can be easily labeled by injection of a bolus of fluorescent dye into the blood stream through either a tail vein or a femoral artery catheter. This approach requires the use of high-molecular-weight carrier molecules, typically 2 MDa dextran, to minimize the excretion of the dye. Such labeling stains only the blood plasma and is ideal for automated particle tracking since flowing red blood cells (RBCs) appear as shadows against a bright background (Fig. 10.4) (Kleinfeld *et al.*, 1998). Either fluorescein- or rhodamine-derivatized dextrans are injected as green or red markers, respectively. Various parameters of microcirculation, including the topological organization of local blood flow and the density, flux, and speed of RBCs, may be quantified. Alternately, RBCs may be labeled directly through the use of a donor animal (Sarelius and Duling, 1982). In this case, a fraction of the RBCs are bright, which is advantageous for calculating speeds in vessels with relatively high velocity as the process of particle tracking is considerably simplified.

Staining of the blood plasma has utility beyond the ability to track RBCs. From an anatomical perspective, high-contrast staining of the microvasculature provides clear landmarks for repeatedly finding the same field of view in long-term imaging studies (Bacskai *et al.*, 2001, 2002). Further, fluorescent staining can help to target experimental manipulations, such as photodisruption or photoinduction of thrombosis, to specific vessels and to verify and visualize subsequent extravasation.

2.2.2. Glial cells

Several methods are available for fluorescence staining of glial cells *in vivo*, including the expression of FPs, discussed below, and the application of synthetic organic dyes. Astrocytes in the neocortex of rats and mice can be readily stained by sulforhodamine 101 (SR101), a water-soluble red fluorescent dye. Brief topical application of SR101 to the exposed surface of the intact neocortex results in bright staining of the astrocyte network (Nimmerjahn *et al.*, 2004). The specificity of the stain for astrocytes verified in three ways: (a) postmortem immunohistochemistry with antibodies against common markers of neurons and glial subtypes indicates that the SR101-labeled cells co-labeled as astrocytes; (2) application of SR101 to transgenic

---

with a gas mixture. Blood pressure is measured with a tail-cuff device. The femoral artery catheter is used to collect blood samples for blood gas measurements, and is also the delivery route for contrast agents that stain the blood plasma. Heart rate and blood oxygen saturation are continuously monitored using a pulse oximeter. (Adapted from Kleinfeld, D., Friedman, B., Lyden, P. D., and Shih, A. Y. (2008). Targeted occlusion to surface and deep vessels in neocortex via linear and nonlinear optical absorption. In: Chen, J., Xu, X.-M., Zhang, J., eds. "Animal Models of Acute Neurological Injuries." Humana Press, Inc., Totowa, pp. 19–185)

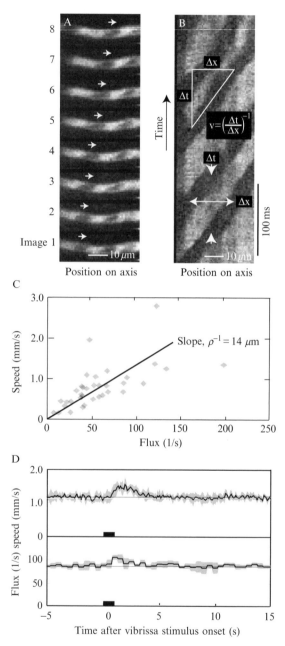

Figure 10.4 Imaging of capillary blood flow in cortex with fluorescently labeled plasma. (A) Successive planar images through a capillary, at a depth of 450 μm below the pial surface, acquired every 60 ms. The change in position of a particular unstained object, interpreted as a RBC, is indicated by the series of arrows ($\rightarrow$); the velocity of the RBC is $+0.18$ mm/s. (B) Characterization of the transport of RBCs in capillaries.

animals with astrocyte-specific expression of enhanced green fluorescent protein (EGFP) results in a complete overlap of exogenous and endogenous markers; and (3) counterstaining the blood plasma with fluorescein/dextran reveals that the entire vasculature system is sheathed by astrocytic endfeet (Fig. 10. 5A). The mechanism of SR101 uptake remains, however, unclear. SR101 can be transported by multi–drug-resistant proteins, but this transport mechanism usually extrudes SR101 from the cytosol. Perhaps astrocytes possess similar transporter proteins that operate in reverse mode, shuffling SR101 inside the cell. Once inside cells, SR101 easily distributes throughout the astrocyte network. Evidence suggests that the distribution is via gap junctions (Nimmerjahn *et al.*, 2004).

There appears to be confusion in the literature with regard to the specificity of glial SR101 uptake in brain slices and in brain regions aside from neocortex. Early studies reported that SR101 stains oligodendroglia in the retina (Ehinger *et al.*, 1994) and labels activated neurons in a turtle brainstem–cerebellum preparation (Keifer *et al.*, 1992). A difficulty with brain slices is that SR101 is apparently taken up by dead or damaged cells near the tissue surface. Further, as shown in a recent report that demonstrated SR101 labeling of astrocytes in hippocampal slices (Kafitz *et al.*, 2008), efficient SR101 uptake requires physiological temperature. An additional issue is that the mode of SR101 application may matter. In the cerebellum, it was originally reported that SR101 fails to stain Bergmann glia (Nimmerjahn *et al.*, 2004). However, in recent experiments bright SR101-staining of Bergmann glia was achieved by directly injecting the dye through a micropipette into the tissue, rather than applying it topically to the surface (unpublished data of W. Göbel in the Helmchen laboratory). Hence, specific properties of ependymal cells and their connections to the parenchymal glial network may be important for dye uptake from the tissue surface. *In toto*, SR101 is a simple and robust stain of astrocytes in the rodent neocortex *in vivo*, but further work is required to fully understand regional differences in staining and the uptake mechanism of the dye.

The vessel was longitudinally scanned in line scan mode at 2 ms/line. The instantaneous velocity is $v_{cap} = \Delta x/\Delta t$; the flux, F_{cap}, is $1/\Delta t$, and the linear density is $\rho_{cap} = 1/\Delta x$. (C) Plot of the speed versus flux. The solid line is a best fit and corresponds to the density (Eq. 10.1). (D) The trial-averaged response of RBC flow to vibrissa stimulation. We recorded from a single capillary in vibrissa S1 cortex at a depth of 255 μm below the dura. The dark line is the average over all six trials, the gray band is the standard deviation of the average. Black bar indicates the whisker stimulation period. (Adapted from Kleinfeld, D., Mitra, P. P., Helmchen, F., and Denk, W. (1998). Fluctuations and stimulus-induced changes in blood flow observed in individual capillaries in layers 2 through 4 of rat neocortex. *Proc. Natl. Acad. Sci. U.S.A.* **95**, 15741–15746.)

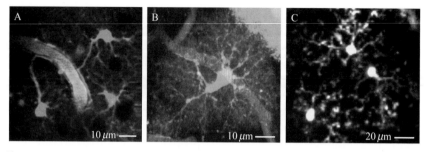

Figure 10.5 Examples of *in vivo* fluorescence staining of vasculature and glial cells. (A) Co-staining of microvessels and astrocytes by injection of fluorescein/dextran into the blood plasma and topic application of SR101 to the brain surface. (B) An EGFP-expressing astrocyte in an hGFAP/EGFP transgenic mouse. The enwrapped blood capillary was visualized by tail-vein injection of rhodamine/dextran. (C) *In vivo* TPLSM image of three EGFP-expressing microglial cells in the neocortex of CX3CR1/EGFP mice. (Images in (A) and (C) are adapted, respectively, from Nimmerjahn, A., Kirchhoff, F., Kerr, J. N., and Helmchen, F. (2004). Sulforhodamine 101 as a specific marker of astroglia in the neocortex *in vivo*. *Nat. Methods* **29**, 31–37; and Nimmerjahn, A., Kirchhoff, F., and Helmchen, F. (2005). Resting microglial cells are highly dynamic surveillants of brain parenchyma *in vivo*. *Science* **308**, 1314–1318.) (See color insert.)

2.2.3. Expression of fluorescent proteins

The alternate approach for specific *in vivo* staining of glial cells is the expression of FPs. While local transfection with viral constructs using an unspecific promoter such as the ubiquitin promoter can result in staining of some glial cells, this labeling is sparse and uncontrolled. Better results are obtained by generation of transgenic animals through the use of specific promoters. Meanwhile, there are a number of transgenic mouse lines with specific glial expression pattern available. For example, astrocytes can be visualized in transgenic mouse lines that express FPs under control of the promoter for human–glial-fibrillary-acid protein (GFAP), which is glia-specific (Zhuo *et al.*, 1997; Nolte *et al.*, 2001) (Fig. 10.5B). Other transgenic mouse lines show FP expression in non-astrocytic glial cells, including oligodendrocytes (Fuss *et al.*, 2000) and microglial cells (Jung *et al.*, 2000). In the latter case a knock-in approach was used to introduce the gene sequence for EGFP into the gene locus of the fractalkine receptor (CX3CR1). The chemokine fractalkine is a transmembrane glycoprotein that is found in endothelial cells and neurons and can be released in a soluble form following proteolytic cleavage (Cook *et al.*, 2001). In the CNS, microglial cells are the only cells that express CX3CR1. As a result, they are selectively labeled with high contrast with respect to background fluorescence (Fig. 10.5C), which make them very well suited for *in vivo* imaging studies.

2.3. Functional imaging

2.3.1. Blood flow

Blood flow can be most readily measured in capillaries, where the RBCs pass in single file. A line-scan through a capillary leads to a sequence of bright pixels, corresponding to labeled plasma, and dark pixels, corresponding to RBCs (Fig. 10.4A). This results in diagonal bands in a space–time image constructed from the line-scan data (Fig. 10.4B). The slope of the bands is the inverse of the velocity, denoted v_{cap}, which may be determined automatically. The linear flux is found by counting the number of RBCs that pass per unit time. The flux, F_{cap}, and velocity are related by

$$F_{cap} = \rho v_{cap} \qquad (10.1)$$

A plot of v_{cap} versus F_{cap} shows a fairly linear relation (Fig. 10.4C), which implies that the linear density, ρ, and thus the hematocrit, is constant. As an application, we consider possible changes in the speed and flux concurrent with sensory stimulation. An example measurement from the vibrissa primary sensory cortex shows that the speed and flux of RBCs in a single capillary both increase concomitant with brief motion of the vibrissae (Kleinfeld *et al.*, 1998) (Fig. 10.4D).

A more interesting case of blood flow dynamics concerns flow in large vessels. The line-scan scheme can be used to measure flow so long as the flow is laminar, that is, the speed along a line parallel to the vessel stays constant (Fig. 10.6A and B). This can be shown self-consistently by measuring the velocity at different radii from the centerline and plotting the speed as a function of the radius (Fig. 10.6C). We expect to get Poiseuille's parabolic curve, that is,

$$v(r) = \frac{\Delta P}{4L\eta} \left(R^2 - r^2 \right) \qquad (10.2)$$

slightly flattened by the non–zero size of the RBC, where ΔP is the pressure drop across a vessel of length L and radius R and η is effective viscosity. In practice, this condition is fulfilled (Fig. 10.6C). This implies that the blood flows as a series of nested cylinders, fastest in the center and slower toward the walls. The average velocity across the cross-section of the vessel is $\langle v \rangle = (\Delta P/8\eta L)R^2$, so that $\langle v \rangle = v(0)/2$ where $v(0)$ is the measured center-line velocity. The volume flux is

$$F_{vol} = \langle v \rangle A = \frac{\Pi}{2} v(0)R^2 \qquad (10.3)$$

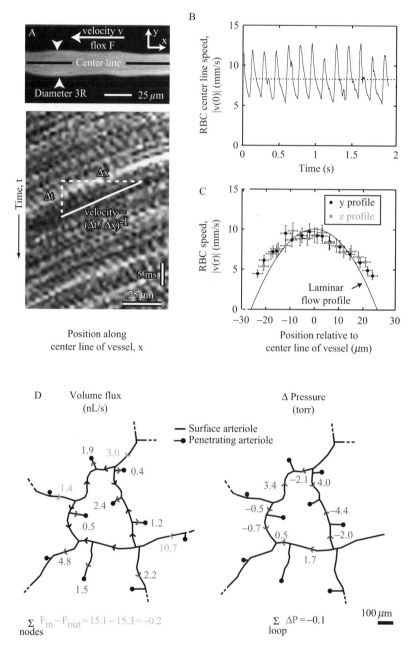

Figure 10.6 Imaging of blood flow in pial arterioles with fluorescently labeled plasma. (A, top) Maximum-intensity projection of a TPLSM image stack through a cortical arteriole. The dark line indicates the location where the line-scan data was taken, and the arrow represents the direction of flow obtained from these scans. (A, bottom) Line-scan data from the vessel in top panel forms a space-time image with time running down the image. The dark streaks running from upper right to lower left are formed by the motion of the

Lastly, the pressure versus velocity relation can be written in the form of Ohm's law, with ΔP playing the role of potential drop, $\langle v \rangle$ playing the role of current, and $8\Pi\eta$ playing the role of resistivity, that is,

$$\Delta P = F_{vol} \frac{8\eta L}{\pi R^4} = \left(8\pi\eta \frac{L}{\pi R^2} \right) \langle v \rangle \qquad (10.4)$$

Plots of the flux into and out of a region of the brain are useful as a means to study RBC utilization. They also act as a methodological control, as the flux into a loop must equal the flux out of that loop. This is illustrated in Fig. 10.6D, for which the total flux balances to within 2%. A similar zero-sum rule holds for the pressure drops along a vascular loop (Fig. 10.6D); here we see that the total pressure drop sums to within 5% of the average drop across a given segment.

2.3.2. Astrocyte dynamics

Beyond issues of determining network structure and cell morphology, fluorescence imaging is a powerful tool for assessing the functional state of cells. The most common functional markers are fluorescent indicators of intracellular calcium, which can be loaded into neuronal and glial cell populations in the neocortex *in vivo*. The simplest method for loading is to apply the membrane-permeable acetoxymethyl(AM)-ester form of indi-cator, which is nonspecifically taken up and trapped inside cells once the ester groups have been cleaved by endogenous esterases. For unknown reason some indicators, such as Fluo-4/AM and Rhod-2/AM, are prefer-entially taken up by astrocytes, at least when topically applied to the brain surface (Hirase *et al.*, 2004; Takano *et al.*, 2006). In contrast, Oregon-Green BAPTA-1/AM (OGB-1) labels both the neuronal and glial networks

---

nonfluorescent RBCs. The RBC speed is given by the inverse of the slope of these streaks, while the direction of flow is discerned from the sign of the slope. (B) RBC speed along the center of the arteriole shown in part (A) as a function of time. The periodic modulation of the RBC speed occurs at the ~6-Hz heart rate. The dotted line represents the temporal aver-age of the speed. (C) RBC speed in a different, larger arteriole, averaged over 40 s, as a func-tion of the transverse position in the vessel along horizontal (y) and vertical (z) directions. The parabolic curve (Eq. 10.2) represents the laminar flow profile that most closely matches the data, that is, $v(r) = (A/R^2) - (R^2 - r^2)$, where v is the velocity of the RBCs, R is the measured vessel radius of 26 μm, and A is a free parameter (A = 10 mm/s). (D) Example data from measurements around a closed loop of pial vessels and the vessels that flow into and out of the loop. We measured the radius and length, L, of each segment, along with the center-line velocity, v(0). The volume flux was calculated according to Eq. 10.3, and the pressure drops were calculated according to Eq. 10.4 with $\eta = 5 \times 10^{-3}$ Pa·s (1 Torr ~ 133 Pa). Ideally, the total flux should be zero and the total pressure drop should be zero; the actual values differ slightly. (Panels (A) to (C) are adapted from Schaffer, C. B., Friedman, B., Nishimura, N., Schroeder, L. F., Tsai, P. S., Ebner, F. F., Lyden, P. D., and Kleinfeld, D. (2006). Two-photon imaging of cortical surface microvessels reveals a robust redistribution in blood flow after vascular occlusion. *Public Library Sci. Biol.* 4, 258–270.)

(Garaschuk *et al.*, 2006a; Stosiek *et al.*, 2003). The green fluorescence of OGB-1 permits counterstaining with red fluorescent SR101 (Nimmerjahn *et al.*, 2004) to yield a crisp discrimination of astrocytic versus neuronal networks (Fig. 10.7A). It is unclear whether microglial cells load calcium indicators because they remain invisible within the diffusely stained background.

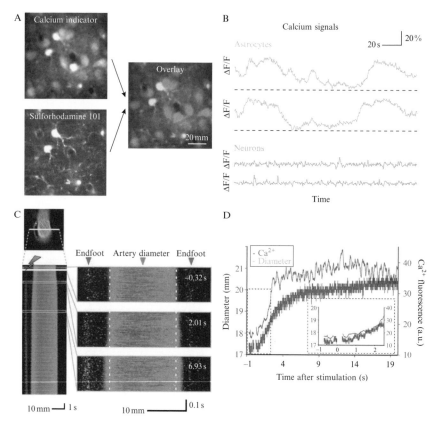

Figure 10.7 *In vivo* measurement of astrocyte function performed with TPLSM. (A) Multicell bolus loading of the calcium indicator OGB-1 into cortical layer 2 stains astrocytes as well as neurons. The astrocyte-specific SR101 stain unambiguously labels only astrocytes. (B) Spontaneous slow intracellular calcium oscillations occur in astrocyte cell bodies. Brief neuronal calcium transients, presumably coupled to action-potential firing, occur in neurons. (C) Images of a vascular astrocyte loaded with the Ca^{2+} indicator dye Rhod-2/AM (red) and DMNP-EDTA/AM, a caged Ca^{2+} compound. Light-induced Ca^{2+} uncaging in the astrocytic endfoot triggered vasodilation. Vasculature was stained with fluorescein/dextran (green). Purple arrow indicates the position of photostimulation. Scale bar, 10 μm. (D) Time course of astrocytic Ca^{2+} increase (red) and arterial vasodilation (green) following photostimulation. (Panel (B) is adapted from Takano, T., Tian, G. F., Peng, W., Lou, N., Libionka, W., Han, X., and Nedergaard, M. (2006). Astrocyte-mediated control of cerebral blood flow. *Nat. Neurosci.* **9**, 260–267.) (See Color Insert.)

Several laboratories have started to investigate astroglial calcium signaling *in vivo*. In anesthetized animals, neocortical astrocytes show spontaneous calcium oscillations and waves (Fig. 10.7B) (Hirase *et al.*, 2004; Nimmerjahn *et al.*, 2004). A caveat is that the rate of these spontaneous events increases with laser illumination power (Wang *et al.*, 2006) so that laser power should be kept as low as possible to avoid artifacts. Astrocytes also communicate with various other tissue components. For example, their endfeet are in tight contact with endothelial cells, enabling astrocytes to participate in regulation of local blood flow. Using a combination of *in vivo* TPLSM and uncaging of caged Ca^{2+}, Takano *et al.* (2006) demonstrated that intracellular calcium elevation in glial endfeet alone could cause a rapid and transient dilation of the associated vessel (Fig. 10.7C and D).

A close functional relationship of astroglia with the surrounding neuronal network is indicated by the recent finding of sensory-evoked calcium signals in astrocytes (Wang *et al.*, 2006). These signals were inhibited by antagonists of metabotropic, but not ionotropic, glutamate receptors. This suggests a direct action of synaptically released glutamate on astrocytes. The functional implications of this finding remain to be elucidated, in particular for cortical processing in awake animals. Finally, *in vivo* imaging of astroglia calcium signaling provides new promising means to investigate functional alterations under various conditions of brain pathology, such as epilepsy (Tian *et al.*, 2005) and neurodegenerative diseases (Eichhoff *et al.*, 2008).

2.3.3. Microglia dynamics

Microglial cells are the primary immune effector cells in the CNS and they are known to be activated in response to many kinds of brain damage and injury. Two recent studies employed *in vivo* TPLSM to directly visualize dynamic changes in microglia cell morphology in the neocortex of CX3CR1/EGFP mice (Davalos *et al.*, 2005; Nimmerjahn *et al.*, 2005). In both studies, time-lapse imaging was performed by repeatedly collecting small image stacks of EGFP-expressing microglial cells over the time course of several minutes to hours. Structural changes were analyzed from movies of the maximum-intensity projections. Surprisingly, microglial cells displayed an extraordinary level of restructuring even in their resting state. In contrast to neurons and astrocytes, which in the adult brain are mostly stable structures on the time scale of days and longer, microglial cells continually expand and retract their processes to interact with as well as screen the surrounding tissue components (Fig. 10.8A). The mean rate of length changes is about 1.5 μm/min (Fig. 10.8B). To get a sense of this rate, consider that the human cortex roughly contains 5×10^9 microglial cells, each possessing at least seven dynamic branches; thus the total restructuring of microglial cell processes in the human brain sums to 80,000 km/day.

Microglial cells are in a position to rapidly react to brain injury. For example, if focal lesions are produced by heating at a spot with the

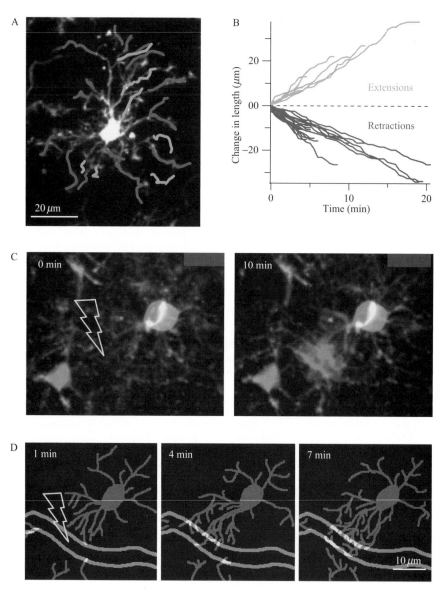

Figure 10.8 *In vivo* measurement of microglia function. (A) EGFP-expressing microglial cell in the neocortex of a CX3CR1/EGFP mouse. Extensions and retractions of cell processes over the time course of 20 min are indicated in green and red, respectively. (B) Length changes of the processes shown in (B) as a function of time. (C) Laser-induced damage to blood vessels causes a rapid focal response in neighboring microglial cells. Overlay of green microglia and SR101-stained astrocytes before and 10 min after a mild laser injury to a blood vessel. (D) Microglia morphology at intermediate time points during this event showing rapid, targeted movement of microglial processes toward the injured blood vessel (outlined in red); yellow flash indicates the site of injury. (Adapted from Nimmerjahn, A., Kirchhoff, F., and Helmchen, F. (2005). Resting microglial cells are highly dynamic surveillants of brain parenchyma *in vivo*. *Science* **308**, 1314–1318.) (See color insert.)

same pulsed laser used for TPLSM, microglial processes are attracted by the injury site, switching their seemingly random patrolling behavior to a targeted response (Fig. 10.8C and D). The molecular mechanism underlying this structural activity involves extracellular ATP signaling (Davalos *et al.*, 2005; Haynes *et al.*, 2006). In the case of severe laser-induced disruptions of blood vessels, microglial cells become progressively activated over hours, exhibiting clear signs of phagocytosis (Nimmerjahn *et al.*, 2005). Quantification of microglia dynamics in the intact brain thus opens new opportunities for revealing molecular mechanisms of microglia activation and for studying their role in a variety of brain pathologies, such as Alzheimer's disease, multiple sclerosis, and ischemia.

3. Photoprocesses for Targeted Disruption of Vascular Flow

The ability to form occlusions in targeted pial arterioles and subsurface microvessels is an essential element in the study of microstrokes. A central question is how blockages in single small blood vessels lead to potential degradation of neuronal viability and the onset of tissue inflammation and necrosis.

3.1. Photoinduced thrombosis

The methods for targeted clot formation fall into two classes. The first makes use of the introduction of a photosensitizer into the blood stream and the subsequent irradiation of the target vessel with actinic light. This will drive photothrombosis and, with judicious adjustment of the intensity, will lead to a localized, focal blockage in surface vessels. This method is not appropriate for subsurface vessels, as clots will form along the entire path of the incident light-cone. Nonetheless, application of this technique led to an understanding of the redundancy in the pial arteriole network (Schaffer *et al.*, 2006) as well as to the identification of penetrating arterioles as the weak-link in the supply of blood from the pial vessels to the subsurface microvasculature (Nishimura *et al.*, 2007).

3.2. Plasma-mediated ablation

The ability to create an occlusion below the pial surface may be achieved solely with light through the use of plasma-mediated photoprocesses. This technique makes use of the dissociation of matter by high-fluence, ultrafast pulses of near-infrared laser light (~100 fs pulse duration). The pulse initially leads to ionization of the material, such as blood plasma or vascular lumen, within a femtoliter-sized focal volume of the incident laser pulse. The interaction photogenerated free electrons and light toward the end of the laser pulse results in a spatially limited release of mechanical energy in the form of a shock wave (Joglekar et al., 2004; Schaffer et al., 2002). In practice, this method allows occlusions to be formed at least down to 500 μm below the pia without disruption of neighboring tissue.

The hardware for plasma-mediated occlusion is readily combined with TPLSM through a thinned skull window to enable the targeting and real-time monitoring of blood vessels. The typical set-up makes use of a standard two-photon microscope in which an additional amplified 100 fs light source is introduced to the beam path through a polarizing beam splitter (Fig. 10.9A). The imaging and photodisruption beams are brought to the same focus so that photodisruption occurs at the center of the imaged field. The energy per pulse of the amplified beam needs to be at least 0.03 μJ at the focus, which corresponds to a near-threshold fluence for damage of approximately 1 J/cm^2 with a 40× dipping objective. Calibration curves are established for each sample. The end result is the occlusion of flow in a targeted microvessel (Fig. 10.9B) (Nishimura et al., 2006).

As an application, we consider a microvessel that lay approximately 250 μm below the pia (Fig. 10.9C). Uniform flow is seen in the vessel prior to irradiation, as evidenced by streaks in the raster-scanned imaged (panel 1 in Fig. 10.9C). After the first burst of amplified laser pulses, there was temporary cessation of RBC motion and swelling of the target vessel (panels 2 and 3 in Fig. 10.9C). Labeling of the vessel wall with trapped fluorescein/dextran is also observed, and flow rapidly returns (panels 2 and 3 in Fig. 10.9C). A second burst of pulses led to limited extravasation and permanent occlusion of the vessel lumen (panel 4 in Fig. 10.9C). Quantitative measurements of the speed and direction of RBC flow before and after formation of the occlusion (Fig. 10.6) indicate that blockage of flow in the microvessel further leads to the cessation of flow in vessels that lie immediately downstream of the occlusion (Fig. 10.9C).

4. Outlook

The technology for measuring blood flow at the level of individual vessels throughout the upper layers of neocortex (Kleinfeld et al., 1998; Nishimura et al., 2007; Schaffer et al., 2006; Zhang and Murphy, 2007; Zhang et al., 2005) and the olfactory bulb (Chaigneau et al., 2003) is largely mature. Open issues on flow per se exist in three areas. The first concerns the absence of a map, or "plumbing diagram," of the connectivity of brain vasculature. Such a map is critical as a means to calculate potential modularity in the vascular system and correlations in flow, as well as the

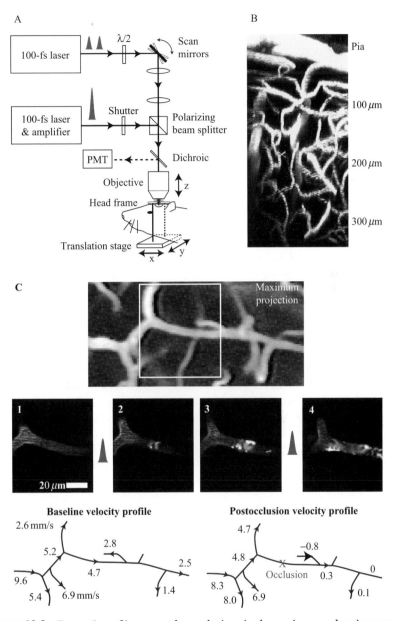

Figure 10.9 Formation of intravascular occlusions in deep microvessels using amplified 100-fs laser pulses. (A) Rough schematic of a TPLSM system modified for delivery of amplified 100-fs pulses. The polarizing beam splitter combined imaging and ablation beams. (B) An xz-projection of a TPLSM image stack that shows deep microvessels (yellow cross) that are routinely targeted for selective occlusion using this technique. (C) Maximum-intensity projection of a TPLSM stack showing a tortuous network of subsurface microvessels. The second row shows planar images taken from a region of

control of flow by neuronal and astrocytic activity. Preliminary anatomical work in this direction is promising (Tsai et al., 2003; Weber et al., 2008). The second issue is the need to measure both resilience and control of flow in subcortical areas, in particular the striatum, where vascularization is relatively sparse and damage from stroke relatively high compared with neocortex (Feekes and Cassell, 2006). Surgical resection and the use of small-diameter, gradient-index lenses for endoscopic imaging may enable such studies (Jung et al., 2004; Levene et al., 2004). The third issue in flow concerns the need to directly measure pressure, which at this time is only inferred and depends on estimates of "effective viscosity." A direct method would enable observation of homeostatic control at the level of microvessels. A viable approach for building an optical-based pressure sensor is unclear, but may involve reporters of the deformation of endothelial cells that comprise vessel walls.

The ability to observe calcium dynamics in astrocytes concomitant with changes in blood flow provides a means to study how astrocytes mediate the coupling of neuronal activity to changes in blood flow (Mulligan and MacVicar, 2004; Takano et al., 2006). A second means of neuronal control involves direct control of vascular dilation or contraction by inhibitory interneurons (Cauli et al., 2004). The advent of functional labeling of subpopulations of inhibitory cells will allow this critical but largely unexplored pathway to be studied in vivo. Of particular relevance for future work are the new Brainbow mice, in which glial cells as well as neuronal cell types are made to express FPs of different color in a combinatorial manner, resulting, for example, in a multicolor stain of the astrocytic network (Livet et al., 2007). Moreover, glia-specific expression of genetically encoded FP-based calcium indicators (Garaschuk et al., 2006b; Kotlikoff, 2007; Miyawaki, 2005) may foster new studies of glial function in the intact brain.

Experimental models of stroke are also poised to gain from measurements of functional changes in neurons and glial cells concomitant to those of changes of blood flow in response to targeted occlusion of vessels. The work of Murphy and colleagues explored the use of transgenic animals with GFP-labeled dendrites as a means to look at cellular changes that track

---

interest (white rectangle) depicting the time-course for intravascular clot formation (frames 1 to 4) in a specific, 10-μm-diameter vessel. The red pulses indicate irradiation with multiple trains of 0.03-μJ pulses delivered at 1 kHz. The third row shows vascular traces with baseline and postocclusion RBC velocity profiles, in millimeters per second, of the vascular network. Arrowheads denote the direction of RBC movement and the red cross marks the occluded microvessel. (Adapted from Nishimura, N., Schaffer, C. B., Friedman, B.,Tsai, P. S., Lyden, P. D., and Kleinfeld, D. (2006).Targeted insult to individual subsurface cortical blood vessels using ultrashort laser pulses: Three models of stroke. Nat. Methods 3, 99–108.)

ischemia (Zhang *et al.*, 2005). Future work will likely involve chronic studies of changes in flow concomitant with changes in cell function. The latter may be measured with endogenous indicators of function, such as FP-based calcium indicators (Garaschuk *et al.*, 2007). Further, stroke-based research will benefit from real-time monitoring of microglia dynamics and direct visualization of invading immune effector cells, along with molecular components of clot formation (Ogawa *et al.*, 1990), such as platelets, fibrin, and thrombin, during ischemia.

A final point is that all measurements to date on blood flow, their control by neuronal and glial activity, and changes in molecular markers of tissue viability, have involved anesthetized animals. While anesthesia does not necessarily block homeostasis, it does affect the extent of modulation of the pial and deep brain vasculature by small neuroactive molecules such as acetylcholine. It is thus important to move towards recording in awake animals (Dombeck *et al.*, 2007). This is particularly important for the case of experimental stroke, where homeostasis may be compromised by anesthesia.

ACKNOWLEDGMENTS

Recent work in the Helmchen laboratory was funded by the Max Planck Society, the University of Zurich, and grants from the Human Frontier Science Program and the Swiss National Science Foundation. Recent work in the Kleinfeld laboratory was funded by grants from the National Science Foundation (Division of Biological Infrastructure) and the National Institutes of Health, National Center for Research Resources (NCRR), National Institute of Neurological Disorders and Stroke (NINDS), National Institute of Biomedical Imaging and Bioengineering (NIBIB). We thank Philbert Tsai and Pablo Blinder for supplying the unpublished data for Fig. 10.2, and Andy Shih for performing the analysis shown in Fig. 10.6D.

REFERENCES

Abbott, N. J., Rönnbäck, L., and Hansson, E. (2006). Astrocyte–endothelial interactions at the blood–brain barrier. *Nat. Rev. Neurosci.* **7,** 41–53.

Bacskai, B. J., Klunk, W. E., Mathis, C. A., and Hyman, B. T. (2002). Imaging amyloid-beta deposits *in vivo. J. Cerebral Blood Flow Metab.* **22,** 1035–1041.

Bacskai, B. J., Kajdasz, S. T., Christie, R. H., Carter, C., Games, D., Seubert, P., Schenk, D., and Hyman, B. T. (2001). Imaging of amyloid-beta deposits in brains of living mice permits direct observation of clearance of plaques with immunotherapy. *Nat. Med.* **7,** 369–372.

Brozici, M., van der Zwain, A., and Hillen, B. (2003). Anatomy and functionality of leptomeningeal anastomoses: A review. *Stroke* **34,** 2750–2762.

Cauli, B., Tong, X. K., Rancillac, A., Serluca, N., Lambolez, B., Rossier, J., and Hamel, E. (2004). Cortical GABA interneurons in neurovascular coupling: Relays for subcortical vasoactive pathways. *J. Neurosci.* **24,** 8940–8949.

Chaigneau, E., Oheim, M., Audinat, E., and Charpak, S. (2003). Two-photon imaging of capillary blood flow in olfactory bulb glomeruli. *Proc. Natl. Acad. Sci. U.S.A.* **100,** 13081–13086.

Cook, D. N., Chen, S. C., Sullivan, L. M., Manfra, D. J., Wiekowski, M. T., Prosser, D. M., Vassileva, G., and Lira, S. A. (2001). Generation and analysis of mice lacking the chemokine fractalkine. *Mol. Cell Biol.* **21,** 3159–3165.

Davalos, D., Grutzendler, J., Yang, G., Kim, J. V., Zuo, Y., Jung, S., Littman, D. R., Dustin, M. L., and Gan, W. B. (2005). ATP mediates rapid microglial response to local brain injury *in vivo*. *Nat. Neurosci.* **8,** 752–758.

Denk, W., Strickler, J. H., and Webb, W. W. (1990). Two-photon laser scanning fluorescence microscopy. *Science* **248,** 73–76.

Denk, W., Delaney, K. R., Kleinfeld, D., Strowbridge, B., Tank, D. W., and Yuste, R. (1994). Anatomical and functional imaging of neurons and circuits using two photon laser scanning microscopy. *J. Neurosci. Methods* **54,** 151–162.

Dombeck, D. A., Khabbaz, A. N., Collman, F., Adelman, T. L., and Tank, D. W. (2007). Imaging large-scale neural activity with cellular resolution in awake, mobile mice. *Neuron* **56,** 43–57.

Ehinger, B., Zucker, C. L., Bruun, A., and Adolph, A. (1994). *In vivo* staining of oligodendroglia in the rabbit retina. *Glia* **10,** 40–48.

Eichhoff, G., Busche, M. A., and Garaschuk, O. (2008). *In vivo* calcium imaging of the aging and diseased brain. *Eur. J. Nucl. Res. Mol. Med.* **35** (Suppl. 1), S99–S106.

Feekes, J. A., and Cassell, M. D. (2006). The vascular supply of the functional compartments of the human striatum. *Brain* **129,** 2189–2201.

Feng, G., Mellor, R. H., Bernstein, M., Keller-Peck, C., Nguyen, Q. T., Wallace, M., Nerbonne, J. M., Lichtman, J. W., and Sanes, J. R. (2000). Imaging neuronal subsets in transgenic mice expressing multiple spectral variants of GFP. *Neuron* **28,** 41–51.

Frostig, R. D., Dory, Y., Kwon, M. C., and Masino, S. A. (1993). Characterization of functional organization within rat barrel cortex using intrinsic signal optical imaging through a thinned skull. *Proc. Natl. Acad. Sci. U.S.A.* **90,** 9998–10002.

Fuss, B., Mallon, B., Phan, T., Ohlemeyer, C., Kirchhoff, F., Nishiyama, A., and Macklin, W. B. (2000). Purification and analysis of *in vivo*-differentiated oligodendrocytes expressing the green fluorescent protein. *Dev. Biol.* **218,** 259–274.

Garaschuk, O., Milos, R. I., and Konnerth, A. (2006a). Targeted bulk-loading of fluorescent indicators for two-photon brain imaging *in vivo*. *Nat. Protoc.* **1,** 380–386.

Garaschuk, O., Milos, R. I., Grienberger, C., Marandi, N., Adelsberger, H., and Konnerth, A. (2006b). Optical monitoring of brain function *in vivo*: From neurons to networks. *Pflugers Arch.* **453,** 385–396.

Garaschuk, O., Griesbeck, O., and Konnerth, A. (2007). Troponin C-based biosensors: A new family of genetically encoded indicators for *in vivo* calcium imaging in the nervous system. *Cell Calcium* **42,** 351–361.

Göbel, W., and Helmchen, F. (2007). *In vivo* calcium imaging of neural network function. *Physiology (Bethesda)* **22,** 358–365.

Grutzendler, J., Kasthuri, N., and Gan, W. B. (2002). Long-term dendritic spine stability in the adult cortex. *Nature* **420,** 812–816.

Haynes, S. E., Hollopeter, G., Yang, G., Kurpius, D., Dailey, M. E., Gan, W. B., and Julius, D. (2006). The P2Y(12) receptor regulates microglial activation by extracellular nucleotides. *Nat. Neurosci.* **9,** 1512–1519.

Helmchen, F., and Denk, W. (2002). New developments in multiphoton microscopy. *Curr. Opin. Neurobiol.* **12,** 593–601.

Helmchen, F., and Denk, W. (2005). Deep tissue two-photon microscopy. *Nat. Methods* **2,** 932–940.

Hirase, H., Qian, L., Bartho, P., and Buzsaki, G. (2004). Calcium dynamics of cortical astrocytic networks *in vivo*. *PLoS Biol*. **2**, e96.

Joglekar, A. P., Liu, H. H., Meyhofer, E., Mourou, G., and Hunt, A. J. (2004). Optics at critical intensity: Applications to nanomorphing. *Proc. Natl. Acad. Sci. U.S.A.* **101**, 5856–5861.

Jung, J. C., Mehta, A. D., Aksay, E., Stepnoski, R., and Schnitzer, M. J. (2004). *In vivo* mammalian brain imaging using one- and two-photon fluorescence microendoscopy. *J. Neurophysiol.* **92**, 3121–3133.

Jung, S., Aliberti, J., Graemmel, P., Sunshine, M. J., Kreutzberg, G. W., Sher, A., and Littman, D. R. (2000). Analysis of fractalkine receptor CX(3)CR1 function by targeted deletion and green fluorescent protein reporter gene insertion. *Mol. Cell Biol.* **20**, 4106–4114.

Kafitz, K. W., Meier, S. D., Stephan, J., and Rose, C. R. (2008). Developmental profile and properties of sulforhodamine 101-labeled glial cells in acute brain slices of rat hippocampus. *J. Neurosci. Methods.* **169**, 84–92.

Keifer, J., Vyas, D., and Houk, J. C. (1992). Sulforhodamine labeling of neural circuits engaged in motor pattern generation in the *in vitro* turtle brainstem-cerebellum. *J. Neurosci.* **12**, 3187–3199.

Kleinfeld, D., and Delaney, K. R. (1996). Distributed representation of vibrissa movement in the upper layers of somatosensory cortex revealed with voltage sensitive dyes. *J. Comp. Neurol.* **375**, 89–108.

Kleinfeld, D., and Griesbeck, O. (2005). From art to engineering? The rise of *in vivo* mammalian electrophysiology via genetically targeted labeling and nonlinear imaging. *Public Library Sci. Biol.* **3**, 1685–1689.

Kleinfeld, D., Mitra, P. P., Helmchen, F., and Denk, W. (1998). Fluctuations and stimulus-induced changes in blood flow observed in individual capillaries in layers 2 through 4 of rat neocortex. *Proc. Natl. Acad. Sci. U.S.A.* **95**, 15741–15746.

Kleinfeld, D., Friedman, B., Lyden, P. D., and Shih, A. Y. (2008). Targeted occlusion to surface and deep vessels in neocortex via linear and nonlinear optical absorption. *In* "Animal Models of Acute Neurological Injuries." (J. Chen, X.-M. Xu, and J. Zhang, eds.). Totowa: The Humana Press Inc.

Kotlikoff, M. I. (2007). Genetically encoded Ca^{2+} indicators: Using genetics and molecular design to understand complex physiology. *J. Physiol.* **578**, 55–67.

Levene, M. J., Dombeck, D. A., Kasischke, K. A., Molloy, R. P., and Webb, W. W. (2004). *In vivo* multiphoton microscopy of deep brain tissue. *J. Neurophysiol.* **91**, 1908–1912.

Livet, J., Weissman, T. A., Kang, H., Draft, R. W., Lu, J., Bennis, R. A., Sanes, J. R., and Lichtman, J. W. (2007). Transgenic strategies for combinatorial expression of fluorescent proteins in the nervous system. *Nat. Neurosci.* **450**, 56–62.

Miyawaki, A. (2005). Innovations in the imaging of brain functions using fluorescent proteins. *Neuron* **48**, 189–199.

Mulligan, S. J., and MacVicar, B. A. (2004). Calcium transients in astrocyte endfeet cause cerebrovascular constrictions. *Nature* **431**, 195–199.

Nimmerjahn, A., Kirchhoff, F., Kerr, J. N., and Helmchen, F. (2004). Sulforhodamine 101 as a specific marker of astroglia in the neocortex *in vivo*. *Nat. Methods* **29**, 31–37.

Nimmerjahn, A., Kirchhoff, F., and Helmchen, F. (2005). Resting microglial cells are highly dynamic surveillants of brain parenchyma *in vivo*. *Science* **308**, 1314–1318.

Nishimura, B., Schaffer, C. B., Friedman, B., Lyden, P. D., and Kleinfeld, D. (2007). Penetrating arterioles are a bottleneck in the perfusion of neocortex. *Proc. Natl. Acad. Sci. U.S.A.* **104**, 365–370.

Nishimura, N., Schaffer, C. B., Friedman, B., Tsai, P. S., Lyden, P. D., and Kleinfeld, D. (2006). Targeted insult to individual subsurface cortical blood vessels using ultrashort laser pulses: Three models of stroke. *Nat. Methods* **3**, 99–108.

Nolte, C., Matyash, M., Pivneva, T., Schipke, C. G., Ohlemeyer, C., Hanisch, U. K., Kirchhoff, F., and Kettenmann, H. (2001). GFAP promoter-controlled EGFP-expressing transgenic mice: A tool to visualize astrocytes and astrogliosis in living brain tissue. *Glia* **33,** 72–86.

Ogawa, S., Lee, T. M., Kay, A. R., and Tank, D. W. (1990). Brain magnetic resonance imaging with contrast dependent on blood oxygenation. *Proc. Natl. Acad. Sci. U.S.A.* **87,** 9868–9872.

Sarelius, I. H., and Duling, B. R. (1982). Direct measurement of microvessel hematocrit, red cell flux, velocity, and transit time. *Am. J. Physiol.* **243,** H1018–H1022.

Schaffer, C. B., Nishimura, N., Glezer, E. N., Kim, A. M. T., and Mazur, E. (2002). Dynamics of femtosecond laser-induced breakdown in water from femtoseconds to microseconds. *Opt. Express* **10,** 196–203.

Schaffer, C. B., Friedman, B., Nishimura, N., Schroeder, L. F., Tsai, P. S., Ebner, F. F., Lyden, P. D., and Kleinfeld, D. (2006). Two-photon imaging of cortical surface microvessels reveals a robust redistribution in blood flow after vascular occlusion. *Public Library Sci. Biol.* **4,** 258–270.

Stosiek, C., Garaschuk, O., Holthoff, K., and Konnerth, A. (2003). In vivo two-photon calcium imaging of neuronal networks. *Proc. Natl. Acad. Sci. U.S.A.* **100,** 7319–7324.

Takano, T., Tian, G. F., Peng, W., Lou, N., Libionka, W., Han, X., and Nedergaard, M. (2006). Astrocyte-mediated control of cerebral blood flow. *Nat. Neurosci.* **9,** 260–267.

Tian, G. F., Azmi, H. T. T., Xu, Q., Peng, W., Lin, J., Oberheim, N., Lou, N., Wang, X., Zielke, H. R., Kang, J., and Nedergaard, M. (2005). An astrocytic basis of epilepsy. *Nat. Med.* **11,** 973–981.

Tsai, P. S., Friedman, B., Schaffer, C. B., Squier, J. A., and Kleinfeld, D. (2005). All-optical, *in situ* histology of neuronal tissue with femtosecond laser pulses. *In* "Imaging in Neuroscience and Development: A Laboratory Manual." (R. Yuste and A. Konnerth, eds.), pp. 815–826. Cold Spring Harbor Laboratory Press, New York.

Tsai, P. S., Friedman, B., Ifarraguerri, A. I., Thompson, B. D., Lev-Ram, V., Schaffer, C. B., Xiong, Q., Tsien, R. Y., Squier, J. A., and Kleinfeld, D. (2003). All-optical histology using ultrashort laser pulses. *Neuron* **39,** 27–41.

Vermeer, S. E., Prins, N. D., den Heijer, T., Hofman, A., Koudstaal, P. J., and Breteler, M. M. (2003). Silent brain infarcts and the risk of dementia and cognitive decline. *N. Engl. J. Med.* **348,** 1215–1222.

Volterra, A., and Meldolesi, J. (2005). Astrocytes, from brain glue to communication elements: The revolution continues. *Nat. Rev. Neurosci.* **6,** 626–640.

Wang, X., Lou, N., Xu, Q., Tian, G. F., Peng, W. G., Han, X., Kang, J., Takano, T., and Nedergaard, M. (2006). Astrocytic Ca^{2+} signaling evoked by sensory stimulation *in vivo*. *Nat. Neurosci.* **9,** 816–823.

Weber, B., Keller, A. L., Reichold, J., and Logothetis, N. K. (2008). The microvascular system of the striate and extrastriate visual cortex of the macaque. *Cerebral Cortex* **18,** 2318–2330.

Zhang, S., and Murphy, T. H. (2007). Imaging the impact of cortical microcirculation on synaptic structure and sensory-evoked hemodynamic responses *in vivo*. *Public Library Sci. Biol.* **5,** e119.

Zhang, S., Boyd, J., Delaney, K. R., and Murphy, T. H. (2005). Rapid reversible changes in dendritic spine structure *in vivo* gated by the degree of ischemia. *J. Neurosci.* **25,** 5333–5228.

Zhuo, L., Sun, B., Zhang, C. L., Fine, A., Chiu, S. Y., and Messing, A. (1997). Live astrocytes visualized by green fluorescent protein in transgenic mice. *Dev. Biol.* **187,** 36–42.

CHAPTER ELEVEN

THE ROLE OF BONE MARROW–DERIVED CELLS IN TUMOR ANGIOGENESIS AND METASTATIC PROGRESSION

Marianna Papaspyridonos* *and* David Lyden*,†

Contents

Abstract

Tumor angiogenesis is orchestrated by a complex set of secreted factors and collaboration between many different cell types. This chapter discusses the role of tumor-secreted angiogenic factors that are responsible for the recruitment of bone marrow–derived cells (BMDCs) at tumor sites and their role in tumor angiogenesis and progression.

* Department of Pediatrics and Department of Cell and Developmental Biology, Weill Cornell Medical College, New York
† Memorial Sloan-Kettering Cancer Center, New York

Methods in Enzymology, Volume 444
ISSN 0076-6879, DOI: 10.1016/S0076-6879(08)02811-5

Furthermore, recent findings indicating a role for BMDCs in initiating metastasis and their interactions with tumor and stromal cells are also discussed with an emphasis on the SDF-1–CXCR4 axis used for the migration of most BMDCs. Finally, therapeutic strategies that manipulate the migrating BMDCs themselves as well as their signaling pathways are discussed in the context of controlling and preventing tumor progression.

1. THE ANGIOGENIC SWITCH

During the premalignant stages of carcinogenesis epithelial cells manifest altered responses to factors present in the extracellular microenvironment, resulting in genetic and epigenetic changes and uncontrolled cell proliferation. However, continued growth of transformed cells requires induction of new vessel formation, a process referred to as tumor angiogenesis, which is dependent on proliferation and migration of endothelial cells (Hanahan and Folkman, 1996). The "angiogenic switch" is the rapid formation of blood vessels thought to occur as a discrete step required for tumor progression. It is thought that the angiogenic switch occurs when the balance between the levels of activators and inhibitors of angiogenesis in the premalignant lesion tips in favor of angiogenesis (Bergers and Benjamin, 2003). Oncogene activation and/or loss of tumor suppressor(s) can alter the expression of both angiogenic activators (including vascular endothelial growth factors [VEGFs], fibroblast growth factors [FGFs], matrix metalloproteinases [MMPs], and hepatocyte growth factor) and inhibitors (including thrombospondin-1 and the statins) in the transformed cells, leading to rapid tumor growth (De Palma and Naldini, 2006).

Although studies of mouse tumor models have suggested that therapeutic intervention by inhibition of factors that control various aspects of angiogenesis would effectively inhibit or delay tumor growth, clinical trials with anti-angiogenic factors for the treatment of human malignancies have been less effective than anticipated (Rafii et al., 2002). These results indicate that tumor angiogenesis might be orchestrated by a complex set of growth factors and collaboration among many different cell types.

2. ORIGIN OF ANGIOGENIC ENDOTHELIAL CELLS

Numerous reports have shown that tumor growth depends on the rapid recruitment of endothelial cells, which then contribute to functional neovasculature. Most tumor vessels differ from normal vasculature in that they are dilated (Carmeliet and Jain, 2000; Folkman, 1995; Folkman et al., 1989), leaky (Dvorak et al., 1988), and are made up of a disorganized array of pericytes (the periendothelial cell layer) and smooth muscle cells (Morikawa *et al.*, 2002). The lack of architectural stability in tumor vasculature occasionally leads to microhemorrhages or vessel collapse (Bergers *et al.*, 1999).

The origin of nascent blood vessels within the expanding tumor tissue has been the subject of intense investigation (Carmeliet and Jain, 2000; Gao *et al.*, 2008; Yancopoulos *et al.*, 2000). New blood vessels may (1) sprout from preexisting mature ones by cooption (Holash *et al.*, 1999) and proliferation or migration of neighboring vessels (Folkman, 1995; Hanahan and Folkman, 1996), or (2) form de novo by recruiting circulating endothelial progenitor cells (CEPs), a process known as adult vasculogenesis (Asahara *et al.*, 1997; De Palma and Naldini, 2006; Gao *et al.*, 2008). Circulating CEPs can originate from the bone marrow (BM) (Lyden *et al.*, 2001), or from other organs, such as the liver (Aicher and Heeschen, 2007). In addition, some tumor cells may sufficiently resemble functional endothelial cells, giving rise to mosaic vessels (Maniotis *et al.*, 1999). The relative contribution of each of these processes to tumor angiogenesis is likely dictated by the cytokine repertoire and matrix components of each tumor cell type (Rafii *et al.*, 2002).

3. BONE-MARROW–DERIVED CELLS: ENDOTHELIAL AND HEMATOPOIETIC PROGENITOR CELLS

3.1. Bone marrow–derived endothelial progenitor cells

Bone marrow–derived EPCs comprise a unique endothelial cell type that has the capacity to repair injured denuded vasculature (Asahara *et al.*, 1997; Lin *et al.*, 2000; Shi *et al.*, 1998). Initial studies showed that EPCs could be identified by their expression pattern of specific cell-surface markers, including VEGFR2 (Gill *et al.*, 2001), AC133 (Peichev *et al.*, 2000), CXCR4 (Peichev *et al.*, 2000), and CD146 (Solovey *et al.*, 1997). The first human EPCs were successfully isolated using monoclonal antibodies against those markers. $CD133^{+}VEGFR2^{+}$ EPCs were found to be present at low frequencies in human umbilical cord blood (Gill *et al.*, 2001), adult bone marrow (Gehling *et al.*, 2000; Peichev *et al.*, 2000; Quirici *et al.*, 2001), cytokine-mobilized peripheral blood (Gehling *et al.*, 2000), and human fetal liver cells (Peichev *et al.*, 2000).

3.2. Bone-marrow–derived hematopoietic stem and progenitor cells

Hematopoietic stem cells (HSCs), which express VEGFR1, Sca1, and c-Kit, are pluripotent cells that have the capacity to undergo self-renewal and differentiate into specific lineages, including erythroid, megakaryocytic, lymphoid, and myeloid progenitors (Rafii *et al.*, 2002). The lineage-specific differentiation of HSCs is determined by the local availability of certain cytokines (Eriksson and Alitalo, 2002; Hattori *et al.*, 2002). For example, granulocyte–monocyte colony-stimulating factor (GM-CSF) and granulocyte colony–stimulating factor (G-CSF) promote the generation of myelomonocytic precursor cells, including CD11b$^+$ monocytes and neutrophils.

Hematopoietic stem cells are stored in a bone marrow microenvironment known as the stem cell niche, where they are maintained in an undifferentiated and quiescent state (Reya *et al.*, 2001). These niches are crucial for regulating the self-renewal and cell fate decisions. Under steady-state conditions, most stem cells are maintained in the G phase of the cell cycle (Cheng *et al.*, 2000). Bone marrow ablation by cytotoxic agents, such as the chemotherapeutic drug 5-fluorouracil (5-FU), activates recruitment of stem cells from their niches, proliferation, differentiation, and mobilization to the circulation, leading to reconstitution of hematopoiesis.

4. BONE MARROW–CELL MOBILIZATION IN RESPONSE TO TUMOR FACTORS

Tumors produce angiogenic factors, such as vascular endothelial growth factor (VEGF) (Bergers *et al.*, 2000), platelet-derived growth factor (PDGF), and angiopoietins (Holash *et al.*, 1999). These factors activate matrix metalloproteinase-9 (MMP-9) within the bone marrow microenvironment, which is thought to result in cleavage of membrane-bound KIT ligand (mKitL) from the surface of the bone marrow stromal cells. Increased levels of soluble KIT ligand (sKitL) within the bone marrow promote proliferation and motility of stem and progenitor cells, including circulating VEGFR2$^+$ endothelial progenitor cells (EPCs) and VEGFR1$^+$ hematopoietic progenitor cells (HPCs) (Hattori *et al.*, 2002; Heissig *et al.*, 2002). VEGFR2$^+$ EPCs and VEGFR1$^+$ HPCs translocate from an osteoblastic zone within the bone marrow to a vascular zone, where they proliferate, differentiate, and are launched into peripheral circulation (Kopp and Rafii, 2007). Additionally, production of MMPs by tumor cells promotes the release of ECM-bound or cell-surface–bound cytokines (Vu and Werb, 2000), which also mobilize EPCs. It remains to be determined whether alterations in the chemokine/cytokine repertoire of the tumor influence the differentiation of EPCs before their release into peripheral circulation.

5. CONTRIBUTION OF BMDCs IN TUMOR ANGIOGENESIS

One of the first studies to demonstrate the importance of BMDCs in the regulation of tumor angiogenesis showed that the angiogenic defect in placental growth factor was alleviated after transplantation of wildtype bone marrow (Carmeliet *et al.*, 2001). Pl-null mice demonstrate impaired tumor angiogenesis due to decreased vessel density, indicating that this growth factor might function to stimulate angiogenesis. When Pl-null mice received bone marrow grafts from wildtype mice, they were able to support tumor angiogenesis, indicating that BMDCs serve as a source of PlGF in tumor vasculature. Thus, BMDCs provide not only the building blocks that are required for tumor angiogenesis, but also the growth factors that modulate blood vessel maturation and stabilization.

Increased numbers of CEPs have been detected in the circulation of cancer patients and lymphoma-bearing mice, and tumor volume and tumor production of VEGF have been found to correlate with CEP mobilization (Mancuso *et al.*, 2001; Monestiroli *et al.*, 2001). To distinguish between BM-derived EPCs and cells recruited from neighboring vessels, many studies have used transplantation techniques in which sex-mismatched or genetically marked donor bone marrow is infused into irradiated recipients. For example, multipotent human adult progenitor cells (MAPCs), a group of highly primitive cells that has the capacity to differentiate into many different cell types, were induced to differentiate into endothelial cells *in vitro* and injected into immunocompromised mice that carried mouse Lewis lung carcinomas (Reyes *et al.*, 2002). After 5 days, the contribution from the human MAPC–derived endothelial cells was detected in 30% of the newly formed tumor vasculature. Human MAPC–derived endothelial cells also contributed to the vasculature of spontaneous lymphomas that commonly develop in aging immunocompromised mice (Reyes *et al.*, 2002).

Finally, Asahara and colleagues engrafted tumor-bearing mice with transgenic bone marrow cells that were designed to express a cell marker under the control of the VEGFR2 and TIE2 endothelial-specific promoters (Asahara *et al.*, 1997). Cells that expressed this marker were detected throughout the implanted colon tumors, indicating that BM-derived cells such as EPCs can home to the tumor vasculature and differentiate into endothelial cells.

6. EPCs AND HPCs ARE REQUIRED FOR TUMOR ANGIOGENESIS

Studying mice deficient in the transcription factors called inhibitors of differentiation (Id)1 and 3 provided further evidence that the recruitment of BMDCs was required for tumor growth. The Id proteins interact with other helix-loop-helix transcription factors to modulate cellular differentiation during early development (Lyden *et al.*, 1999). Id1/3 double knock-out mouse embryos have vascular malformations in the forebrain, leading to fatal hemorrhage. Adult mice with reduced Id gene copies are viable but cannot support tumor-induced angiogenesis (Lyden *et al.*, 1999). The mechanism for the underlying angiogenic defect is believed to be mainly due to impaired recruitment of EPCs from the bone marrow to the tumor.

Both B6RV2 lymphoma and Lewis lung carcinoma cells, which generate slow poorly vascularized tumors in Id-deficient mice, are able to form fully vascularized tumors in Id-deficient mice that have received wildtype bone marrow transplants (Lyden *et al.*, 2001). This result indicates that BM-derived EPCs and HPCs might be sufficient to support tumor angiogenesis. Analysis of the tumors that developed in these mice revealed that BM-derived VEGFR2$^+$ EPCs were incorporated into vessels in close proximity with VEGFR1$^+$ HPCs. In the early phases of tumor growth, most vessels within both types of tumors were derived from the transplanted bone marrow cells. However, by day 14, only 50% of the vessels in Lewis lung carcinoma contained the donor cells, whereas most of the vessels in B6RV2 lymphomas contained the donor cells. These findings demonstrate that the long-term dependence of a particular tumor on BM-derived precursors might be dictated by additional factors, such as the constitution of the ECM, or the chemokine/cytokine repertoire of a particular tumor, as well as the tumor type itself.

The role of Id1 in the mobilization of EPCs and their contribution in the tumor vasculature and progression were further explored in a recent study by Gao *et al.* (2008), in which EPCs were labeled and tracked with green fluorescent protein. Using the metastatic mouse Lewis lung carcinoma and spontaneous MMTVPyMT breast cancer model, the authors demonstrated that about 12% of the endothelial cells within macrometastases were derived from the bone marrow. When the expression levels of Id1 in this small progenitor cell population were reduced, their mobilization from bone marrow decreased by 96%, angiogenesis was blocked, tumor formation decreased, and the animal's survival improved. Interestingly, the inhibition of Id1 did not affect tumor cell dissemination or the initial colonization of organs by malignant cells; instead, Id1 loss shut off the mobilization and recruitment of particular endothelial progenitor cells (cells expressing Id1, VE-cadherin, and low amounts of CD31) to the tumor site. These specific progenitor cells infiltrated micrometastatic lesions and produced proangiogenic growth factors before the initiation of macrometastases. These data suggest that BM-derived endothelial progenitors are unique in providing both instructive (paracrine) and structural (vessel incorporation) roles to promote tumor macrometastasis.

7. COOPERATION BETWEEN BM-DERIVED HEMATOPOIETIC AND ENDOTHELIAL PRECURSORS

Recruitment of EPCs and HPCs from the bone marrow is a dynamic process that requires sequential activation of a number of molecular switches, including activation of metalloproteinases and release of cytokines from the bone marrow stromal cells. Angiogenic factors that are released by tumor cells induce co-mobilization of VEGFR1[+] HPCs, terminally differentiated monocytic precursor cells, and VEGFR2[+] EPCs from the bone marrow (Lyden *et al.*, 2001).

It appears that recruitment of both EPCs and HPCs is required for the functional incorporation of differentiated EPCs into the tumor vasculature. The co-localization of VEGFR1[+] HPCs and VEGFR2[+] EPCs in the tumor vasculature provides evidence that they cooperate in forming functional tumor vessels. Both HPCs and tumor cells release proangiogenic factors including VEGF and MMPs, whereas co-recruited EPCs can release growth factors that induce the maturation of subsets of hematopoietic progenitor cells into proangiogenic monocytes and macrophages. Of note, functional studies involving antibody inhibition of either VEGFR1 or VEGFR2 showed that inhibition of either receptor alone was not sufficient to induce tumor regression (Luttun *et al.*, 2002; Lyden *et al.*, 2001). However, inhibition of both VEGFR1 and VEGFR2 was sufficient to block tumor growth and induce necrosis, indicating that signaling through both receptors is required for tumor angiogenesis. More specifically, inhibition of VEGFR2 signaling alone resulted in decreased vessel density and diffuse hemorrhage formation, whereas inhibition of VEGFR1 alone diminished the number of perivascular hematopoietic cells typically found in mouse tumors. These data indicate that corecruitment of VEGFR1[+] hematopoietic cells facilitate the incorporation of EPCs into tumor vasculature.

More recently, using lentiviral transduction of a suicide gene (De Palma *et al.*, 2003a), BMDCs were eliminated during the early phases of tumor growth. The resultant impairment in the recruitment of these cells substantially inhibited tumor angiogenesis and growth (De Palma *et al.*, 2003b). This study is intriguing as it suggests possible new avenues for gene therapy in tumors, involving transplantation of genetically modified bone marrow–derived progenitor cells as a vehicle for selective cell elimination.

8. VEGFR1[+] HPCs DEFINE THE PREMETASTATIC NICHE

Exploring the earliest steps in metastasis before the arrival of metastatic tumor cells at distant organ sites has recently revealed a key role for BMDCs in initiating a microenvironment that fosters the recruitment of disseminating tumor cells (Kaplan *et al.*, 2005). At these sites, termed the premetastatic niche, clusters of BM-derived HPCs prime distant tissues for the influx of tumor cells and the establishment of metastatic lesions. When implanted intradermal tumors and spontaneous transgenic c-Myc lymphoma murine models were examined, minimal numbers of BMDCs were observed at future metastatic sites before the onset of carcinogenesis. However, VEGFR1$^+$ HPC clusters in organs of future metastasis were apparent by 14 days after tumor implantation. Furthermore, the appearance of VEGFR1$^+$ HPCs clusters predated the arrival of VEGFR2$^+$ EPCs, which were found to migrate to HPC clusters in parallel with tumor cells.

The premetastatic sites described above appear to possess niche characteristics. Within the local tissue parenchyma, the VEGFR1$^+$ cells maintained their progenitor status, expressing markers of immature hematopoietic lineage including CD34, CD11b, c-Kit, and stem cell antigen-1 (Sca-1). As a result of multiple ongoing events in response to primary tumor chemokines, VEGFR1$^+$ HPCs proliferate and circulate in the bloodstream but also preferentially adhere to areas of increased fibronectin, newly synthesized by resident fibroblasts and fibroblast-like cells that often reside at the subcapsular region of an organ (Kaplan *et al.*, 2006). VEGFR1$^+$ HPCs express the integrin VLA-4 ($_4\beta_1$), allowing them to adhere specifically to fibronectin and initiate cellular cluster formation. MMP-9 expression by HPCs allows vascular remodeling, accelerating the extravasation of more VEGFR1$^+$ cells into the niche. In addition, Id3 is expressed within the premetastatic niche, and may be involved in the proliferation and mobilization of HPCs from the bone marrow to distinct sites, as well as maintaining an activated progenitor state in the cells within these cellular clusters. Together with fibronectin and associated stromal cells, VEGFR1$^+$ HPCs alter the local microenvironment, leading to the activation of other integrins and chemokines, such as SDF-1, that promote attachment, survival, and growth of tumor cells. Soon after the implantation of tumor cells, VEGFR2$^+$ progenitors are recruited to promote vasculogenesis, enabling maturation to a fully developed metastatic lesion at the site of the premetastatic niche. The dependence of the evolving metastatic process on these changes is further illustrated when monoclonal antibodies are used. Using neutralizing antibodies to inhibit VEGFR1$^+$ HPCs eliminates the premetastatic niche, while targeting VEGFR2$^+$ EPCs results in the formation of small micrometastases without vascularization, preventing formation of full metastatic lesions. Similar to carcinoma-associated fibroblasts and tumor-associated macrophages, which promote tumor progression via the creation of a supportive microenvironment, the VEGFR1$^+$ BMDCs could also foster inflammation and sustain tumor cell growth at distant organ sites (Kaplan *et al.*, 2005; Lin *et al.*, 2001; Orimo *et al.*, 2005).

These findings provide evidence for alterations at distant metastatic sites, including newly synthesized fibronectin, VEGFR1$^+$ cellular infiltration, and MMP production, and suggest that research focusing exclusively on the intrinsic properties of tumor cells insufficient. Insights into key interactions between tumor cells, EPCs, HPCs, and stromal cells will yield important clues to aid in our understanding of the complex processes behind cancer progression.

9. STROMAL CELLS: CARCINOMA-ASSOCIATED FIBROBLASTS (CAFs)

Fibroblasts comprise the majority of stromal cells within the primary tumor bed in various types of human carcinomas (Sappino *et al.*, 1988). Until recently, the role of these cells in tumor progression was unknown. As with fibroblasts associated with wound healing, carcinoma-associated fibroblasts (CAFs) are referred to as "activated fibroblasts" or myofibroblasts (Olumi *et al.*, 1999), and are characterized by the production of alpha-smooth muscle actin. The role of CAFs in tumor progression was elucidated by several studies that demonstrated the unique cancer-promoting properties of tumor stroma–associated fibroblasts. Early studies established the role of fibroblasts in tumor progression by analyzing grafts of tumorigenic epithelial cells mixed with normal fibroblasts or fibroblasts that were immortalized, transformed, or tumor associated (Atula *et al.*, 1997; Camps *et al.*, 1990; Gleave *et al.*, 1991). More recent studies in which CAFs were co-implanted with nontumorigenic, prostate epithelial cells showed that these activated fibroblasts could induce tumorigenesis in immunocompromised mice (Hayward *et al.*, 2001). Additionally, CAFs isolated from invasive human breast carcinomas were shown to be more competent than normal fibroblasts in promoting growth of breast cancer cells in a murine model of breast carcinoma (Orimo *et al.*, 2005). Understanding the molecular events for the generation of CAFs from normal fibroblasts in either distant or local environments could hold the key to unraveling the crucial mechanisms underlying metastasis and tumor angiogenesis.

10. ACTIVATION OF THE SDF-1–CXCR4 PATHWAY IN MOBILIZATION OF PROANGIOGENIC CELLS

It was recently shown that carcinoma-associated fibroblasts (CAFs) isolated from breast cancers, but not normal tissue fibroblasts, secrete high levels of SDF-1, a molecule that functions as a potent chemoattractant for endothelial cells (ECs) and hematopoietic cells (HCs) and enhances tumor angiogenesis (Orimo *et al.*, 2005). SDF-1 can also enhance tumor progression by directly stimulating the growth of carcinoma cells expressing its cognate receptor, CXCR4. Additionally, the SDF-1–CXCR4 axis used by activated fibroblasts may directly promote tumor cell motility based on chemokine gradients of SDF-1 and CXCR4 expression on most tumor cells.

SDF-1 signals exclusively through the receptor CXCR4, which is expressed by a broad range of cell types, including cancer cells, as well as

ECs, HCs, and their progenitors. In a recent study, Keshet and coworkers showed that an increase in VEGF expression in these cells upregulated the local expression levels of SDF-1 and induced robust angiogenesis (Grunewald et al., 2006). VEGF efficiently mobilized VEGFR1$^+$CXCR4$^+$ myeloid cells from the BM and recruited them to the target organs. Locally expressed SDF-1 was sufficient to retain the recruited BMDCs, which produced MMP-9 and other proangiogenic factors. The latter study contributes to the emerging view that proangiogenic factors (such as VEGF) not only directly stimulate the local proliferation of vascular cells, but also recruit proangiogenic BMDCs to the sites of angiogenesis.

Direct evidence for the role of SDF-1 in regulating mobilization of proangiogenic BM cells *in vivo* has also been demonstrated by plasma elevation of SDF-1, which stimulates mobilization of CXCR4$^+$ BM cells, including HPCs and EPCs (Hattori et al., 2001; Heissig et al., 2002). In ischemic models, increased levels of SDF-1 in peripheral blood concurrent with decreased levels in the BM have been documented, resulting in the mobilization of the proangiogenic cells (Askari et al., 2003; De Falco et al., 2004; Togel et al., 2005).

Recently, the precise mechanism by which stress-induced release of SDF-1 mediates mobilization of BM-derived proangiogenic cells was described (Dar et al., 2005). SDF-1 released within the intravascular compartment is translocated from the plasma to the BM through a complex vesicular transport system, a process referred to as transcytosis (Dar et al., 2005). When SDF-1 enters the BM microenvironment, it induces the activation of matrix metalloproteinase-9 (MMP-9) and the release of soluble kit ligand (sKitL). Subsequently, sKitL induces the release of additional SDF-1, enhancing mobilization of the CXCR4$^+$ and c-Kit$^+$ cells to the circulation (Heissig et al., 2002). Another participating pathway in the amplification of the SDF-1 response is the nitric oxide (NO) pathway. SDF-1 induces the release of NO by endothelial cells, resulting in the upregulation of MMP-9. Activation of MMP-9 and subsequent release of sKitL and mobilization of proangiogenic cells were found to be impaired in eNOS-null mice, delineating a new role for the NO signaling pathway in SDF-1–induced BM cell mobilization (Aicher et al., 2005).

Activation of osteoclasts and cathepsins has also been implicated in the mobilization of the hematopoietic and endothelial progenitor cells (Kollet et al., 2006; Urbich et al., 2005). Finally, additional molecules such as extracellular matrix (ECM) compounds, matrix degradation products, and activated complement proteins have crucial roles in priming CXCR4$^+$ cells and modulating the amplitude of SDF-1 signals (Reca et al., 2003; Wysoczynski et al., 2005). These data suggest that SDF-1 mediated mobilization of CXCR4$^+$ cells is a complex process driven by activation of several molecular pathways, including protease activation, release of cytokines and matrix remodeling.

11. FUTURE DIRECTIONS

Early trials of antiangiogenic therapies in malignancy, such as the VEGF-A neutralizing antibody bevacizumab (Avastin), have demonstrated an increase in initial survival rates in colorectal cancer patients treated with the drug (Fernando and Hurwitz, 2003). Impairment of the cell recruitment for primary tumor neoangiogenesis, subsequent priming of distant premetastatic tissues, and direct cell–cell communication might be accomplished by targeting VEGF. However, resistance to selective antiangiogenic targeting can develop, and the early survival benefit of bevacizumab therapy in colon cancer is subsequently lost (Jubb et al., 2006). This might be due to synthesis of alternative angiogenic mediators such as FGF (Casanovas et al., 2005, Miller et al., 2005) and PlGF (Carmeliet et al., 2001). Thus, the utility of specific immunotherapeutic agents might be hampered by the redundancy that exists in most signaling pathways. Instead, a multimodal approach, such as that which incorporates anti-VEGFR1 and anti-VEGFR2 therapeutic agents, in combination with traditional chemotherapies might prove useful.

Furthermore, therapeutic strategies that manipulate the migrating BMDCs themselves as well as their signaling pathways might prove quite successful in preventing tumor progression. BMDCs that home to the tumor neovasculature and premetastatic sites could be used as "magic bullets," vehicles for the delivery of anticancer agents (Arafat et al. 2000). The feasibility of integrating a suicide gene into BM-derived progenitors to reduce tumor size and vascularity has already been confirmed by several animal studies (Komarova et al. 2006; Lotem and Sachs 2006). Not only HPCs and EPCs, but also fibroblasts and stromal progenitors that migrate to tumor sites might prove useful as carriers of oncolytic adenoviruses or as direct targets of "activated" or genetically altered stromal cells.

As described above, one emerging concept is the invasive and migratory phenotype of a cell may be a function of the niche that it inhabits (Scadden et al., 2006). Dissecting the precise interactions of homing stem cells and their niches in normal physiology and in disease might lead to the development of specific therapeutic targets that prevent aberrant niche formation. Alternatively, identifying and inhibiting the cytokines and growth factors that promote cellular migration, as recently demonstrated with antibodies to PlGF (Fischer et al. 2007), may provide an additional arsenal for the abrogation of the early and late processes of tumor growth and metastasis. Perhaps it is in this respect that therapeutic strategies targeting BMDC mechanisms will prove most beneficial.

REFERENCES

Aicher, A., and Heeschen, C. (2007). Nonbone marrow–derived endothelial progenitor cells: What is their exact location? *Circ. Res.* **101,** e102(comment).

Aicher, A., Zeiher, A. M., and Dimmeler, S. (2005). Mobilizing endothelial progenitor cells. *Hypertension* **45,** 321–325.

Asahara, T., Murohara, T., Sullivan, A., Silver, M., van der Zee, R., Li, T., Witzenbichler, B., Schatteman, G., and Isner, J. M. (1997). Isolation of putative progenitor endothelial cells for angiogenesis. *Science* **275,** 964–967.

Askari, A. T., Unzek, S., Popovic, Z. B., Goldman, C. K., Forudi, F., Kiedrowski, M., Rovner, A., Ellis, S. G., Thomas, J. D., and DiCorleto, P. E. (2003). Effect of stromal-cell–derived factor 1 on stem-cell homing and tissue regeneration in ischaemic cardiomyopathy. *Lancet* **362,** 697–703(see comment).

Atula, S., Grenman, R., and Syrjanen, S. (1997). Fibroblasts can modulate the phenotype of malignant epithelial cells *in vitro. Exp. Cell Res.* **235,** 180–187.

Bergers, G., and Benjamin, L. E. (2003). Tumorigenesis and the angiogenic switch. *Nat. Rev. Cancer* **3,** 401–410.

Bergers, G., Brekken, R., McMahon, G., Vu, T. H., Itoh, T., Tamaki, K., Tanzawa, K., Thorpe, P., Itohara, S., and Werb, Z. (2000). Matrix metalloproteinase-9 triggers the angiogenic switch during carcinogenesis. *Nat. Cell Biol.* **2,** 737–744.

Bergers, G., Javaherian, K., Lo, K. M., Folkman, J., and Hanahan, D. (1999). Effects of angiogenesis inhibitors on multistage carcinogenesis in mice. *Science* **284,** 808–812.

Camps, J. L., Chang, S. M., Hsu, T. C., Freeman, M. R., Hong, S. J., Zhau, H. E., von Eschenbach, A. C., and Chung, L. W. (1990). Fibroblast-mediated acceleration of human epithelial tumor growth *in vivo. Proc. Natl. Acad. Sci. USA* **87,** 75–79.

Carmeliet, P., and Jain, R. K. (2000). Angiogenesis in cancer and other diseases. *Nature* **407,** 249–257.

Carmeliet, P., Moons, L., Luttun, A., Vincenti, V., Compernolle, V., De Mol, M., Wu, Y., Bono, F., Devy, L., and Beck, H. (2001). Synergism between vascular endothelial growth factor and placental growth factor contributes to angiogenesis and plasma extravasation in pathological conditions. *Nat. Med.* **7,** 575–583.

Casanovas, O., Hicklin, D. J., Bergers, G., and Hanahan, D. (2005). Drug resistance by evasion of antiangiogenic targeting of VEGF signaling in late-stage pancreatic islet tumors. *Cancer Cell* **8,** 299–309.

Cheng, T., Rodrigues, N., Shen, H., Yang, Y., Dombkowski, D., Sykes, M., and Scadden, D. T. (2000). Hematopoietic stem cell quiescence maintained by p21cip1/waf1. *Science* **287,** 1804–1808.

Dar, A., Goichberg, P., Shinder, V., Kalinkovich, A., Kollet, O., Netzer, N., Margalit, R., Zsak, M., Nagler, A., and Hardan, I. (2005). Chemokine receptor CXCR4-dependent internalization and resecretion of functional chemokine SDF-1 by bone marrow endothelial and stromal cells. *Nat. Immunol.* **6,** 1038–1046.

De Falco, E., Porcelli, D., Torella, A. R., Straino, S., Iachininoto, M. G., Orlandi, A., Truffa, S., Biglioli, P., Napolitano, M., and Capogrossi, M. C. (2004). SDF-1 involvement in endothelial phenotype and ischemia-induced recruitment of bone marrow progenitor cells. *Blood* **104,** 3472–3482.

De Palma, M., and Naldini, L. (2006). Role of haematopoietic cells and endothelial progenitors in tumour angiogenesis. *Biochim. Biophys. Acta* **1766,** 159–166.

De Palma, M., Venneri, M. A., and Naldini, L. (2003a). *In vivo* targeting of tumor endothelial cells by systemic delivery of lentiviral vectors. *Hum. Gene Ther.* **14,** 1193–1206.

De Palma, M., Venneri, M. A., Roca, C., and Naldini, L. (2003b). Targeting exogenous genes to tumor angiogenesis by transplantation of genetically modified hematopoietic stem cells. *Nat. Med.* **9,** 789–795.

Dvorak, H. F., Nagy, J. A., Dvorak, J. T., and Dvorak, A. M. (1988). Identification and characterization of the blood vessels of solid tumors that are leaky to circulating macromolecules. *Am. J. Pathol.* **133,** 95–109.

Eriksson, U., and Alitalo, K. (2002). VEGF receptor 1 stimulates stem-cell recruitment and new hope for angiogenesis therapies. *Nat. Med.* **8,** 775–777(comment).

Fernando, N., and Hurwitz, H. (2003). Inhibition of vascular endothelial growth factor in the treatment of colorectal cancer. *Semin. Oncol.* **30,** 39–50.

Fischer, C., Jonckx, B., Mazzone, M., Zacchigna, S., Loges, S., Pattarini, L., Chorianopoulos, E., Liesenborghs, L., Koch, M., and De Mol, M. (2007). Anti-PlGF inhibits growth of VEGF(R)-inhibitor-resistant tumors without affecting healthy vessels. [see comment]. *Cell* **131,** 463–475.

Folkman, J. (1995). Angiogenesis in cancer, vascular, rheumatoid and other disease. *Nat. Med.* **1,** 27–31.

Folkman, J., Watson, K., Ingber, D., and Hanahan, D. (1989). Induction of angiogenesis during the transition from hyperplasia to neoplasia. *Nature* **339,** 58–61.

Gao, D., Nolan, D. J., Mellick, A. S., Bambino, K., McDonnell, K., and Mittal, V. (2008). Endothelial progenitor cells control the angiogenic switch in mouse lung metastasis. *Science* **319,** 195–198(see comment).

Gehling, U. M., Ergun, S., Schumacher, U., Wagener, C., Pantel, K., Otte, M., Schuch, G., Schafhausen, P., Mende, T., and Kilic, N. (2000). *In vitro* differentiation of endothelial cells from AC133-positive progenitor cells. *Blood* **95,** 3106–3112.

Gill, M., Dias, S., Hattori, K., Rivera, M. L., Hicklin, D., Witte, L., Girardi, L., Yurt, R., Himel, H., and Rafii, S. (2001). Vascular trauma induces rapid but transient mobilization of VEGFR2(+)AC133(+) endothelial precursor cells. *Circ. Res.* **88,** 167–174.

Gleave, M., Hsieh, J. T., Gao, C. A., von Eschenbach, A. C., and Chung, L. W. (1991). Acceleration of human prostate cancer growth *in vivo* by factors produced by prostate and bone fibroblasts. *Cancer Res.* **51,** 3753–3761.

Grunewald, M., Avraham, I., Dor, Y., Bachar-Lustig, E., Itin, A., Jung, S., Chimenti, S., Landsman, L., Abramovitch, R., and Keshet, E. (2006). VEGF-induced adult neovascularization: Recruitment, retention, and role of accessory cells. *Cell* **124,** 175–189(see comment) (erratum appears in *Cell* 2006;126(4):811).

Hanahan, D., and Folkman, J. (1996). Patterns and emerging mechanisms of the angiogenic switch during tumorigenesis. *Cell* **86,** 353–364.

Hattori, K., Heissig, B., Tashiro, K., Honjo, T., Tateno, M., Shieh, J. H., Hackett, N. R., Quitoriano, M. S., Crystal, R. G., and Rafii, S. (2001). Plasma elevation of stromal cell–derived factor-1 induces mobilization of mature and immature hematopoietic progenitor and stem cells. *Blood* **97,** 3354–3360.

Hattori, K., Heissig, B., Wu, Y., Dias, S., Tejada, R., Ferris, B., Hicklin, D. J., Zhu, Z., Bohlen, P., and Witte, L. (2002). Placental growth factor reconstitutes hematopoiesis by recruiting VEGFR1(+) stem cells from bone-marrow microenvironment. *Nat. Med.* **8,** 841–849(see comment).

Hayward, S. W., Wang, Y., Cao, M., Hom, Y. K., Zhang, B., Grossfeld, G. D., Sudilovsky, D., and Cunha, G. R. (2001). Malignant transformation in a nontumorigenic human prostatic epithelial cell line. *Cancer Res.* **61,** 8135–8142.

Heissig, B., Hattori, K., Dias, S., Friedrich, M., Ferris, B., Hackett, N. R., Crystal, R. G., Besmer, P., Lyden, D., and Moore, M. A. (2002). Recruitment of stem and progenitor cells from the bone marrow niche requires MMP-9 mediated release of kit-ligand. *Cell* **109,** 625–637.

Holash, J., Maisonpierre, P. C., Compton, D., Boland, P., Alexander, C. R., Zagzag, D., Yancopoulos, G. D., and Wiegand, S. J. (1999). Vessel cooption, regression, and growth in tumors mediated by angiopoietins and VEGF. *Science* **284,** 1994–1998.

Jubb, A. M., Oates, A. J., Holden, S., and Koeppen, H. (2006). Predicting benefit from anti-angiogenic agents in malignancy. *Nat. Rev. Cancer* **6,** 626–635.

Kaplan, R. N., Rafii, S., Lyden, D., *et al.* (2006). Preparing the "soil": The premetastatic niche. *Cancer Res.* **66,** 11089–11093.

Kaplan, R. N., Riba, R. D., Zacharoulis, S., Bramley, A. H., Vincent, L., Costa, C., MacDonald, D. D., Jin, D. K., Shido, K., and Kerns, S. A. (2005). VEGFR1-positive haematopoietic bone marrow progenitors initiate the pre-metastatic niche. *Nature* **438,** 820–827(see comment).

Kollet, O., Dar, A., Shivtiel, S., Kalinkovich, A., Lapid, K., Sztainberg, Y., Tesio, M., Samstein, R., Goichberg, P., and Spiegel, A. (2006). Osteoclasts degrade endosteal components and promote mobilization of hematopoietic progenitor cells. *Nat. Med.* **12,** 657–664.

Komarova, S., Kawakami, Y., Stoff-Khalili, M. A., Curiel, D. T., and Pereboeva, L. (2006). Mesenchymal progenitor cells as cellular vehicles for delivery of oncolytic adenoviruses. *Mol. Cancer Ther.* **5,** 755–766.

Kopp, H. G., and Rafii, S. (2007). Thrombopoietic cells and the bone marrow vascular niche. *Ann. N. Y. Acad. Sci.* **1106,** 175–179.

Lin, E. Y., Nguyen, A. V., Russell, R. G., and Pollard, J. W. (2001). Colony-stimulating factor 1 promotes progression of mammary tumors to malignancy. *J. Exp. Med.* **193,** 727–740.

Lin, Y., Weisdorf, D. J., Solovey, A., and Hebbel, R. P. (2000). Origins of circulating endothelial cells and endothelial outgrowth from blood. *J. Clin. Invest.* **105,** 71–77(see comment).

Lotem, M., Zhao, Y., Riley, J., Hwu, P., Morgan, R. A., Rosenberg, S. A., and Parkhurst, M. R. (2006). Presentation of tumor antigens by dendritic cells genetically modified with viral and nonviral vectors. *J. Immunother.* **29,** 616–627.

Luttun, A., Tjwa, M., Moons, L., Wu, Y., Angelillo-Scherrer, A., Liao, F., Nagy, J. A., Hooper, A., Priller, J., and De Klerck, B. (2002). Revascularization of ischemic tissues by PlGF treatment, and inhibition of tumor angiogenesis, arthritis and atherosclerosis by anti-Flt1. *Nat. Med.* **8,** 831–840(see comment).

Lyden, D., Hattori, K., Dias, S., Costa, C., Blaikie, P., Butros, L., Chadburn, A., Heissig, B., Marks, W., and Witte, L. (2001). Impaired recruitment of bone-marrow-derived endothelial and hematopoietic precursor cells blocks tumor angiogenesis and growth. *Nat. Med.* **7,** 1194–1201.

Lyden, D., Young, A. Z., Zagzag, D., Yan, W., Gerald, W., O'Reilly, R., Bader, B. L., Hynes, R. O., Zhuang, Y., and Manova, K. (1999). Id1 and Id3 are required for neurogenesis, angiogenesis and vascularization of tumour xenografts. *Nature* **401,** 670–677(see comment).

Mancuso, P., Burlini, A., Pruneri, G., Goldhirsch, A., Martinelli, G., and Bertolini, F. (2001). Resting and activated endothelial cells are increased in the peripheral blood of cancer patients. *Blood* **97,** 3658–3661.

Maniotis, A. J., Folberg, R., Hess, A., Seftor, E. A., Gardner, L. M. G., Pe'er, J., Trent, J. M., Meltzer, P. S., and Hendrix, M. J. C. (1999). Vascular channel formation by human melanoma cells *in vivo* and *in vitro*: Vasculogenic mimicry. *Am. J. Pathol.* **155,** 739–752.

Miller, K. D., Sweeney, C. J., and Sledge, G. W., Jr. (2005). Can tumor angiogenesis be inhibited without resistance? *EXS* **94,** 95–112.

Monestiroli, S., Mancuso, P., Burlini, A., Pruneri, G., Dell'Agnola, C., Gobbi, A., Martinelli, G., and Bertolini, F. (2001). Kinetics and viability of circulating endothelial cells as surrogate angiogenesis marker in an animal model of human lymphoma. *Cancer Res.* **61,** 4341–4344.

Morikawa, S., Baluk, P., Kaidoh, T., Haskell, A., Jain, R. K., and McDonald, D. M. (2002). Abnormalities in pericytes on blood vessels and endothelial sprouts in tumors. *Am. J. Pathol.* **160,** 985–1000.

Olumi, A. F., Grossfeld, G. D., Hayward, S. W., Carroll, P. R., Tlsty, T. D., and Cunha, G. R. (1999). Carcinoma-associated fibroblasts direct tumor progression of initiated human prostatic epithelium. *Cancer Res.* **59,** 5002–5011.

Orimo, A., Gupta, P. B., Sgroi, D. C., Arenzana-Seisdedos, F., Delaunay, T., Naeem, R., Carey, V. J., Richardson, A. L., and Weinberg, R. A. (2005). Stromal fibroblasts present in invasive human breast carcinomas promote tumor growth and angiogenesis through elevated SDF-1/CXCL12 secretion. *Cell* **121,** 335–348.

Peichev, M., Naiyer, A. J., Pereira, D., Zhu, Z., Lane, W. J., Williams, M., Oz, M. C., Hicklin, D. J., Witte, L., and Moore, M. A. (2000). Expression of VEGFR-2 and AC133 by circulating human CD34(+) cells identifies a population of functional endothelial precursors. *Blood* **95,** 952–958.

Quirici, N., Soligo, D., Caneva, L., Servida, F., Bossolasco, P., and Deliliers, G. L. (2001). Differentiation and expansion of endothelial cells from human bone marrow CD133(+) cells. *Br. J. Haematol.* **115,** 186–194.

Rafii, S., Lyden, D., Benezra, R., Hattori, K., and Heissig, B. (2002). Vascular and haematopoietic stem cells: Novel targets for anti-angiogenesis therapy? *Nat. Rev. Cancer* **2,** 826–835.

Reca, R., Mastellos, D., Majka, M., Marquez, L., Ratajczak, J., Franchini, S., Glodek, A., Honczarenko, M., Spruce, L. A., and Janowska-Wieczorek, A. (2003). Functional receptor for C3a anaphylatoxin is expressed by normal hematopoietic stem/progenitor cells, and C3a enhances their homing-related responses to SDF-1. *Blood* **101,** 3784–3793.

Reya, T., Morrison, S. J., Clarke, M. F., and Weissman, I. L. (2001). Stem cells, cancer, and cancer stem cells. *Nature* **414,** 105–111.

Reyes, M., Dudek, A., Jahagirdar, B., Koodie, L., Marker, P. H., and Verfaillie, C. M. (2002). Origin of endothelial progenitors in human postnatal bone marrow. *J. Clin. Invest.* **109,** 337–346(see comment).

Sappino, A. P., Skalli, O., Jackson, B., Schurch, W., and Gabbiani, G. (1988). Smooth-muscle differentiation in stromal cells of malignant and non-malignant breast tissues. *Int. J. Cancer* **41,** 707–712.

Scadden, D. T. (2006). The stem-cell niche as an entity of action. *Nature* **441,** 1075–1079.

Shi, Q., Rafii, S., Wu, M. H., Wijelath, E. S., Yu, C., Ishida, A., Fujita, Y., Kothari, S., Mohle, R., and Sauvage, L. R. (1998). Evidence for circulating bone marrow-derived endothelial cells. *Blood* **92,** 362–367.

Solovey, A., Lin, Y., Browne, P., Choong, S., Wayner, E., and Hebbel, R. P. (1997). Circulating activated endothelial cells in sickle cell anemia. *N. Engl. J. Med.* **337,** 1584–1590(see comment).

Togel, F., Isaac, J., Hu, Z., Weiss, K., and Westenfelder, C. (2005). Renal SDF-1 signals mobilization and homing of CXCR4-positive cells to the kidney after ischemic injury. *Kidney Int.* **67,** 1772–1784.

Urbich, C., Aicher, A., Heeschen, C., Dernbach, E., Hofmann, W., Zeiher, A., and Dimmeler, S. (2005). Soluble factors released by endothelial progenitor cells promote migration of endothelial cells and cardiac resident progenitor cells. *J. Mol. Cell Cardiol.* **39,** 733–742.

Vu, T. H., and Werb, Z. (2000). Matrix metalloproteinases: Effectors of development and normal physiology. *Genes Dev.* **14,** 2123–2133.

Wysoczynski, M., Reca, R., Ratajczak, J., Kucia, M., Shirvaikar, N., Honczarenko, M., Mills, M., Wanzeck, J., Janowska-Wieczorek, A., and Ratajczak, M. Z. (2005). Incorporation of CXCR4 into membrane lipid rafts primes homing-related responses of hematopoietic stem/progenitor cells to an SDF-1 gradient. *Blood* **105,** 40–48.

Yancopoulos, G. D., Davis, S., Gale, N. W., Rudge, J. S., Wiegand, S. J., and Holash, J. (2000). Vascular-specific growth factors and blood vessel formation. *Nature* **407,** 242–248.

CHAPTER TWELVE

STRUCTURE OF MICROVASCULAR NETWORKS IN GENETIC HYPERTENSION

Walter L. Murfee* *and* Geert W. Schmid-Schönbein[†]

Contents

Abstract

Microvascular rarefaction, defined by a loss of terminal arterioles, small venules, and/or capillaries, is a common characteristic of the hypertension syndrome. While rarefaction has been associated with vessel-specific free radical production, deficient leukocyte adhesion, and cellular apoptosis, the relationships of rarefaction with structural alterations at the network and cellular level remain largely unexplored. The objective of this study was to examine the architecture and perivascular cell phenotypes along microvascular networks in hypertensive versus normotensive controls in the context of imbalanced angiogenesis. Mesenteric tissues from age-matched adult male spontaneously hypertensive (SHR) and Wistar-Kyoto (WKY) rats were harvested and immunolabeled for PECAM and neuron-glia antigen 2 (NG2). Evaluation of intact rat mesenteric microvascular networks rats suggests that network alterations associated with hypertension are more complex than just a loss of vessels. Typical SHR versus WKY networks

* Assistant Professor Department of Biomedical Engineering, Tulane University, New Orleans, Louisiana
† Professor Department of Bioengineering, University of California, San Diego, La Jolla, California

Methods in Enzymology, Volume 444
ISSN 0076-6879, DOI: 10.1016/S0076-6879(08)02812-7

demonstrate a reduced branching architecture marked by more proximal arteriole/venous anastomoses and an absence of NG2 labeling along arterioles. Although less frequent, larger SHR microvascular networks display regions of dramatically increased vascular density. SHR and WKY lymphatic networks demonstrate increased vessel diameters and vascular density compared to networks in normotensive Wistar rats (the strain from which both the SHR and WKY originated). These observations provide a rationale for investigating the presence of local angiogenic factors and response of microvascular networks to therapies aimed at reversing rarefaction in genetic hypertension.

1. INTRODUCTION

Hypertension is associated with an increase of microvascular resistance due to two major mechanisms: arteriolar wall remodeling and microvessel rarefaction. Microvascular rarefaction can be classified as functional or structural. Functional rarefaction in hypertension refers to nonperfusion of a vessel with blood cells and can be reversed by application of a vasodilator. Structural rarefaction is caused by the actual anatomical loss of microvessels, including small terminal arterioles, small venules, and/or capillaries. Given that the development of elevated blood pressure is thought to be accompanied and in some cases preceded by a loss of microvessels (Antonios et al., 1999, 2003; Hutchins and Darnell, 1974; Noon et al., 1997), therapies aimed at reversing microvascular rarefaction potentially represent candidate treatments of hypertension.

In the spontaneously hypertensive rat (SHR), which represents a model of hypertension with a genetic background, the microcirculation is characterized by smooth muscle medial hyperplasia, smooth muscle hypertrophy, enhanced smooth muscle longitudinal coverage, enhanced microvessel specific oxidative stress, elevated counts of activated circulating leukocytes, enhanced capillary flow resistance, impairment of selectin-mediated leukocyte adhesion, and extensive nonuniform endothelial cell apoptosis (Fukuda et al., 2004; Suematsu et al., 2002). These characteristics represent a complex remodeling scenario across multiple scales. In order to develop a new rationale to therapeutically manipulate microvascular rarefaction, a full understanding of whether these vessel–specific alterations in hypertension drive vessel loss or are resultants of impaired vessel angiogenesis is warranted. Angiogenesis, defined as the growth of new vessels from existing vessels, is itself dependent on the coordination of multiple mechanical, cellular, and molecular players (Peirce and Skalak, 2003). For the purpose of this chapter, the term "angiogenesis" will be used in the general sense, and will also encompass the remodeling of existing vessels described by capillary arterialization or arteriogenesis (Peirce and Skalak, 2003).

The objective of this chapter is to examine the architecture and perivascular cell phenotypes along microvascular networks in the SHR and its normotensive control, the Wistar Kyoto (WKY) rat. We present evidence for disturbed angiogenesis in rat mesenteric tissue in hypertension. Evaluation of microvascular architectural patterns and cell alterations in hypertensive animal models serve as the background for future investigation of microvascular dysfunction in this disease.

2. METHODS

2.1. Tissue collection

All experimental protocols were reviewed and approved by the University of California-San Diego Animal Care and Use Committee. Age-matched adult (16 to 20 weeks) male, spontaneous hypertensive rats (SHR) ($n = 4$) and Wistar-Kyoto (WKY) ($n = 4$) rats were given general anesthesia with pentobarbital (50 mg /kg b.w., i.m.). Approximately 30 min after the initial anesthetic injection, rats were euthanized by direct intravenous injection of pentobarbital via the femoral vein. The mesentery was exteriorized, and vascularized sectors, defined as the thin translucent connective tissues in between the mesenteric arterial/venous vessels feeding the bowel, were harvested from each rat. Tissues were preferably harvested from the ileum. The mesenteric tissues were immediately placed in 10 mM of phosphate-buffered saline (PBS), mounted on positively charged glass slides, and fixed for 1 h in 4% paraformaldehyde at 4 °C. The rat mesentery was selected for this study because it allows for two-dimensional visualization of entire intact microvascular networks down to the single cell level.

2.2. In situ immunohistochemistry

Similar to methods previously described (Murfee et al., 2005, 2007), tissue specimens were labeled with one of the following immunolabeling protocols using primary antibodies against platelet endothelial adhesion molecule (PECAM, CD31), or PECAM + neuron-glia antigen 2 (NG2). For the PECAM group, two tissues were labeled per animal. For the PECAM + NG2 group, seven tissues from three SHR animals and four tissues from three WKY animals were labeled. All labeling steps were followed by at least three rinses with PBS + 0.1% saponin (10 min each), except after the developing steps, which was followed by incubation in either distilled water (Vector Red) or PBS (Vector SG) for 5 min.

2.2.1. PECAM labeling

1. Incubated for 1 h at room temperature with 1:200 mouse monoclonal biotinylated CD31 (BD PharMingen) antibody, diluted in antibody buffer (0.1% saponin in PBS + 2% BSA).
2. Incubated for 1 h at room temperature with streptavidin-peroxidase secondary antibody solution (VECTASTAIN Elite ABC, Vector Laboratories).
3. Incubated for 15 to 20 min at room temperature with Vector Nova Red (Vector Laboratories) substrate.

2.2.2. PECAM + NG2 labeling

1. Incubated for 20 min in 2.5% normal horse blocking serum (ImmPRESS kit, Vector Laboratories).
2. Incubated for 1.5 h at room temperature with 1:100 or 1:200 rabbit polyclonal NG2 antibody (a gift from Bill Stallcup, Burnham Institute, La Jolla, CA) diluted in antibody buffer.
3. Incubated for 1 h at room temperature with a peroxidase horse anti-rabbit secondary solution (ImmPRESS kit, Vector Laboratories).
4. Incubated for 5 to 10 min at room temperature with Vector Nova Red substrate (Vector Laboratories).
5. Incubated overnight at 4 °C with 1:200 mouse monoclonal biotinylated CD31 antibody (BD Pharmingen) diluted in antibody buffer.
6. Incubated for 1 h at room temperature with streptavidin-peroxidase secondary antibody solution (Vectastain Elite ABC, Vector Laboratories).
7. Incubated for 10 to 20 min at room temperature with Vector SG substrate (Vector Laboratories).

2.3. Microvascular network evaluation and image acquisition

Mesenteric microvascular networks were examined with an inverted microscope (Olympus IX71) with an Olympus 10×/numerical aperture (NA) equal to 0.3 and Olympus 20×/NA equal to 0.4. Images were captured using a similar microscope system (Olympus IX70), a Cooke 5×/NA equal to 0.15, CCD camera and a Scion frame grabber or an upright microscope (Olympus BX60), an Olympus 10×/NA equal to 0.3 or Olympus 40×/NA equal to 0.75, and a SPOT Insight camera. For the purpose of this study, microvascular networks were defined as having at least one arteriole/venule vessel pair originating from the periphery of the mesenteric connective tissue connecting to a branching vasculature and being isolated from another network. Small networks were classified as spanning an area less than a 10× field of view. Larger networks were classified as spanning an area greater than a $10\times$ field of view, and in some cases included multiple subnetworks interconnected at the capillary level. Mesenteric tissues were excluded from analysis if they were avascular or did not contain an isolated microvascular network. Arterial and venous vessel identity was determined based on position within the microvascular network, relative vessel diameter, and endothelial cell morphology. An arterial/venous (A/V) anastomosis was defined by direct connection between an arteriole and a venule. Capillaries were identified based on either vessel diameter (<10 μm) or blind-ended location. Pericyte-specific, NG2-positive cell labeling along capillaries was confirmed by identification of the typical elongated pericyte morphology (Murfee et al., 2005). Finally, for the evaluation of lymphatic vessels in the SHR and WKY tissues, additional mesenteric sectors were harvested from adult Wistar rats and immunolabeled for PECAM following the protocol as described above.

3. RESULTS

In comparison to tissues harvested from WKY rats, SHR mesenteric tissues contained smaller microvascular networks, consistent with previously documented microvascular rarefaction in other organs of the SHR (Chen et al., 1981; Greene et al., 1990; Hutchins and Darnell, 1974; Kobayashi et al., 2005; Prewitt et al., 1982). Fifteen of fifteen SHR isolated networks from six tissues were classified as small defined by being contained in a $10\times$ field of view. A seventh tissue contained five larger networks, atypical of the architectures associated with the other tissues and spanning vascular areas greater than a $10\times$ field of view. In contrast to the small SHR networks, five of eight WKY networks spanned multiple fields of view. Five of six WKY tissues contained these dramatically larger networks. Overall, fewer WKY networks were observed because, based on our definition, an isolated network could consist of multiple interconnected subnetworks.

Inspection of microvascular architectures within SHR and WKY networks of comparable area served to identify alterations more complex than just a loss of microvessels (Fig. 12.1). Typical small SHR networks had arterial/venous (A/V) anastomoses located proximal to the capillary level and prior to multiple arterial or venule branching (Figs. 12.1A and 12.2A). As a consequence, these networks had less of a tree-stucture as compared to WKY networks (Fig. 12.1A and B). Individual arterial and venous vessels were determined based on position within the microvascular network, relative vessel diameter, and endothelial cell morphology. The larger, but less frequent, SHR microvascular networks displayed an increased capillary density per vascular area (Fig. 12.1C) compared to WKY networks (Fig. 12.1D).

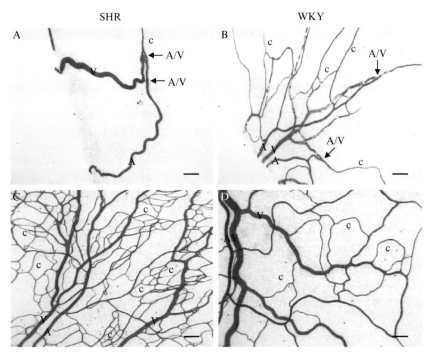

Figure 12.1   Representative examples of SHR and WKY adult mesenteric microvascular networks identified by PECAM labeling. (A) Typical SHR microvascular networks display arterial/venous anastomoses and a nonbranching or non–tree-like architecture. (B) In comparison, similarly sized WKY networks display a more tree-like structure and increased vascular density per tissue area. (C) Atypical large SHR microvascular networks demonstrate increased vessel density and loop formation when compared to WKY networks (D) of comparable area. A, arteriole; V, venule; c, capillary; A/V, arterial/venous anastomoses; L, lymphatic vessel. Scale bars, 100 μm.

SHR mesenteric microvascular networks also displayed an altered perivascular cell NG2 expression pattern. In typical small SHR networks, NG2 labeling did not identify smooth muscle cells, wrapping around arterioles or venules (Fig. 12.2A). In comparably sized WKY networks, NG2 expression identified smooth muscle cells wrapping around arterioles and pericytes along most, if not all, capillaries (Fig. 12.2B). In the SHR networks, NG2-positive labeling was present only on a subset of capillaries (Fig. 12.2A). Pericyte-specific, NG2-positive cell labeling along capillaries was confirmed by identification of the characteristic elongated, wrapping morphology (Murfee et al., 2005) (Fig. 12.3). These NG2 labeling patterns seemed to be associated with the SHR and WKY architectural differences. In one example of a typical SHR small network observed in a WKY tissue, NG2 expression patterns were similar to the SHR.

SHR   WKY

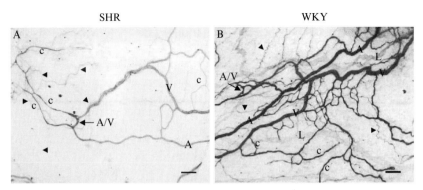

Figure 12.2 Representative examples of SHR and WKY adult mesenteric microvascular networks labeled with PECAM (blue) and NG2 (dark red). (A) In typical SHR networks, perivascular cells along both arterioles and venules lack NG2-positive labeling. At the capillary level, vascular pericyte expression of NG2 is vessel-specific. (B) In comparable WKY microvascular networks, positive NG2 labeling identifies arterioles versus venules and is present along most, if not all, capillaries. A, arteriole; V, venule; c, capillary; A/V, arterial/venous anastomoses; L, lymphatic vessel. Arrowheads identify NG2-positive nerve cells running along blood vessels or within the interstitial space. Scale bars, 100 μm. (See Color Insert.)

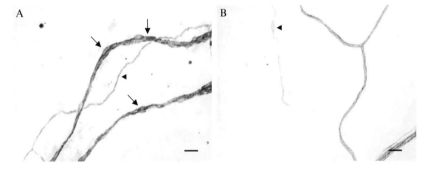

Figure 12.3 Vascular pericyte identification along capillaries in SHR microvascular networks. (A) NG2-positive (dark red) labeling identifies pericytes (arrows), which wrap and elongate along the PECAM-positive (blue) endothelial lined capillaries. (B) Along PECAM-positive (blue) capillaries lacking pericytes, no observable NG2 labeling is observed. Note that both images represent higher-magnification rotated views of vessels in Figure 12.2A. Arrowheads identify NG2-positive nerve cells. Scale bars, 20 μm. (See Color Insert.)

In addition to the blood vascular endothelial cells, PECAM labeling in rat mesenteric tissues also identified lymphatic endothelial cells. Lymphatic vessels were distinguishable from blood vessels by diameter, decreased PECAM labeling intensity, and endothelial cell morphology (Murfee et al., 2007). We observed significant structural differences in initial lymphatic vessels, including increased diameters and altered branching patterns, in SHR and WKY mesenteric tissues compared to tissues harvested from adult Wistar rats (Fig. 12.4). The Wistar strain represents an alternative normotensive control given that it is the origin for both the SHR and the WKY and that the WKY, in spite of its lower blood pressure relative to the SHR, is still slightly hypertensive (DeLano *et al.*, 2005).

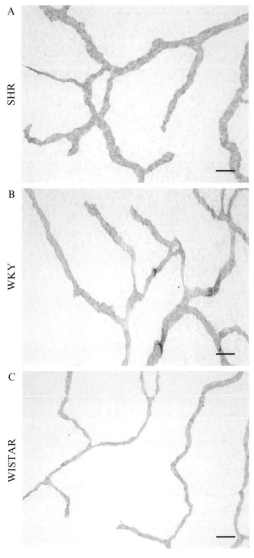

Figure 12.4 Representative examples of terminal lymphatic vessels labeled with PECAM in adult SHR, WKY, and WISTAR mesenteric tissues. In general, lymphatic vessels in SHR (A) and WKY (B) networks demonstrate increased diameters and increased branching patterns as compared to Wistar (C) vessels. Scale bars, 100 μm.

4. DISCUSSION

4.1. Microvascular rarefaction in hypertension

Structural microvascular rarefaction, defined as an anatomical loss of microvessels, has been well documented in various tissues harvested from hypertensive animals (le Noble *et al.*, 1998; Sane *et al.*, 2004; Seumatsu *et al.*, 2002), and is mechanistically supported by evidence for enhanced endothelial cell apoptosis along capillaries in hypertensive networks (Gobe *et al.*, 1997; Suematsu *et al.*, 2002; Tran and Schmid-Schönbein, 2007; Vogt and Schmid-Schönbein, 2001). However, recent evidence suggests that microvascular rarefaction might not be characteristic of all tissues. For example, no changes were found in the number of capillaries in human quadriceps muscle samples harvested from hypertensive patients (Hernández *et al.*, 1999). In left ventricular tissue, absent numbers of capillaries were found only in microareas in NO-deficient hypertensive Wistar rats (Okruhlicova *et al.*, 2005). In SHR animals, rarefaction of arterioles was reported not to occur in either the pial vasculature (Werber *et al.*, 1990) or the intestine (Bohlen, 1983). Finally, the length of arcade arterioles in spinotrapezius muscle harvested from SHR animals was higher compared to WKY animals (Engelson *et al.*, 1986). Thus, although microvascular rarefaction and cellular apoptosis exist in hypertension, this phenomenon may be in part dependent on local tissue environments.

4.2. Signs for disturbed angiogenesis in hypertension

The association of microvascular rarefaction with genetic forms of hypertension implicates angiogenesis as a potential treatment option. However, rarefaction is not characteristic of all tissues (see above). Based on recent evidence of elevated levels of angiogenic growth factors in certain models of hypertension (Sane *et al.*, 2004), evaluation of angiogenesis aimed therapies for treating hypertension and, more broadly, a full understanding of the microcirculation's role in hypertension requires (1) a continued characterization of microvascular architectures in different tissues, (2) characterization of how microvascular network patterns change over the time course of the disease compared to the maturation of networks in tissues harvested normotensive animals, and (3) evaluation of angiogenic responses in hypertensive animals at different ages.

As evidence for an altered angiogenic activity during hypertension, hypertensive patients have been reported to have elevated circulating levels of VEGF, bFGF, and TGF-β (Sane *et al.*, 2004). In 18-week-old SHR versus WKY animals, VEGF mRNA levels and capillary density were significantly higher (Gu *et al.*, 2004). Patients with essential hypertension were found to have elevated plasma VEGF levels when compared to healthy controls (Belgore *et al.*, 2001). In 4-week-old SHR animals, implanted fibrin chambers contained more vessels than those implanted into WKY animals (Hudlett *et al.*, 2005). Finally, endothelial cell proliferation and increased angiogenesis are common characteristics of pulmonary hypertension (Tuder and Voelkel, 2002).

Disturbed angiogenesis in hypertensive tissues is supported by our observations of local increased vessel density in larger SHR microvascular networks compared to WKY networks of similar area. From our observations, we hypothesize that hypertensive networks in the rat mesentery are hypersensitive to angiogenic stimuli. Investigation of microvascular network structural remodeling during hypertension from an angiogenesis perspective will serve to provide valuable insight for assessing the potential for angiogenesis therapies aimed at reversing microvascular rarefaction.

4.3. Microvascular pattern alterations in hypertension

Signs of microvascular network architectural alterations in tissues of hypertensive animals support the concept that microvascular rarefaction is more multivariate than just a loss of vessels. In typical SHR mesenteric networks, we observed non–tree-like architectures marked by arterial/venous anastomoses. Whether these structural features are characteristic of younger normotensive networks and merely represent impaired microvascular network maturation remains to be determined, yet their presence do represent an altered network pattern that could influence, blood flow, capillary stasis, cellular apoptosis, and the ability of the network to respond to angiogenic stimuli. More importantly, our observations of altered network patterns are supported by the characterization of microvascular structure in other tissues and harvested from adult SHR animals (Engelson *et al.*, 1986). In spinotrapezius muscle, Engelson *et al.* (1986) reported almost twice as many transverse arterioles with shorter branches per unit tissue volume. In adult SHR retinas, microvascular networks display abnormal patterning characterized by increased arteriovenous crossings and loop formation (Bhutto *et al.*, 1997). Even in their hallmark paper identifying rarefaction in the cremaster muscle of the SHR, Hutchins and Darnell (1974) documented an increased arteriolar tortuosity and an increase in the number of smallest venules.

Our observations in rat mesenteric networks also highlight a potential for increased lymphatic vessel diameter and density in hypertension. The mechanisms for these alterations remain currently unknown. Enhanced lymph flow in myocardial and intestinal tissue has been reported for the one-kidney, one-clip hypertension model (Laine, 1988; Laine and Granger, 1983) and in adult SHR animals mesenteric lymphatic vessels have been reported to exhibit extensive oxygen free radical formation (DeLano *et al.*, 2005). Although future investigation is required to compare lymphatic function in the SHR, WKY, and alternate normotensive strains, the potential association of altered lymphatic network patterns emphasizes a need to better understand the structural regulation of the microcirculation in various models of hypertension.

4.4. Role for perivascular cells in hypertension

Perivascular cells include smooth muscle cells and vascular pericytes, which elongate along endothelial-lined capillaries. Pericytes, which make direct membrane contact with endothelial cells (Gerhardt and Betsholtz, 2003) and regulate vessel permeability, vessel diameter, and endothelial cell proliferation, have been identified as critical players in angiogenesis (Gerhardt and Betsholtz, 2003; Hellstrom et al., 2001).

Thus, not surprisingly, a role of pericytes has been implicated in hypertension. The number of SM α-actin–positive pericytes is increased in SHR versus WKY brains (Herman and Jacobson, 1988). In the retina of SHR rats, Wallow et al. (1993) report a decreased number of vascular pericytes with increased coverage along capillaries and increased contractility. To our knowledge, a comprehensive description of perivascular cell presence, phenotype, and function across multiple tissues harvested from hypertensive animals has not been completed.

Common markers for pericytes include smooth muscle (SM) α-actin, desmin, and PDGFR-β, yet the expression of these markers is not pericyte-specific (Gerhardt and Betsholtz, 2003). For example, SM α-actin can be expressed by pericytes, immature smooth muscle cells, which express SM α-actin and not SM myosin heavy chain, and mature smooth muscle cells, which express both contractile proteins (Nehls and Drenckhahn, 1991; Price et al., 1994; Van Gieson et al., 2003; Yoshida and Owens, 2005). Desmin and PDGF-β can also be expressed by smooth muscle cells and interstitial fibroblasts (Nehls et al., 1992). Further complicating the issue of pericyte identification, expression of the various markers differs depending on species, tissue, and even developmental stage. Thus, characterization of perivascular cells in hypertensive microvascular networks entails a multifaceted approach involving morphological, phenotypic, and function descriptions.

Recently, NG2 (neuron-glia antigen 2), the rat homologue of the human melanoma proteoglycan (HMP), has been characterized as a pericyte maker (Ozerdem et al., 2001, 2002). In quiescent rat mesenteric tissues, NG2 labeling identifies smooth muscle cells along arterioles and vascular pericytes along all capillary vessels, yet not by smooth muscle cells along venules (Murfee et al., 2005). Our observations in typical SHR mesenteric networks indicate that a subset of capillaries are covered by NG2-positive pericytes, and that pre- and post-capillary vessels positive for SM α-actin expression (data not shown) lack NG2 arterial/venous identification. Further investigation is needed to determine the functional implications of altered NG2-labeling patterns and arterial/venous identity on microvascular network development in the adult.

5. SUMMARY

In summary, the results presented in this chapter indicate that mesenteric microvascular networks in spontaneously hypertensive rats are characterized by architectural and perivascular cell-phenotype alterations in comparison to normotensive networks. These qualitative observations support an emerging paradigm that microvascular rarefaction is more complex than just a loss of vessels or apoptosis of endothelial cells and is associated with imbalanced angiogenesis. In the context of designing therapies to promote angiogenesis aimed at reversing vessel loss, this study highlights the importance for future quantitative description of the local tissue environments in microvascular networks over the time course of aging in hypertensive animals.

REFERENCES

Antonios, T. F., Rattray, F. M., Singer, D. R., Markandu, N. D., Mortimer, P. S., and MacGregor, G. A. (2003). Rarefaction of skin capillaries in normotensive offspring of individuals with essential hypertension. *Heart* **89,** 175–178.

Antonios, T. F., Singer, D. R., Markandu, N. D., Mortimer, P. S., and MacGregor, G. A. (1999). Rarefaction of skin capillaries in borderline essential hypertension suggests an early structural abnormality. *Hypertension* **34,** 655–658.

Belgore, F. M., Blann, A. D., Li-Saw-Hee, F. L., Beevers, D. G., and Lip, G. Y. (2001). Plasma levels of vascular endothelial growth factor and its soluble receptor (SFlt-1) in essential hypertension. *Am. J. Cardiol.* **87,** 805–807, A9.

Bhutto, I. A., and Amemiya, T. (1997). Vascular changes in retinas of spontaneously hypertensive rats demonstrated by corrosion casts. *Ophthalmic Res.* **29,** 12–23.

Bohlen, H. G. (1983). Intestinal microvascular adaptation during maturation of spontaneously hypertensive rats. *Hypertension* **5,** 739–745.

Chen, I. I., Prewitt, R. L., and Dowell, R. F. (1981). Microvascular rarefaction in spontaneously hypertensive rat cremaster muscle. *Am. J. Physiol.* **241,** H306–H310.

DeLano, F. A., Balete, R., and Schmid-Schönbein, G. W. (2005). Control of oxidative stress in the microcirculation of the spontaneously hypertensive rat. *Am. J. Physiol. Heart Circ. Physiol.* **288,** H805–H812.

Engelson, E. T., Schmid-Schönbein, G. W., and Zweifach, B. W. (1986). The microvasculature in skeletal muscle. II. Arteriolar network anatomy in normotensive and spontaneously hypertensive rats. *Microvasc. Res.* **31,** 356–374.

Fukuda, F., Yasu, T., Kobayashi, N., Ikeda, N., and Schmid-Schönbein, G. W. (2004). Contribution of fluid shear response in leukocyte to hemodynamic resistance in the spontaneously hypertensive rat. *Circ. Res.* **95,** 100–108.

Gerhardt, H., and Betsholtz, C. (2003). Endothelial-pericyte interactions in angiogenesis. *Cell Tissue Res.* **314,** 15–23.

Gobé, G., Browning, J., Howard, T., Hogg, N., Winterford, C., and Cross, R. (1997). Apoptosis occurs in endothelial cells during hypertension-induced microvascular rarefaction. *J. Struct. Biol.* **118,** 63–72.

Greene, A. S., Lombard, J. H., Cowley, A. W., Jr., and Hansen-Smith, F. M. (1990). Microvessel changes in hypertension measured by Griffonia simplicifolia I lectin. *Hypertension* **15**(6 Pt 2), 779–783.

Gu, J. W., Fortepiani, L. A., Reckelhoff, J. F., Adair, T. H., Wang, J., and Hall, J. E. (2004). Increased expression of vascular endothelial growth factor and capillary density in hearts of spontaneously hypertensive rats. *Microcirculation* **11,** 689–697.

Hellström, M., Gerhardt, H., Kalén, M., Li, X., Eriksson, U., Wolburg, H., and Betsholtz, C. (2001). Lack of pericytes leads to endothelial hyperplasia and abnormal vascular morphogenesis. *J. Cell Biol.* **153**(3), 543–553.

Herman, I. M., and Jacobson, S. (1988). *In situ* analysis of microvascular pericytes in hypertensive rat brains. *Tissue Cell.* **20**(1), 1–12.

Hernández, N., Torres, S. H., Finol, H. J., and Vera, O. (1999). Capillary changes in skeletal muscle of patients with essential hypertension. *Anat. Rec.* **256,** 425–432.

Hudlett, P., Neuville, A., Miternique, A., Griffon, C., Weltin, D., and Stephan, D. (2005). Angiogenesis and arteriogenesis are increased in fibrin gel chambers implanted in prehypertensive spontaneously hypertensive rats. *J. Hypertens.* **23,** 1559–1564.

Hutchins, P. M., and Darnell, A. E. (1974). Observations of a decreased number of small arterioles in spontaneously hypertensive rats. *Circ. Res.* **34–35**(Suppl 1), 161–165.

Kobayashi, N., DeLano, F. A., and Schmid-Schönbein, G. W. (2005). Oxidative stress promotes endothelial cell apoptosis and loss of microvessels in the spontaneously hypertensive rats. *Arterioscler. Thromb. Vasc. Biol.* **25,** 2114–2121.

Laine, G. A. (1988). Microvascular changes in the heart during chronic arterial hypertension. *Circ. Res.* **62,** 953–960.

Laine, G. A., and Granger, H. J. (1983). Permeability of intestinal microvessels in chronic arterial hypertension. *Hypertension* **5,** 722–727.

le Noble, F. A., Stassen, F. R., Hacking, W. J., and Struijker Boudier, H. A. (1998). Angiogenesis and hypertension. *J. Hypertens.* **16,** 1563–1572.

Murfee, W. L., Rappleye, J. W., Ceballos, M., and Schmid-Schönbein, G. W. (2007). Discontinuous expression of endothelial cell adhesion molecules along initial lymphatic vessels in mesentery: The primary valve structure. *Lymphat. Res. Biol.* **5,** 81–89.

Murfee, W. L., Skalak, T. C., and Peirce, S. M. (2005). Differential arterial/venous expression of NG2 proteoglycan in perivascular cells along microvessels: Identifying a venule-specific phenotype. *Microcirculation* **12,** 151–160.

Nehls, V., Denzer, K., and Drenckhahn, D. (1992). Pericyte involvement in capillary sprouting during angiogenesis *in situ*. *Cell Tissue Res.* **270,** 469–474.

Nehls, V., and Drenckhahn, D. (1991). Heterogeneity of microvascular pericytes for smooth muscle alpha-actin. *J. Cell Biol.* **113,** 147–154.

Noon, J. P., Walker, B. R., Webb, D. J., Shore, A. C., Holton, D. W., Edwards, H. V., and Watt, G. C. (1997). Impaired microvascular dilatation and capillary rarefaction in young adults with a predisposition to high blood pressure. *J. Clin. Invest.* **99,** 1873–1879.

Okruhlicova, L., Tribulova, N., Weismann, P., and Sotnikova, R. (2005). Ultrastructure and histochemistry of rat myocardial capillary endothelial cells in response to diabetes and hypertension. *Cell Res.* **15,** 532–538.

Ozerdem, U., Grako, K. A., Dahlin-Huppe, K., Monosov, E., and Stallcup, W. B. (2001). NG2 proteoglycan is expressed exclusively by mural cells during vascular morphogenesis. *Dev. Dyn.* **222,** 218–227.

Ozerdem, U., Monosov, E., and Stallcup, W. B. (2002). NG2 proteoglycan expression by pericytes in pathological microvasculature. *Microvasc. Res.* **63,** 129–134.

Peirce, S. M., and Skalak, T. C. (2003). Microvascular remodeling: A complex continuum spanning angiogenesis to arteriogenesis. *Microcirculation* **10**(1), 99–111.

Prewitt, R. L., Chen, I. I., and Dowell, R. (1982). Development of microvascular rarefaction in the spontaneously hypertensive rat. *Am. J. Physiol.* **243**, H243–H251.

Price, R. J., Owens, G. K., and Skalak, T. C. (1994). Immunohistochemical identification of arteriolar development using markers of smooth muscle differentiation. Evidence that capillary arterialization proceeds from terminal arterioles. *Circ. Res.* **75**, 520–527.

Sane, D. C., Anton, L., and Brosnihan, K. B. (2004). Angiogenic growth factors and hypertension. *Angiogenesis* **7**, 193–201.

Suematsu, M., Suzuki, H., Delano, F. A., and Schmid-Schönbein, G. W. (2002). The inflammatory aspect of the microcirculation in hypertension: Oxidative stress, leukocytes/endothelial interaction, apoptosis. *Microcirculation* **9**, 259–276.

Tran, E. D., and Schmid-Schönbein, G. W. (2007). An *in-vivo* analysis of capillary stasis and endothelial apoptosis in a model of hypertension. *Microcirculation* **14**, 793–804.

Tuder, R. M., and Voelkel, N. F. (2002). Angiogenesis and pulmonary hypertension: A unique process in a unique disease. *Antioxid. Redox. Signal.* **4**, 833–843.

Van Gieson, E. J., Murfee, W. L., Skalak, T. C., and Price, R. J. (2003). Enhanced smooth muscle cell coverage of microvessels exposed to increased hemodynamic stresses *in vivo*. *Circ. Res.* **92**, 929–936.

Vogt, C. J., and Schmid-Schönbein, G. W. (2001). Microvascular endothelial cell death and rarefaction in the glucocorticoid-induced hypertensive rat. *Microcirculation* **8**, 129–139.

Wallow, I. H., Bindley, C. D., Reboussin, D. M., Gange, S. J., and Fisher, M. R. (1993). Systemic hypertension produces pericyte changes in retinal capillaries. *Invest. Ophthalmol. Vis. Sci.* **34**, 420–430.

Werber, A. H., Fitch-Burke, M. C., Harrington, D. G., and Shah, J. (1990). No rarefaction of cerebral arterioles in hypertensive rats. *Can. J. Physiol. Pharmacol.* **68**, 476–479.

Yoshida, T., and Owens, G. K. (2005). Molecular determinants of vascular smooth muscle cell diversity. *Circ. Res.* **96**, 280–291.

CHAPTER THIRTEEN

Oxygen as a Direct and Indirect Biological Determinant in the Vasculature

Yan Huang *and* Frank J. Giordano

Contents

Abstract

A fundamental function of the vasculature is to deliver oxygen to tissues and organs. The cells that make up the vasculature also require oxygen, and are acted upon by oxygen in direct and indirect ways that can have significant effects on acute and chronic vascular function and morphology. The role that oxygen, or its absence, plays in defining the biology of the vasculature is thus of critical importance, yet remains an area about which there are many gaps in knowledge and understanding. Oxygen-associated paracrine mechanisms can drive vascular processes such as angiogenesis. The vasculature can also directly sense blood oxygen levels and differentially translate this information into rapid vasoconstriction responses in some vascular beds, and vasodilation

Cardiovascular Medicine, Department of Medicine, and Vascular Biology and Translation Program, Yale University School of Medicine, New Haven, Connecticut

Methods in Enzymology, Volume 444
ISSN 0076-6879, DOI: 10.1016/S0076-6879(08)02813-9

in others. Furthering our understanding of how oxygen and hypoxia affect the vasculature may lead to greater insights into the mechanisms and pathogenesis of disease processes involving the vasculature, and lead to new therapeutic paradigms.

1. INTRODUCTION

One of the most fundamental and crucial functions of the vasculature is to deliver oxygen to organs and tissues. In this role distinct components of the vasculature are exposed to different levels of oxygen. Large arterial vessels are normally exposed to blood oxygen saturation levels in the 95 to 99% range. Large veins that carry oxygen-spent mixed-venous blood back to the heart are exposed to a mean O_2 saturation of ~68%, although this can vary markedly depending on cardiac output and a variety of other factors. In the microcirculation, O_2 saturation can vary even more, with saturations below 20% occurring, and saturations in the 20 to 40% range common. Despite this wide range of oxygen exposures experienced by the vasculature, relatively little is known about how oxygen directly influences vascular biology. It is clear, however, that the vasculature is more than a passive conduit for the oxygen it carries in red blood cells and dissolved in the fluid compartment of blood.

In this chapter of the series, the manner in which oxygen-associated events influence physiological and pathophysiological vascular biology *in vivo*, directly and indirectly, is explored. Elsewhere in this issue, M. Celese Simon addresses the basic biological effects of hypoxia and hyperoxia on vascular biology, with an emphasis on the role of the HIF family of hypoxia-inducible transcription factors. In addressing the mechanisms whereby oxygen influences vascular biology *in vivo*, this chapter divides oxygen-associated effects categorically into those mediated via the HIF pathway, and those mediated via other mechanisms.

2. MECHANISMS OF OXYGEN SENSING AND TRANSDUCTION INTO VASCULAR BIOLOGICAL RESPONSES

The effects of oxygen on the vasculature can be broadly divided into two major categories: (1) *indirect* induction of responses by the vasculature initiated by factors produced and/or released by nonvascular cells in reaction to altered oxygen levels, and (2) *direct* induction of responses by the vasculature initiated by oxygen-sensing within the cells of the vasculature. In a practical sense, indirect and direct effects of oxygen on the vasculature often contribute simultaneously. Examples of indirect effects include promotion of angiogenesis by growth factors produced by hypoxic nonvascular parenchymal cells in ischemic tissues, and acute vasodilation mediated by vasoactive substances released by ischemic/hypoxic tissue beds. Examples of direct effects include pulmonary vasoconstriction in response to hypoxemia, and constriction of the ductus arteriosis in response to postnatal normoxia. Oxygen sensing by the vasculature can also be involved in the regulation of non-vascular biological responses, as exemplified by carotid body–mediated activation of central nervous system (CNS) respiratory centers in response to hypoxemia. Both direct and indirect vascular responses to oxygen can be acute or delayed, and lead to either short-term or longer-term vascular adaptations. Whereas acute vascular responses can occur independent of altered gene expression, longer-term responses often involve complex changes that include alterations in the vascular transcriptosome and/or proteome.

Several mechanistic pathways linking oxygen to specific vascular responses have been identified. Among the least complex is the modulation of local vasomotor tone by vasoactive substances elaborated from adjacent metabolically active tissue beds (Delp and O'Leary, 2004; Feigl, 1983; Tune, 2007). When the rate of energy catabolism is high, and the availability of oxygen limited (e.g., as occurs in skeletal muscle beds during vigorous exercise), a number of vasoactive byproducts of energy metabolism/catabolism are released that can act on the local microcirculation and mediate changes in vascular tone (e.g., vasodilation to promote greater substrate and oxygen delivery to these actively metabolizing tissue beds). This particular example of a mechanistic link between oxygen availability and a consequent vascular response does not involve a specific oxygen-sensing function, *per se*, and is dependent instead on the effects of oxygen availability on local tissue metabolism.

Several other mechanistic pathways linking oxygen to vascular biology do involve a more definitive oxygen-sensing function(Giaccia et al., 2004; Giordano, 2005; Lahiri et al., 2006; Paffett and Walker, 2007; Safran and Kaelin, 2003; Semenza, 2004a; Weir et al., 2002, 2005; Wolin, 2000; Wolin et al., 2005). Oxygen-sensing is complex and, at best, incompletely understood. The field is also rapidly evolving. In that context, an extensive review of oxygen-sensing is beyond the scope and focus of this chapter, but a few representative oxygen-sensing pathways relevant to vascular biology will be addressed herein.

2.1. Oxygen-sensitive potassium channels

As in other excitatory cell types, an outward potassium current maintains the negative resting membrane potential in vascular smooth muscle cells. When this outward current is attenuated, membrane potential becomes less negative and the L-type calcium channels allow calcium entry into the cells, mediating a contractile response and thus vasoconstriction

(Weir *et al.*, 2005). Interestingly, in the pulmonary resistance vessels hypoxia decreases the outward potassium current and thus promotes smooth muscle cell contraction and consequent vasoconstriction. This phenomenon has been linked to a specific voltage-activated potassium channel, Kv1.5, which exhibits oxygen-sensitive potassium conductance (Archer *et al.*, 2008; Weir *et al.*, 2005). In this context, Kv1.5 acts as an oxygen sensor in the pulmonary vasculature. Alterations in Kv1.5 expression and/or activity have been postulated as playing an important role in the pathogenesis of some forms of pulmonary hypertension (Archer *et al.*, 2008; Weir *et al.*, 2005). Other potassium channels have also been implicated, including Kv2.1, 3.1, and TASK (two-pore, acid-sensitive potassium channel) (Gurney *et al.*, 2003; Hogg *et al.*, 2002; Osipenko *et al.*, 2000). Underscoring the complexity of the biology of oxygen sensing and its role in vascular biology is that in the smooth muscle of the ductus arteriosus, hypoxia does not inhibit the outward potassium current, but normoxia does; thus, normoxia induces vasoconstriction of the ductus arteriosus (discussed in further detail below) (Archer *et al.*, 2008; Cornfield *et al.*, 1996). A calcium-sensitive potassium channel has been shown to mediate dilation of the fetal pulmonary vasculature in response to oxygen (Cornfield *et al.*, 1996).

2.2. Reactive oxygen species, NAD (P)H oxidases, and mitochondria-associated oxygen sensing

The role of reactive oxygen species (ROS) in vascular biology has been shown in multiple contexts (Finkel, 2000; Griendling *et al.*, 2000; Seshiah *et al.*, 2002; Sundaresan *et al.*, 1995; Tang *et al.*, 2004b; Ushio-Fukai, 2006), and ROS are also purported to play a role in vascular oxygen sensing and transduction (Paffett and Walker, 2007; Waypa *et al.*, 2006; Waypa and Schumacker, 2005; Weir *et al.*, 2002; Wolin *et al.*, 2005). Most relevant to the connection between oxygen and vascular tone, ROS have been implicated in the modulation of potassium channel activity and in the modulation of sarcoplasmic reticulum calcium release, a crucial determinant of smooth muscle cell contraction and thus vascular tone. One source of ROS generation in the vasculature is the NAD (P)H oxidases. NAD (P)H oxidases NOX1,2 and 4 are present in the vascular wall and have the capacity to generate ROS, including superoxide and H_2O_2 (Ushio-Fukai, 2006; Waypa and Schumacker, 2005). ROS generation by the NOXs involves flavin-associated electron transfer and appears to vary relative to oxygen levels, thus supporting the ostensible role of the NOXs in oxygen sensing. Complicating this, however, are various competing hypotheses purporting that the NOXs either increase, or conversely decrease ROS formation in the vasculature in response to hypoxia (Wolin *et al.*, 2005). The reason for this inconsistency is unclear, but may reflect the complexity of ROS generation and turnover, and the difficulties inherent in studying these systems.

Mitochondria are also a significant source of ROS generation, and their purported role in vascular oxygen sensing has been linked to this function. ROS generation in mitochondria is, however, also quite complex, and it has been proposed in competing converse hypotheses that it is either a hypoxia-induced decrease in ROS generation by the mitochondria, or a hypoxia-induced increase in mitochondrial ROS generation that mediates pulmonary hypoxic vasoconstriction (Moudgil et al., 2005; Waypa et al., 2001, 2002; Waypa and Schumacker, 2002, 2008). Experimental data support each of these competing hypotheses; additional data also suggest an alternative calcium-related mechanism whereby mitochondria regulate hypoxic pulmonary vasoconstriction (Kang et al., 2002, 2003).

2.3. Heme oxygenases

Of three well-described heme oxygenases (HO-1, -2, and -3), the expression of only one (HO-1) has been definitively shown to be inducible (HO-1 is inducible by hypoxia, and potentially by increased oxidative stress or shear stress) (Paffett and Walker, 2007; Shibahara et al., 2007). HO-1 and -2 participate in the degradation of heme, a process that results in the formation of carbon monoxide, biliverdin, and ferrous iron in equimolar concentrations (Paffett and Walker, 2007). This process is dependent on oxygen, and on this basis the heme oxygenases have been implicated as oxygen sensors. Supporting the contention that heme oxygenases are oxygen sensors in the vasculature are the findings that HO-2 is highly expressed in the carotid body, and that mice null for HO-2 demonstrate abnormal ventilatory reflexes to hypoxia, presumably related to abnormal carotid body function (Adachi et al., 2004; Shibahara et al., 2007). In yet another study, HO-2 was shown to be an oxygen sensor for a calcium-sensitive potassium channel (Williams et al., 2004). As with several other purported oxygen sensors, however, there are conflicting data that show no specific effect of HO-2 on carotid body function and suggest that it is not a commonly used acute oxygen sensor (Ortega-Saenz et al., 2006).

HO-1 is, however, a definitive hypoxia-responsive gene, and its expression is regulated transcriptionally via the hypoxia-inducible factor (HIF)-pathway (the HIFs are discussed in detail below). As such, HO-1 is linked to a separate oxygen-sensing pathway, irrespective of the intrinsic oxygen sensing functions of HO-1 itself. This provides an additional link between oxygen and vasomotor tone, given that heme oxygenase–mediated breakdown of heme results in the production of carbon monoxide. Carbon monoxide interacts with soluble guanylate cyclase, stimulating the production of cGMP, ultimately promoting vasodilation via protein kinase G (Morita et al., 1995; Paffett and Walker, 2007). A further consideration is that heme oxgenases, via their production of biliverdin as a breakdown product of heme and the subsequent conversion of biliverdin to bilirubin by biliverdin reductase, contribute to the production of a powerful antioxidant (bilirubin). Thus, heme oxygenases may impact on the functions of other oxygen sensing pathways, and vascular biology in general, on the basis of this antioxidant effect.

2.4. The HIF pathway

One of the best-characterized general biological pathways involved in the sensing and transduction of hypoxia to changes in gene expression is the hypoxia-inducible factor (HIF) pathway (Giaccia *et al.*, 2003; Giordano, 2005; Gordan and Simon, 2007; Huang *et al.*, 2004; Lahiri *et al.*, 2006; Lei *et al.*, 2008; Semenza, 2004b, 2006). At the heart of the HIF pathway are a group of basic helix-loop-helix proteins, HIF1, 2, and 3α, that participate in the transcriptional regulation of an extensive repertoire of genes involved in a wide variety of cellular processes (Giordano, 2005; Rankin and Giaccia, 2008). Of those HIF-responsive genes most relevant to vascular biology are included genes encoding growth factors and cytokines involved in angiogenesis (e.g., VEGF-A and PDGF-B), genes encoding growth factor receptors (e.g., flt-1), genes encoding proteins that modulate vascular tone (e.g., heme oxygenase, iNOS, and endothelins 1 and 2), genes encoding proteins involved in vascular transport of nutrients (e.g., glucose transporter Glut-1), genes encoding or regulating vascular cell surface expression of adhesion molecules, genes encoding factors involved in progenitor cell recruitment or mobilization (e.g., erythropoietin, $SDF1\alpha$), and others (Hickey and Simon, 2006). A partial list of hypoxia-responsive genes (not all documented as HIF responsive) that are particularly relevant to the vasculature appears in Table 13.1.

Transcriptional control of genes by the HIF pathway requires dimerization of an HIF-alpha subunit to the aryl hydrocarbon nuclear transferase protein (ARNT), and binding of this dimer to specific hypoxia response elements (HRE) within the regulatory sequences of HIF-responsive genes. Although regulation of HIF1-3α subunit expression at the mRNA level does occur, hypoxia induces increases in these proteins primarily via a post-translational mechanism. During normoxia these subunits undergo prolyl hydroxylation by specific enzymes (Jaakkola *et al.*, 2001; Kaelin, 2005), and this hydroxylation promotes binding to the von Hippel–Lindau protein, a component of an E3 ubiquitin ligase complex that polyubiquitylates the HIF-alpha subunits and targets them for rapid proteolytic degradation by the proteosome (Haase *et al.*, 2001; Maxwell *et al.*, 1999). When oxygen availability decreases, prolyl hydroxylation commensurately decreases, VHL-associated ubiquitylation decreases, and the stability of the HIF alpha subunits rises, increasing their levels. As HIF-alpha–subunit protein levels rise, dimerization with ARNT occurs and these HIF heterodimers

Table 13.1 Hypoxia-responsive genes relevant to vascular biology (partial list)

	HIF-regulated
Vascular endothelial growth factor A (VEGF A)	Yes
Vascular endothelial growth factor C (VEGF C)	Yes
Vascular endothelial growth factor D (VEGF D)	Yes
Platelet derived growth factor-B (PDGF-B)	Yes
Stromal cell–derived factor-1α (SDF1α)	Yes
Endothelin 1 (ET1)	Yes
Endoglin	Yes
aFGF	No
bFGF	No
Leptin (LEP)	Yes
Transforming growth factor-3 (TGF-3)	Yes
Adrenomedulin (ADM)	Yes
Angiopoietin 2	Yes
Inducible nitric oxide synthase (iNOS)	Yes
Plasmin activator inhibitor-1 (PAI-1)	Yes
Fibronectin	Yes
Matrix metalloproteinase 2 (MMP2)	Yes
Flt-1	Yes
Tie-2	Yes
c-MET	Yes
Chemokine receptor CXCR4	Yes
Survivin	Yes
Integrin-linked kinase	?
Integrin αv	Possibly via VEGF
Integrin β3	Possibly via VEGF
Integrin β5	Possibly via VEGF

the bind to the HREs of HIF-responsive genes and transcriptionally regulate their expression. HIF stabilization can occur via other mechanisms as well, including pathways involving reactive oxygen species. The significance of these alternative HIF-associated pathways to *in vivo* vascular biology has not yet been determined, however. Also, as has been shown by Simon, the profile of transcriptional responses mediated by the HIF pathway may vary considerably depending on which of the HIF alpha subunits is primarily involved. The repertoires of genes transcriptionally activated by HIF-1α versus HIF-2α appear to be distinct. Further, HIF-3α has, in fact, been previously implicated as playing a dominant-negative modulating role that is quite different than the roles HIFs 1 and 2α play, primarily as transcriptional activators.

How the HIF pathway is involved in determining *in vivo* vascular biology may be significantly more complex than might be appreciated on initial consideration. In the most straightforward and widely accepted role of the HIF pathway as a determinant of vascular biology, HIF-mediated transcriptional activation of gene expression in ischemic tissues induces the production and secretion of paracrine factors from tissue parenchymal cells that then in turn act on the adjacent vascular beds. Examples include HIF-mediated expression of endothelin 1, iNOS, and heme oxygenase as mediators of vascular tone; and HIF-mediated expression of growth factors and other proteins that participate in angiogenesis. It is this induction of paracrine angiogenic growth factor expression by HIF-1α that has led to clinical testing of HIF-1α gene therapy for treatment of advanced coronary and peripheral vascular disease (Rajagopalan et al., 2007; Vincent et al., 2000). These examples can also be used to illustrate the complexity of HIF pathway effects on the vasculature. In the case of the HIF pathway and angiogenesis, it is interesting that the HIF pathway appears to regulate both angiogenesis promoting proteins as well as angiogenesis inhibitors. Further, in addition to HIF-mediated expression of paracrine angiogenic factors in parenchymal cells, HIF also mediates gene expression in vascular cells, including the expression of specific growth factor receptors. The role of the HIF pathway in angiogenesis and arteriogenesis is discussed more extensively below.

In the case of HIF-pathway involvement in the regulation of vascular tone, the story is equally complex. Some of the genes regulated by the HIF pathway promote vasoconstriction, whereas others promote vasodilation. Further, some HIF-responsive genes, such as heme oxygenase and iNOS, have roles in the regulation of ROS. Also, the HIF pathway has major effects on the transcription of many genes involved in cellular energy metabolism, a function that might affect the local accumulation of vasoactive substances from actively metabolizing tissue beds, and that may also impact ROS generation. Still further, ROS have been shown to have the capacity to regulate HIF stability independent of hypoxia, thus ROS could have transcriptional effects on vascular tone via the HIF pathway (Giordano, 2005; Goyal et al., 2004). In fact, it has been hypothesized that abnormalities in pulmonary vascular tone are mediated by a complex interplay among mitochondrial metabolism, mitochondrial ROS generation, HIF-mediated gene expression, and KV1.5 potassium channel expression and function (Archer et al., 2008). In vivo studies in mice heterozygous for HIF-1α appear to support HIF involvement in determining pulmonary vascular resistance and constriction in response to hypoxia. Specifically, global HIF-1α+/− mice demonstrated attenuated development of pulmonary hypertension, and attenuated pulmonary vascular remodeling after exposure to a hypoxic environment (FIO_2 10%) for a period of 6 weeks (Semenza, 2004b; Yu et al., 1999).

Another consideration is that HIFs are transcription factors and thus the HIF pathway might be thought of as playing a role most predominantly in delayed vascular responses to oxygen; that is, after HIF-associated changes in gene expression have occurred, the HIF pathway also determines the basal level of the expression of many genes important in the vasculature. Thus, by defining basal levels of gene expression, the HIF pathway may have a major influence on what are considered more acute vascular responses. An example that demonstrates how basal changes in HIF-associated gene expression can alter acute biological responses is the finding that mice globally heterozygous for HIF-1α exhibit abnormal carotid body–mediated systemic reflexes (Lahiri *et al.*, 2006; Peng *et al.*, 2006). These findings also further exemplify the role of HIF as a determinant of vascular oxygen sensing.

3. Physiological, Pathophysiological, and Clinical Settings in Which Oxygen Levels Influence the Vasculature

Acute/short-term direct effects of oxygen on the vasculature are involved in several crucial physiological responses. Immediately post-birth, for example, when mammalian lungs are exposed to oxygen and thus the pulmonary circulation begins carrying oxygenated blood, there is a progressive decrease in pulmonary vascular resistance due to vasodilation of the pulmonary vasculature in response to normoxia. This markedly decreases the pressure within the pulmonary circulation to normal physiological levels. As a result, oxygen-spent blood that was shunted from the high-pressure fetal pulmonary circulation to the aorta through the fetal ductus arteriosus no longer flows through the ductus. In fact, blood flow through the ductus can reverse, carrying oxygenated blood from the aorta to the pulmonary artery. As a consequence, the oxygen levels to which the ductus is exposed rise, and this normoxia contributes to a direct constriction of the ductus, which closes permanently in the normal postnatal mammalian circulation. Thus, oxygen sensing and transduction to direct vascular effects play an essential role in the crucial switching from fetal to adult circulation. Also exemplified here is that different vessels and vascular beds can respond quite differently to oxygen. Whereas normoxia causes constriction of the ductus arteriosus and vasodilation in this vascular bed, hypoxia causes constriction of the pulmonary circulation (Moudgil *et al.*, 2005; Weir *et al.*, 2005).

In the adult, hypoxia-induced pulmonary vasoconstriction is essential to maintain normal lung function and prevent deleterious right-to-left shunting of oxygen-spent blood. Whenever aeration of a segment of lung decreases, oxygenation of the blood that circulates through that segment also decreases. Thus, a risk exists that continued blood flow through that segment will result in oxygen-spent blood being mixed with oxygenated blood (right-to-left shunting), leading to a drop in the oxygen saturation of peripheral blood. Physiologically, this is prevented by hypoxia-mediated vasoconstriction of the pulmonary vascular bed subserving the hypoventilated lung segment. Clinically, this process plays an important role in a variety of diverse settings in which segmental lung aeration is altered, including pneumonia, atelectatic collapse of alveolae, heart failure, and plugging of segments of the bronchial tree in various contexts. Also, the ability of the pulmonary vasculature to vasodilate in response to inhaled oxygen is an important provocative test in patients with elevated pulmonary pressures.

Other settings in which direct oxygen-sensing by components of the vasculature plays a crucial role includes O_2 sensing by vascular "chemoreceptors" that transduce sensed changes in blood oxygen levels to physiological adaptations. For example, oxygen sensing in the carotid body can alter the respiratory rate by stimulation of the CNS respiratory center (Hoshi and Lahiri, 2004; Lahiri et al., 2006; Ortega-Saenz et al., 2006; Peng et al., 2006; Semenza, 2006). In the systemic circulation, a mix of direct and indirect oxygen-associated effects contributes to the determination of vascular tone. Some of these mediate acute effects, and some longer-term effects. As an indirect effect, ischemia or hypoxia of parenchymal tissue beds can lead to the acute release of a variety of substances that mediate vasodilation, including adenosine, inosine, hypoxanthine, carbon monoxide, and nitric oxide. As a direct effect, specific components of the systemic vasculature, such as the coronary arteries, can acutely vasodilate in response to hypoxia (Duncker and Merkus, 2005; Tune, 2007). Longer-term effects on vascular tone can be mediated by oxygen-associated changes in gene expression. For example, hypoxia induces the expression of heme oxygenase, an enzyme involved in determining the levels of carbon monoxide to which the vasculature is exposed. Hypoxia also induces the expression of iNOS, and less directly, eNOS, both involved in production of the vasodilating molecule nitric oxide. Conversely, hypoxia induces the expression of endothelins 1 and 2, both potent vasoconstrictors. Oxygen-associated gene regulation is, in fact, a crucial determinant of vascular morphology and function.

In a variety of other clinical settings, direct oxygen-associated vascular effects may be important contributors. An obvious example in which the oxygen levels to which a vessel is exposed are rather abruptly changed is when vein segments are used as bypass conduits in the arterial circulation. In addition to the marked changes in intraluminal pressure and shear forces that vein segments are exposed to when used in this context, they are also exposed to much higher oxygen levels (Kudo et al., 2007). Whether this altered oxygen exposure contributes significantly to the remodeling and maturation of these arterialized veins, or to the pathogenesis of vein graft failure, remains unknown. Another clinically important setting in which oxygen levels could be reasonably postulated as playing a role is in the ontogeny of atherosclerosis, although again this has not been established. A developing atheroma, like a tumor, may reach a point at which it is too large for oxygen and nutrients to reach the cells in its core without the development of a neovasculature. Indeed, there are published data suggesting that angiogenesis plays a role in atherosclerotic plaque progression (Moulton et al., 1999, 2003), and it is reasonable to postulate that this plaque neovascularization is driven in some measure by local hypoxia within the plaque. Hypoxia has been shown to increase lipid accumulation within the vessel wall of ApoE knockout mice exposed to hypoxia (Nakano et al., 2005), and to increase lipid uptake into smooth muscle cells. Further, HIF-1α expression was shown to be increased after arterial injury in ApoE knockout mice, and that neointima formation was inhibited in these mice by siRNA-mediated HIF-1α knock-down (Karshovska et al., 2007). Whether hypoxia is involved directly in the atherosclerotic process in patients, or in the genesis of vulnerable plaque and plaque rupture remains another unresolved question.

3.1. Angiogenesis: Indirect and direct oxygen-associated effects

Arguably, the best-established indirect oxygen-associated effect on the vasculature is in the induction of angiogenesis. In this context, parenchymal cells in ischemic or hypoxic tissue beds produce and/or release growth factors and cytokines that stimulate, in a paracrine fashion, the proliferation, migration, and patterning of vascular cells in the genesis of a neovasculature. This type of ischemia/hypoxia-associated angiogenesis has been shown to be important in wound healing, tumor angiogenesis, angiogenesis in response to obstructive/occlusive vascular disease, and in a number of other settings. Many paracrine mediators of this process have been identified, although undoubtedly many more remain to be delineated. Among the known paracrine factors induced by tissue hypoxia/ischemia are VEGFs-A, -C, and -D; PDGF-B; FGF2; angiopoietin 2; and several others. Some of these, although not all, are transcriptionally regulated by the HIF pathway. Interestingly, as mentioned above, several factors that may act in a paracrine fashion to inhibit angiogenesis are also inducible by hypoxia. How these hypoxia/ischemia-responsive pro- and anti-angiogenic factors balance out and determine a vascular response in the in vivo setting remains unclear.

To investigate the role of the HIF pathway in mediating indirect/paracrine effects on angiogenesis, we deleted HIF-1α specifically from cardiac myocytes in mice (Huang et al., 2004). Previously, we had shown that deletion of VEGF from cardiac myocytes using the same gene targeting strategy resulted in a profound cardiac phenotype with variable embryonic lethality, cardiac contractile dysfunction in the surviving adult mice, and marked hypovascularity in the heart (Giordano *et al.*, 2001). Interestingly, deleting HIF-1α from the heart in the same manner resulted in a much milder phenotype, with no embryonic lethality, milder reductions in myocardial vascularity, and milder cardiac dysfunction, despite significant changes in basal expression of many HIF-regulated genes (Huang *et al.*, 2004). Furthering these studies, we again used the same gene targeting strategy and deleted the von Hippel–Lindau protein (VHL) from the myocardium (Lei *et al.*, 2008). This strategy increased myocardial HIF levels, as expected, and was used as a method to achieve chronic activation of the HIF pathway in the heart. Although these mice were born at expected Mendelian frequencies, and were healthy and viable at birth, between 3 and 4 months of age they developed severe dilated cardiomyopathy and died prematurely. Examination of the hearts of these mice demonstrated severe abnormalities, with rarefaction of myofibrils and dedifferentiation of cardiac myocytes, loss of cardiac muscle with areas of replacement fibrosis, and surprisingly no significant increase in vascularity as assessed by microvessel counts and analysis of vascular casts (Lei *et al.*, 2008). In fact, there was evidence of regionally reduced vascularity in these hearts. These abnormalities did not occur when VHL and HIF-1α were concomitantly deleted, strongly suggesting that they are dependent upon HIF-1α. The reasons for these unexpected effects on cardiac vascularity are unclear, and some of these effects may be attributable to a loss of vessels in conjunction with myocardial degeneration. Taken together, however, the results of these conditional gene deletion studies in cardiac muscle demonstrate the complexity of paracrine regulation of angiogenesis and the role of the HIF pathway in this process, and that HIF-associated effects on angiogenesis are not identical to those of VEGF, the expression of which it regulates.

Also unclear at this point is how paracrine factors derived from ischemic/hypoxic tissue beds might influence levels of circulating bone marrow–derived mononuclear cells, and induce changes in the adhesive properties of adjacent vascular beds to promote tethering and recruitment of these cells. It has been shown, for example, that the expression of SDF-1a and its receptor CXCR4 are hypoxia-inducible (Schioppa *et al.*, 2003). SDF-1α the CXCR4 receptor play an important role in the regulation of progenitor cell release from the bone marrow and a purported important role in the recruitment of progenitor cells to tissue beds (Jo *et al.*, 2000). Hypoxia, in part via the HIF pathway, induces the expression of VEGF, and VEGF has also been shown to have effects on vascular progenitor cell mobilization from the bone marrow and recruitment to tissue beds. Thus, hypoxia-induced gene expression could conceivably play an important role in the mobilization and recruitment of endothelial progenitor cells to sites of angiogenesis. We have previously shown that forced expression of SDF-1α in the normoxic uninjured heart does not increase the recruitment of bone marrow–derived mononuclear cells to the myocardium.

However, concomitant expression of SDF-1a and induction of ischemic myocardial injury acted synergistically to increase progenitor cell recruitment to the heart significantly above levels observed with injury alone (Abbott et al., 2004). This suggests that perhaps a combination of several hypoxia/ischemia-associated effects is involved in progenitor cell recruitment to injured tissue beds and sites of angiogenesis.

In addition to the paracrine indirect effects of oxygen on angiogenesis, the role that direct effects of oxygen on the vasculature play in defining angiogenesis must also be considered. As related above, hypoxia regulates the expression of several growth factor receptors (e.g., flt-1 and cMET), and also plays a role in the expression of adhesion complexes on the endothelium. Further, hypoxia has important effects on cell cycle progression and apoptosis, all of which could have marked effects on angiogenesis. To begin addressing the direct vascular effects of oxygen on angiogenesis we used conditional gene targeting to investigate the role of HIF-1α in the endothelium. Interestingly, mice null for HIF-1α specifically on the endothelium had only a mild nonsignificant reduction in vascularity at baseline. However, these mice showed markedly abnormal tumor angiogenesis with a concomitant increase in tumor necrosis, and delayed cutaneous wound healing with a concomitant reduction in wound-associated capillary sprouting (Tang et al., 2004a). HIF-1α–null endothelial cells from these mice demonstrated altered gene expression, reduced ATP content, decreased proliferation during hypoxia, abnormal chemotaxis, and formed defective capillary structures on matrigel substrates as compared to control endothelial cells with an intact HIF-1α gene derived from the same mice (Tang et al., 2004a). These findings are consistent with a critical role for the HIF pathway in the vasculature itself, and strongly suggest that direct oxygen-associated effects on the vasculature play an essential role in regulating angiogenesis.

3.2. Arteriogenesis, flow sensing, shear sensing, and oxygen

Arteriogenesis is defined generally as the growth and maturation of pre-existing arteriolar networks into collateral vessels. It has been the focus of significant investigational interest as arteriogenesis is deemed to be responsible for the development of the collateral vessels that often develop in response to occlusive atherosclerotic vascular disease, and the possibility that this process could be exploited in biological revascularization strategies has marked clinical implications. Arteriogenesis is thought to be quite distinct from angiogenesis not only in the events and results of the process itself, but also in its biological triggers and mechanisms. Whereas hypoxia is a well-established biological trigger for angiogenesis, it is widely held that arteriogenesis is not induced or regulated by hypoxia.

The primary trigger for induction of arteriogenesis is thought to be fluid shear stress (FSS) caused by increased flow through pre-existing arteriolar networks (Schaper and Scholz, 2003; Scholz et al., 2001). There is also a purported important contribution of altered circumferential wall stress (CWS) caused by increased pressure within the arteriolar network (Schaper and Scholz, 2003; Scholz et al., 2001). Arteriogenesis is additionally thought to involve monocyte recruitment to the developing collateral circulation, in part mediated by shear-induced expression of MCP-1 in these vascular segments. These unique mechanistic aspects of arteriogenesis, in conjunction with the fact that this process generally occurs in a region of the vasculature "upstream" of the frankly ischemic/hypoxic tissue bed, all support the conclusion that arteriogenesis occurs independent of classic hypoxia-induced biological events. Whereas this continues to be the classic view of arteriogenesis, there are hypothetical models that suggest a context in which oxygen sensing within the vessel wall could play a role in arteriogenesis. In these models, oxygen sensing within the vessel wall acts as part of a mechanotransduction cascade that converts the mechanical signals of FSS and CWS into the biological events that support collateral development. The schematic in Fig. 13.1 depicts an arterial occlusion that diminishes blood flow to the lower limb, leading to an ischemic zone in the downstream skeletal muscle tissue bed and a non-ischemic region around the area of arterial occlusion where collateral development via arteriogenesis takes place. In this nonischemic zone, the mechanical effects of fluid shear stress and circumferential wall stress could alter vascular wall stress, and this in turn might alter oxygen tension and oxygen-associated signaling within the vessel wall. Interestingly, in this context, it was recently shown that stretch can induce increases in HIF-1α levels, providing evidence that the HIF pathway can in some manner participate in mechanotransduction.

3.3. Aneurysms, stents, and oxygen

There are a number of additional clinically relevant settings in which oxygen sensing and oxygen-associated biological events could play an important role in the vascular response to mechanical factors. One of the most common procedures performed clinically worldwide is vascular stenting. In this procedure, a cylindrical metal scaffolding is inserted into the narrowed lumen of a blood vessel and expanded at relatively high pressure (e.g., 12 to 14 atmospheres) circumferentially against the vessel wall. Stents prevent elastic recoil of the vessel and thus induce a chronic mechanical circumferential wall stress on the stented vessel segment until biological remodeling of the vessel diminishes this stress. Unfortunately, vascular remodeling in response to stenting is a two-edged sword, and can lead to a prolific ingrowth of cells and matrix from the vessel wall through the fenestrations in the stent such that flow-limiting vascular narrowing is again created. Interestingly, it was

Occlusion of a large artery branch leads to shunted flow through pre-existing arteriolar networks, increasing fluid shear stress (FSS) and circumferential wall stress (CWS) within these networks. This increase in wall stress may lead to altered microenvironmental oxygen tension within the walls of these vessels, leading to oxygen-associated changes in gene expression and induction of vascular biological responses that promote arteriogenesis and development of mature collateral vessels.

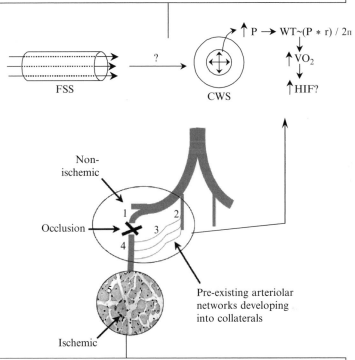

Ischemia/hypoxia leads to changes in gene expression within the underperfused skeletal muscle bed leading to the production of paracrine factors that promote angiogenesis.

Ischemia/hypoxia may also act directly on the vascular cells participating in the angiogenic process, inducing changes in the expression of genes encoding adhesion molecules, growth factor receptors, apotosis-related factors, cell cycle control proteins, and others.

Figure 13.1 Paracrine and direct vascular responses to altered oxygen tension after arterial occlusion. Depicted in this schematic are the ischemic and nonischemic zones resulting from arterial occlusion, and the potential role played by oxygen-associated vascular responses in these zones. (1) Preocclusion arterial segment. (2–3) Feeder vessels and pre-existing arteriolar networks that will participate in arteriogenesis to develop a mature collateral circulation. (4) Postocclusion arterial segment. (5) Downstream ischemic/hypoxic tissue bed in which angiogenesis will occur. Also depicted is a schematic showing how the LaPlace relationship altered fluid shear stress (FSS) and circumferential wall stress (CWS) may lead to changes in oxygen tension, and as a consequence gene expression, in the wall of developing collateral vessels. WT, wall tension; P, pressure; r, radius of curvature; Π, wall thickness; VO_2, oxygen consumption; HIF, hypoxia-inducible factor.

recently shown that stenting of arterial segments results in local hypoxia in the vessel wall and activation of the HIF pathway, and that this can persist for 4 weeks after stent deployment (Cheema et al., 2006). These findings document that mechanical stress can induce alterations in oxygen tension within the vessel wall, and that the HIF pathway can be induced in the vessel wall in this context. Whether HIF-mediated gene expression contributes directly to the remodeling process in this setting remains unclear, but this possibility has significant clinical implications.

As vascular aneurysms enlarge their wall tension and risk of rupture commensurately increase. As applied to aneurysms the LaPlace relationship holds that wall tension is directly proportional to the radius of curvature of the aneurysm and the pressure within the vessel, and inversely proportional to the thickness of the vessel WT $\sim$ ([Radius $\times$ Pressure] / [Wall thickness $\times$ 2]). In other biological contexts, wall tension is proportional to oxygen consumption, and it is clearly reasonable to hypothesize that this also holds in the wall of an expanding vascular aneurysm. In fact, it has been shown that areas in the wall of aortic aneurysms are indeed hypoxic (Vorp et al., 2001), and increased levels of HIF-1α have been shown in the aneurysm wall at sites of rupture (Choke et al., 2006). Given that hypoxia and the HIF pathway can alter the expression of genes involved in extracellular matrix synthesis and turnover, including metalloproteinases, it is reasonable to hypothesize that oxygen-associated biological events within the wall of vascular aneurysms contributes to their pathobiology.

4. SUMMARY

Oxygen-associated molecular processes have a profound effect on the vasculature, directly and indirectly. Various components of the vasculature are exposed to a wide range of oxygen levels, in normal physiological and pathophysiological settings. Oxygen availability can alter the identity and amount of paracrine factors acting on the vasculature, alter vascular levels of reactive oxygen species and thus also alter downstream redox-related molecular events, and elicit short-term biological responses of the vasculature independent of changes in gene expression, but can also profoundly alter gene expression. Oxygen and alterations in its availability should therefore be considered as central in any systematic approach to vascular biology.

REFERENCES

Abbott, J. D., Huang, Y., Liu, D., Hickey, R., Krause, D. S., and Giordano, F. J. (2004). Stromal cell-derived factor-1alpha plays a critical role in stem cell recruitment to the heart after myocardial infarction but is not sufficient to induce homing in the absence of injury. *Circulation* **110,** 3300–3305.

Adachi, T., Ishikawa, K., Hida, W., Matsumoto, H., Masuda, T., Date, F., Ogawa, K., Takeda, K., Furuyama, K., Zhang, Y., Kitamuro, T., Ogawa, H., *et al.* (2004). Hypoxemia and blunted hypoxic ventilatory responses in mice lacking heme oxygenase-2. *Biochem. Biophys. Res. Commun.* **320,** 514–522.

Archer, S. L., Gomberg-Maitland, M., Maitland, M. L., Rich, S., Garcia, J. G., and Weir, E. K. (2008). Mitochondrial metabolism, redox signaling, and fusion: A mitochondria-ROS-HIF-1alpha-Kv1.5 O2-sensing pathway at the intersection of pulmonary hypertension and cancer. *Am. J. Physiol.* **294,** H570–H578.

Cheema, A. N., Hong, T., Nili, N., Segev, A., Moffat, J. G., Lipson, K. E., Howlett, A. R., Holdsworth, D. W., Cole, M. J., Qiang, B., Kolodgie, F., Virmani, R., *et al.* (2006). Adventitial microvessel formation after coronary stenting and the effects of SU11218, a tyrosine kinase inhibitor. *J. Am. Coll. Cardiol.* **47,** 1067–1075.

Choke, E., Cockerill, G. W., Dawson, J., Chung, Y. L., Griffiths, J., Wilson, R. W., Loftus, I. M., and Thompson, M. M. (2006). Hypoxia at the site of abdominal aortic aneurysm rupture is not associated with increased lactate. *Ann. N. Y. Acad. Sci.* **1085,** 306–310.

Cornfield, D. N., Reeve, H. L., Tolarova, S., Weir, E. K., and Archer, S. (1996). Oxygen causes fetal pulmonary vasodilation through activation of a calcium-dependent potassium channel. *Proc. Natl. Acad. Sci. USA* **93,** 8089–8094.

Delp, M. D., and O'Leary, D. S. (2004). Integrative control of the skeletal muscle microcirculation in the maintenance of arterial pressure during exercise. *J. Appl. Physiol. (Bethesda)* **97,** 1112–1118.

Duncker, D. J., and Merkus, D. (2005). Acute adaptations of the coronary circulation to exercise. *Cell Biochem. Biophys.* **43,** 17–35.

Feigl, E. O. (1983). Coronary physiology. *Physiol. Rev.* **63,** 1–205.

Finkel, T. (2000). Redox-dependent signal transduction. *FEBS Lett.* **476,** 52–54.

Giaccia, A., Siim, B. G., and Johnson, R. S. (2003). HIF-1 as a target for drug development. *Nat. Rev.* **2,** 803–811.

Giaccia, A. J., Simon, M. C., and Johnson, R. (2004). The biology of hypoxia: The role of oxygen sensing in development, normal function, and disease. *Genes Dev.* **18,** 2183–2194.

Giordano, F. J. (2005). Oxygen, oxidative stress, hypoxia, and heart failure. *J. Clin. Invest.* **115,** 500–508.

Giordano, F. J., Gerber, H. P., Williams, S. P., VanBruggen, N., Bunting, S., Ruiz-Lozano, P., Gu, Y., Nath, A. K., Huang, Y., Hickey, R., Dalton, N., Peterson, K. L., *et al.* (2001). A cardiac myocyte vascular endothelial growth factor paracrine pathway is required to maintain cardiac function. *Proc. Natl. Acad. Sci. USA* **98,** 5780–5785.

Gordan, J. D., and Simon, M. C. (2007). Hypoxia-inducible factors: Central regulators of the tumor phenotype. *Curr. Opin. Genet. Dev.* **17,** 71–77.

Goyal, P., Weissmann, N., Grimminger, F., Hegel, C., Bader, L., Rose, F., Fink, L., Ghofrani, H. A., Schermuly, R. T., Schmidt, H. H., Seeger, W., and Hänze, J. (2004). Upregulation of NAD(P)H oxidase 1 in hypoxia activates hypoxia-inducible factor 1 via increase in reactive oxygen species. *Free Radic. Biol. Med.* **36,** 1279–1288.

Griendling, K. K., and Ushio-Fukai, M. (2000). NAD(P)H oxidase: Role in cardiovascular biology and disease. *Circ Res.* **86,** 494–501.

Gurney, A. M., Osipenko, O. N., MacMillan, D., McFarlane, K. M., Tate, R. J., and Kempsill, F. E. (2003). Two-pore domain K channel, TASK-1, in pulmonary artery smooth muscle cells. *Circ. Res.* **93,** 957–964.

Haase, V. H., Glickman, J. N., Socolovsky, M., and Jaenisch, R. (2001). Vascular tumors in livers with targeted inactivation of the von Hippel-Lindau tumor suppressor. *Proc. Natl. Acad. Sci. USA* **98,** 1583–1588.

Hickey, M. M., and Simon, M. C. (2006). Regulation of angiogenesis by hypoxia and hypoxia-inducible factors. *Curr. Top. Dev. Biol.* **76,** 217–257.

Hogg, D. S., Davies, A. R., McMurray, G., and Kozlowski, R. Z. (2002). K(V)2.1 channels mediate hypoxic inhibition of I(KV) in native pulmonary arterial smooth muscle cells of the rat. *Cardiovas. Res.* **55,** 349–360.

Hoshi, T., and Lahiri, S. (2004). Cell biology. Oxygen sensing: It's a gas *Science* **306,** 2050–2051.

Huang, Y., Hickey, R. P., Yeh, J. L., Liu, D., Dadak, A., Young, L. H., Johnson, R. S., and Giordano, F. J. (2004). Cardiac myocyte-specific HIF-1alpha deletion alters vascularization, energy availability, calcium flux, and contractility in the normoxic heart. *FASEB J.* **18,** 1138–1140.

Jaakkola, P., Mole, D. R., Tian, Y. M., Wilson, M. I., Gielbert, J., Gaskell, S. J., Kriegsheim, A. V., Hebestreit, H. F., Mukherji, M., Schofield, C. J., Maxwell, P. H., Pugh, C. W., *et al.* (2001). Targeting of HIF-alpha to the von Hippel-Lindau ubiquitylation complex by O2-regulated prolyl hydroxylation. *Science* **292,** 468–472.

Jo, D. Y., Rafii, S., Hamada, T., and Moore, M. A. (2000). Chemotaxis of primitive hematopoietic cells in response to stromal cell-derived factor-1. *J. Clin. Invest.* **105,** 101–111.

Kaelin, W. G. (2005). Proline hydroxylation and gene expression. *Annu. Rev. Biochem.* **74,** 115–128.

Kang, T. M., Park, M. K., and Uhm, D. Y. (2002). Characterization of hypoxia-induced [Ca2+]i rise in rabbit pulmonary arterial smooth muscle cells. *Life Sci.* **70,** 2321–2333.

Kang, T. M., Park, M. K., and Uhm, D. Y. (2003). Effects of hypoxia and mitochondrial inhibition on the capacitative calcium entry in rabbit pulmonary arterial smooth muscle cells. *Life Sci.* **72,** 1467–1479.

Karshovska, E., Zernecke, A., Sevilmis, G., Millet, A., Hristov, M., Cohen, C. D., Schmid, H., Krotz, F., Sohn, H. Y., Klauss, V., Weber, C., and Schober, A. (2007). Expression of HIF-1alpha in injured arteries controls SDF-1alpha mediated neointima formation in apolipoprotein E deficient mice. *Arterioscler. Thromb. Vasc. Biol.* **27,** 2540–2547.

Kudo, F. A., Muto, A., Maloney, S. P., Pimiento, J. M., Bergaya, S., Fitzgerald, T. N., Westvik, T. S., Frattini, J. C., Breuer, C. K., Cha, C. H., Nishibe, T., Tellides, G., *et al.* (2007). Venous identity is lost but arterial identity is not gained during vein graft adaptation. *Arteriosclerosis, thrombosis, and vascular biology* **27,** 1562–1571.

Lahiri, S., Roy, A., Baby, S. M., Hoshi, T., Semenza, G. L., and Prabhakar, N. R. (2006). Oxygen sensing in the body. *Prog. Biophys. Mol. Biol.* **91,** 249–286.

Lei, L., Mason, S., Liu, D., Huang, Y., Marks, C., Hickey, R., Jovin, I. S., Pypaert, M., Johnson, R. S., and Giordano, F. J. (2008). Hypoxia inducible factor-dependent degeneration, failure, and malignant transformation of the heart in the absence of the von Hippel-Lindau protein. *Mol. Cell Biol.* **28,** 3790–3803.

Maxwell, P. H., Wiesener, M. S., Chang, G. W., Clifford, S. C., Vaux, E. C., Cockman, M. E., Wykoff, C. C., Pugh, C. W., Maher, E. R., and Ratcliffe, P. J. (1999). The tumour suppressor protein VHL targets hypoxia-inducible factors for oxygen-dependent proteolysis. *Nature* **399,** 271–275.

Morita, T., Perrella, M. A., Lee, M. E., and Kourembanas, S. (1995). Smooth muscle cell-derived carbon monoxide is a regulator of vascular cGMP. *Proc. Natl. Acad. Sci. USA* **92,** 1475–1479.

Moudgil, R., Michelakis, E. D., and Archer, S. L. (2005). Hypoxic pulmonary vasoconstriction. *J. Appl. Physiol. (Bethesda)* **98,** 390–403.

Moulton, K. S., Heller, E., Konerding, M. A., Flynn, E., Palinski, W., and Folkman, J. (1999). Angiogenesis inhibitors endostatin or TNP-470 reduce intimal neovascularization and plaque growth in apolipoprotein E-deficient mice. *Circulation* **99,** 1726–1732.

Moulton, K. S., Vakili, K., Zurakowski, D., Soliman, M., Butterfield, C., Sylvin, E., Lo, K. M., Gillies, S., Javaherian, K., and Folkman, J. (2003). Inhibition of plaque neovascularization reduces macrophage accumulation and progression of advanced atherosclerosis. *Proc. Natl. Acad. Sci. USA* **100,** 4736–4741.

Nakano, D., Hayashi, T., Tazawa, N., Yamashita, C., Inamoto, S., Okuda, N., Mori, T., Sohmiya, K., Kitaura, Y., Okada, Y., and Matsumura, Y. (2005). Chronic hypoxia accelerates the progression of atherosclerosis in apolipoprotein E-knockout mice. *Hypertens. Res.* **28,** 837–845.

Ortega-Saenz, P., Pascual, A., Gómez-Díaz, R., and López-Barneo, J. (2006). Acute oxygen sensing in heme oxygenase-2 null mice. *J. Gen. Physiol.* **128,** 405–411.

Osipenko, O. N., Tate, R. J., and Gurney, A. M. (2000). Potential role for kv3.1b channels as oxygen sensors. *Circ. Res.* **86,** 534–540.

Paffett, M. L., and Walker, B. R. (2007). Vascular adaptations to hypoxia: Molecular and cellular mechanisms regulating vascular tone. *Essays Biochem.* **43,** 105–119.

Peng, Y. J., Yuan, G., Ramakrishnan, D., Sharma, S. D., Bosch-Marce, M., Kumar, G. K., Semenza, G. L., and Prabhakar, N. R. (2006). Heterozygous HIF-1alpha deficiency impairs carotid body-mediated systemic responses and reactive oxygen species generation in mice exposed to intermittent hypoxia. *J. Physiol.* **577,** 705–716.

Rajagopalan, S., Olin, J., Deitcher, S., Pieczek, A., Laird, J., Grossman, P. M., Goldman, C. K., McEllin, K., Kelly, R., and Chronos, N. (2007). Use of a constitutively active hypoxia-inducible factor-1alpha transgene as a therapeutic strategy in no-option critical limb ischemia patients: Phase I dose-escalation experience. *Circulation* **115,** 1234–1243.

Rankin, E. B., and Giaccia, A. J. (2008). The role of hypoxia-inducible factors in tumorigenesis. *Cell Death Differ.* **15,** 678–685.

Safran, M., and Kaelin, W. G., Jr. (2003). HIF hydroxylation and the mammalian oxygen-sensing pathway. *J. Clin. Invest.* **111,** 779–783.

Schaper, W., and Scholz, D. (2003). Factors regulating arteriogenesis. *Arterioscler. Thromb. Vasc. Biol.* **23,** 1143–1151.

Schioppa, T., Uranchimeg, B., Saccani, A., Biswas, S. K., Doni, A., Rapisarda, A., Bernasconi, S., Saccani, S., Nebuloni, M., Vago, L., Mantovani, A., Melillo, G., et al. (2003). Regulation of the chemokine receptor CXCR4 by hypoxia. *J. Exp. Med.* **198,** 1391–1402.

Scholz, D., Cai, W. J., and Schaper, W. (2001). Arteriogenesis, a new concept of vascular adaptation in occlusive disease. *Angiogenesis* **4,** 247–257.

Semenza, G. L. (2004a). Hydroxylation of HIF-1: Oxygen sensing at the molecular level. *Physiology (Bethesda)* **19,** 176–182.

Semenza, G. L. (2004b). O2-regulated gene expression: Transcriptional control of cardiorespiratory physiology by HIF-1. *J. Appl. Physiol.* **96,** 1173–1117; discussion 1170-1172.

Semenza, G. L. (2006). Regulation of physiological responses to continuous and intermittent hypoxia by hypoxia-inducible factor 1. *Exp. Physiol.* **91,** 803–806.

Seshiah, P. N., Weber, D. S., Rocic, P., Valppu, L., Taniyama, Y., and Griendling, K. K. (2002). Angiotensin II stimulation of NAD(P)H oxidase activity: Upstream mediators. *Circ. Res.* **91,** 406–413.

Shibahara, S., Han, F., Li, B., and Takeda, K. (2007). Hypoxia and heme oxygenases: oxygen sensing and regulation of expression. *Antioxid. Redox Signal.* **9,** 2209–2225.

Sundaresan, M., Yu, Z. X., Ferrans, V. J., Irani, K., and Finkel, T. (1995). Requirement for generation of H2O2 for platelet-derived growth factor signal transduction. *Science* **270,** 296–299.

Tang, N., Wang, L., Esko, J., Giordano, F. J., Huang, Y., Gerber, H. P., Ferrara, N., and Johnson, R. S. (2004). Loss of HIF-1alpha in endothelial cells disrupts a hypoxia-driven VEGF autocrine loop necessary for tumorigenesis. *Cancer Cell.* **6,** 485–495.

Tang, X. D., Garcia, M. L., Heinemann, S. H., and Hoshi, T. (2004). Reactive oxygen species impair Slo1 BK channel function by altering cysteine-mediated calcium sensing. *Nat. Struct. Mol. Biol.* **11,** 171–178.

Tune, J. D. (2007). Control of coronary blood flow during hypoxemia. *Adv. Exp. Med. Biol.* **618,** 25–39.

Ushio-Fukai, M. (2006). Localizing NADPH oxidase-derived ROS. *Sci. STKE* **2006,** re8.

Vincent, K. A., Shyu, K. G., Luo, Y., Magner, M., Tio, R. A., Jiang, C., Goldberg, M. A., Akita, G. Y., Gregory, R. J., and Isner, J. M. (2000). Angiogenesis is induced in a rabbit model of hindlimb ischemia by naked DNA encoding an HIF-1alpha/VP16 hybrid transcription factor. *Circulation* **102,** 2255–2261.

Vorp, D. A., Lee, P. C., Wang, D. H., Makaroun, M. S., Nemoto, E. M., Ogawa, S., and Webster, M. W. (2001). Association of intraluminal thrombus in abdominal aortic aneurysm with local hypoxia and wall weakening. *J. Vasc. Surg.* **34,** 291–299.

Waypa, G. B., Chandel, N. S., and Schumacker, P. T. (2001). Model for hypoxic pulmonary vasoconstriction involving mitochondrial oxygen sensing. *Circ. Res.* **88,** 1259–1266.

Waypa, G. B., Guzy, R., Mungai, P. T., Mack, M. M., Marks, J. D., Roe, M. W., and Schumacker, P. T. (2006). Increases in mitochondrial reactive oxygen species trigger hypoxia-induced calcium responses in pulmonary artery smooth muscle cells. *Circ. Res.* **99,** 970–978.

Waypa, G. B., Marks, J. D., Mack, M. M., Boriboun, C., Mungai, P. T., and Schumacker, P. T. (2002). Mitochondrial reactive oxygen species trigger calcium increases during hypoxia in pulmonary arterial myocytes. *Circ. Res.* **91,** 719–726.

Waypa, G. B., and Schumacker, P. T. (2002). O(2) sensing in hypoxic pulmonary vasoconstriction: The mitochondrial door re-opens. *Respir. Physiol. Neurobiol.* **132,** 81–91.

Waypa, G. B., and Schumacker, P. T. (2005). Hypoxic pulmonary vasoconstriction: redox events in oxygen sensing. *J. Appl. Physiol. (Bethesda)* **98,** 404–414.

Waypa, G. B., and Schumacker, P. T. (2008). Oxygen sensing in hypoxic pulmonary vasoconstriction: Using new tools to answer an age-old question. *Exp. Physiol.* **93,** 133–138.

Weir, E. K., Hong, Z., Porter, V. A., and Reeve, H. L. (2002). Redox signaling in oxygen sensing by vessels. *Resp. Physiol. Neurobiol.* **132,** 121–130.

Weir, E. K., López-Barneo, J., Buckler, K. J., and Archer, S. L. (2005). Acute oxygen-sensing mechanisms. *N. Engl. J. Med.* **353,** 2042–2055.

Williams, S. E., Wootton, P., Mason, H. S., Bould, J., Iles, D. E., Riccardi, D., Peers, C., and Kemp, P. J. (2004). Hemoxygenase-2 is an oxygen sensor for a calcium-sensitive potassium channel. *Science* **306,** 2093–2097.

Wolin, M. S. (2000). Interactions of oxidants with vascular signaling systems. *Arterioscler. Thromb. Vasc. Biol.* **20,** 1430–1442.

Wolin, M. S., Ahmad, M., and Gupte, S. A. (2005). Oxidant and redox signaling in vascular oxygen sensing mechanisms: Basic concepts, current controversies, and potential importance of cytosolic NADPH. *Am. J. Physiol.* **289,** L159–L173.

Yu, A. Y., Shimoda, L. A., Iyer, N. V., Huso, D. L., Sun, X., McWilliams, R., Beaty, T., Sham, J. S., Wiener, C. M., Sylvester, J. T., and Semenza, G. L. (1999). Impaired physiological responses to chronic hypoxia in mice partially deficient for hypoxia-inducible factor 1alpha. *J. Clin. Invest.* **103,** 691–696.

CHAPTER FOURTEEN

Measuring Intratumoral Microvessel Density

Noel Weidner

Contents

Abstract

For a tumor to grow beyond a limited volume of $1-2 \text{ mm}^3$, the tumor cells must not only proliferate, but they must be able to induce the growth of new capillary blood vessels from the host. As early as 1971, it was proposed that tumor growth was dependent on angiogenesis; and, that tumor cells and blood vessels composed a highly integrated ecosystem, that endothelial cells could be switched from a resting state to one of rapid growth by a diffusible signal from tumor cells, and that anti-angiogenesis may become an effective anti-cancer therapy. Indeed, now there is considerable indirect and direct evidence to show that tumor growth is angiogenesis dependent, that tumor cells can produce diffusible angiogenic regulatory molecules, and that angiogenesis inhibitors can slow or prevent tumor growth, and that angiogenesis is a relevant target for anti-cancer therapy. Measuring intratumoral microvessel density (iMVD) in vascular "hot spots" has been shown to correlate with aggressive tumor behavior. This chapter reviews the techniques available for measuring iMVD.

1. Introduction

A multitude of published studies now show that intratumoral microvessel density (iMVD) correlates with aggressive tumor behavior, such as higher stage at presentation, greater incidence of metastases, and/or

Department of Pathology, University of California, San Diego, San Diego, California

Methods in Enzymology, Volume 444

ISSN 0076-6879, DOI: 10.1016/S0076-6879(08)02814-0

decreased patient survival. This technique was first widely applied to patients with breast carcinoma (Barbareschi et al., 1995; Bevilacqua et al., 1995; Bosari et al., 1992; Charpin et al., 1995; Fox et al., 1994, 1995; Gasparini et al., 1994a, 1995a,b,c; Horak et al., 1992; Inada et al., 1995; Lipponen et al., 1994; Obermair et al., 1994a,b, 1995; Ogawa et al., 1995; Scopinaro et al., 1994; Toi et al., 1993, 1994a,b, 1995; Visscher et al., 1993; Weidner et al., 1991, 1992), and a recent meta-analysis of 87 published studies on breast carcinoma patients concluded that high MVD significantly predicted poor survival in node-negative and node-positive patients (Uzzan et al., 2004). These studies involved breast carcinoma patients wherein iMVD, reflecting tumor angiogenesis, was related to relapse-free survival and overall survival. iMVD was assessed by immunohistochemistry, using antibodies against factor VIII (27 studies, $n + 5262$), CD31 (10 studies, $n + 2296$), or CD34 (8 studies, $n + 1726$). Of additional importance, the authors concluded that iMVD might be a better prognostic factor when assessed by CD31 or CD34 versus factor VIII ($p + 0.11$). Indeed, CD34 and CD31 are both more sensitive than factor VIII, although less specific, markers for endothelium. Between CD34 and CD31, CD34 appears more sensitive, although again less specific than CD31 for endothelium.

Subsequently, iMVD has been found to be a predictor of aggressive tumor behavior in those with prostate carcinoma (Brawer et al., 1994; Fregene et al., 1993; Hall et al., 1994; Vesalainen et al., 1994; Wakui et al., 1992; Weidner et al., 1993), head-and-neck squamous carcinoma (Alcalde et al., 1995; Albo et al., 1994; Gasparini et al., 1993; Mikami et al., 1991; Williams et al., 1994), non–small-cell lung carcinoma (Fontanini et al., 1995, 1996; Macchiarini et al., 1992, 1994; Yamazaki et al., 1994; Yuan et al., 1995), malignant melanoma (Barnhill and Levy, 1993; Depasquale et al., 2005; Fallowfield and Cook, 1991; Graham et al., 1994; Smolle et al., 1989; Srivastava et al., 1986, 1988; Vacca et al., 1993), gastrointestinal carcinoma (Maeda et al., 1995a,b; Saclarides et al., 1994; Takebayashi et al., 1996), testicular germ-cell malignancies (Olivarez et al., 1994), multiple myeloma (Vacca et al., 1994), central nervous system tumors (Leon et al., 1996; Li et al., 1994), ovarian carcinoma (Hollingsworth et al., 1995; Palmer et al., 2007; Volm et al., 1996), cervical squamous carcinoma (Bremer et al., 1996; Schlenger et al., 1995; Wiggins et al., 1995), endometrial carcinoma (Abulafia et al., 1995; Ozalp et al., 2003; Salvesen et al., 2003), transitional-cell carcinoma of the bladder (Bochner et al., 1995; Dickinson et al., 1994; Jaeger et al., 1995; Kohno et al., 1993), renal carcinoma (Yoshino et al., 1995), and hepaticellular, gallbladder, and hepatobiliary carcinomas (Chen et al., 2004; Nanashima et al., 2004; Shirabe et al., 2004; Tian et al., 2003).

In these studies iMVD not only correlated with aggressive tumor behavior, but it was also often an independent prognosticator for aggressive behavior when compared to other traditional prognostic markers by multivariate analysis. Yet, iMVD may be more than a prognostic factor, but also a predictive factor—that is, a test that predicts response to specific therapies. For example, Gasparini and coworkers (1995b) found that iMVD was an independent predictive indicator of response in node-positive breast cancer patients treated with either adjuvant chemotherapy or adjuvant hormone therapy, and that assessment of tumor angiogenesis may be useful in selection of those patients who are more likely to benefit from conventional adjuvant therapies (i.e., those with minimally vascularized tumors) from those with highly vascularized and more aggressive tumors, for whom novel forms of systemic therapy are advocated.

Moreover, in many of the previously cited studies, iMVD was linked to other important tumor and vessel growth factors, such as vascular endothelial cell growth factor (VEGF) (Stefanou et al., 2004; Tsuji et al., 2002). Anti-VEGF immunotherapies have now achieved Food and Drug Administration (FDA) approval for the treatment of selected malignancies, such as colon carcinoma. Also, iMVD has been correlated with other biological modulators, such as p53, maspin, heparanase, and p-glycoprotein (Shinyo et al., 2003; Song et al., 2002; Tian et al., 2003).

It is important that assessment of microvessel counting is not restricted to tumor biology. For example, microvessel counting has been shown to be valuable in the study of vascular malformative diseases (Starnes et al., 2002), atherosclerosis (Mofidi et al., 2001), inflammatory bowel disease (Hanabata et al., 2005), and even in infectious disease, such as syphilis (Macaron et al., 2003).

Clearly, during the last two decades microvessel counting has become mainstay research technique in vascular biology and pathophysiology. Mastering the technique(s) has become important. At present the histological assessment of tumor vascularity is mainly used in the research setting, but it may also have applications in the clinic, if appropriate methodology and trained observers perform the studies.

2. MEASURING INTRATUMORAL MICROVESSEL DENSITY

In 1990, my colleagues and I asked if the extent of angiogenesis, as measured by iMVD in human breast carcinoma, correlated with metastasis (Weidner et al., 1991). If so, such information might prove valuable in selecting subsets of breast carcinoma patients for aggressive adjuvant therapies. To be true, it is important that a spectrum of intratumoral microvessel densities exist within the spectrum of invasive breast carcinomas. When the microvessel counts in a number of invasive breast carcinomas are sorted in ascending order on a log scale, the spectrum of low to high microvessel densities becomes apparent. The densities are an evenly distributed continuum, extending from about 10 to 200 microvessels per 0.74 mm^2 (200×) microscopic field.

My colleagues and I examined primary tumor specimens from randomly selected patients with invasive breast carcinoma (Weidner et al., 1991). Hematoxylin and eosin (H&E)–stained sections of the breast tumor were used to choose one paraffin-embedded tissue block clearly representative of a generous cross-section of the invasive carcinoma, and to highlight the endothelial cells lining the blood vessels, one 5-micron–thick section from this block was immunostained for factor VIII-related antigen/von Willebrand's factor (F8RA/vWF). iMVD was assessed by light microscopic analysis for areas of the tumor that contained the most capillaries and small venules (microvessels)—that is, microvessel "hot spots." Only tumors that produced a high-quality and distinct microvessel immunoperoxidase staining pattern with low background staining were included in this or subsequent studies. This is very important, because the quality of immuno-peroxidase staining can vary considerably between laboratories, and before measuring iMVD, high-quality immunoperoxidase staining must be consistently achieved.

Finding these neovascular "hot spots" is critical to accurately assess a particular tumor's angiogenic potential. This is to be expected since there is considerable evidence that, like tumor proliferation rate, tumor angiogene-sis is heterogeneous within tumors (Folkman et al., 1985, 1994; Kandel et al., 1991; Weidner et al., 1991). The technique for identifying neovas-cular "hot spots" is very similar to that for finding mitotic "hot spots" for assessing mitotic figure content and is subject to the same kinds of inter- and intra-observer variability. In our study, sclerotic hypocellular areas within tumors (i.e., usually found toward the tumor center) and immediately adjacent to benign breast tissue were not considered in iMVD determina-tions. In fact, there is a positive correlation between neovascular "hot spots" and tumor-cell mitotic activity. This correlation is underscored by Begum and coworkers (2003), who studied the distributions of intratumoral micro-vessels and tumor-cell proliferation (i.e., defined by Ki-67 immunostain-ing). They found that the total number of microvessels correlated closely with areas also having highest tumor-cell proliferative activity. iMVD increased toward the invasive periphery of the tumor, which also had higher tumor-cell proliferative indices.

Areas of highest neovascularization were found by scanning the tumor sections at low power (40× and 100× total magnification) and selecting those areas of invasive carcinoma with the greatest density of distinct F8RA/vWF-staining microvessels. These highly neovascular areas could occur anywhere within the tumor, but most frequently appeared at the tumor margins or invasive periphery. After the area of highest neovascularization was identified, individual microvessel counts were made on a 200× field (20× objective and 10× ocular, Olympus BH-2 microscope, 0.74 mm^2 per field with the field size measured with an ocular micrometer). Any highlighted endothelial cell or endothelial-cell cluster clearly separate from adjacent microvessels, tumor cells, and other connective-tissue elements was considered a single, countable microvessel. Even those distinct clusters of brown-staining endothelial cells, which might be from the same microvessel "snaking" its way in and out of the section, were considered distinct and countable as separate microvessels. Vessel lumens, although usually present, were not necessary for a structure to be defined as a microvessel, and red cells were not used to define a vessel lumen. Results were expressed as the highest number of microvessels in any single 200× field. An average of multiple fields was not performed.

Invasive breast carcinomas from patients with metastases (either lymph nodal or distant site) had a mean microvessel count of 101 per 200× field. For those carcinomas from patients without metastases the corresponding value was 45 per 200× field ($p + 0.003$). We plotted the percent of patients with metastatic disease in whom a vessel count was carried out within progressive 33 vessel increments (Weidner *et al.*, 1991). The plot showed that the incidence of metastatic disease increased with the number of microvessels, reaching 100% for patients having invasive carcinomas with more than 100 microvessels per 200× field.

To further define the relationship of iMVD to overall and relapse-free survival and to other reported prognostic indicators in breast carcinoma, a blinded study of 165 consecutive carcinoma patients was performed using identical techniques to measure intratumoral microvessel density (Weidner *et al.*, 1992). The other prognostic indicators evaluated were metastasis to axillary lymph nodes, patient age, menopausal status, tumor size, histologic grade (i.e., modified Bloom-Richardson criteria), peritumoral lymphatic-vascular invasion (LVI), flow DNA ploidy analysis, flow S-phase fraction, growth fraction by Ki-67 binding, c-erbB2 oncoprotein expression, pro-cathepsin-D content, estrogen-receptor content, progesterone-receptor content, and EGFR expression. We found a highly significant association of iMVD with overall survival and relapse-free survival in all patients, including node-negative and node-positive subsets. All patients with breast carcinomas having more than 100 microvessels per 200× field experienced tumor recurrence within 33 months of diagnosis, compared to less than 5% of patients who had fewer than 33 microvessels per 200× field. Moreover, iMVD was the only significant predictor of overall and relapse-free survival among node-negative women.

When highlighting microvessels, it is important that previously published protocols for measuring iMVD be followed carefully, and considerable experience is needed at the senior staff pathologist level for assessing iMVD. It is necessary not only for supervising the immunostaining of endothelial cells, but also for selecting generous sections of representative invasive tumor and for localizing the neovascular "hot spot." Counting microvessels has been shown to be reproducible (Weidner, 1992), especially following a period of training. Brawer *et al.* (1994) compared manual iMVD determinations with those determined by automated counting (i.e., Optimas Image Analysis) and found a very high correlation ($r^2 + 0.98$, $p<0.001$). Finally, accurate staging and adequate patient follow-up are needed to identify patients who have metastases or will experience recurrent tumor, and proper technique must be supplemented with unbiased case selection and proper statistical analysis of data. These reasons may explain why some investigators have not found this association between iMVD and prognosis in some solid tumors, including breast and others (Axelsson *et al.*, 1995; Barnhill *et al.*, 1994; Carnochan *et al.*, 1991; Costello *et al.*, 1995; Dray *et al.*, 1995; Goulding *et al.*, 1995; Hall *et al.*, 1992; Kainz *et al.*, 1995; Leedy *et al.*, 1994; MacLennan and Bostwick, 1995; Miliaras *et al.*, 1995; Rutger *et al.*, 1995; Siitonen *et al.*, 1995; Tahan and Stein, 1995; van Diest *et al.*, 1995; Van Hoef *et al.*, 1993). Why these reports are contradictory to other reports is not clear; however, Leedy *et al.* (1994) noted that the tongue was already a highly vascular organ. They implied that tumor growth and spread may be facilitated by pre-existing vessels in highly vascular organs, and it may be that iMVD will prove not as useful in predicting outcome in patients with tumors developing in such highly vascular organs such the tongue, liver, skin, kidney, or gastrointestinal tract. Clearly, it should not prove useful in solid tumor systems, wherein there is no spectrum from low to high iMVD (MacLennan and Bostwick, 1995). Paradoxically, Kainz *et al.* (1995) found that patients with cervical cancers showing relatively low iMVD had a significantly poorer recurrence-free survival than those with higher densities. Tumor progression is determined by the net balance between tumoral and patient-dependent processes, and both are highly complex biological processes that remain incompletely understood.

3. HIGHLIGHTING MICROVESSELS FOR MICROVESSEL COUNTING

Although no endothelial marker is perfect, when applied properly, anti-F8RA/vWF remains the most specific endothelial marker, providing very good contrast between microvessels and other tissue components. Unfortunately, anti-F8RA/vWF may not highlight all intratumoral microvessels. Although apparently more sensitive, CD31 strongly cross-reacts with plasma cells (DeYoung *et al.*, 1993; Longacre and Rouse, 1994). This complication can markedly obscure the microvessels in those tumors with a prominent plasma cellular inflammatory background. CD34 is an acceptable alternative and the most reproducible endothelial-cell highlighter in many laboratories, but CD34 will highlight perivascular stromal cells and has been noted to stain a wide variety of stromal neoplasms (Traweek *et al.*, 1991; van de Rijn and

Rouse, 1994). Like antibodies to F8RA/vWF, antibodies to CD31 and CD34 also do not immunostain all intratumoral microvessels (Schlingemann *et al.*, 1991). *Ulex europeus* lectin will stain many tumor cells, seriously decreasing its specificity; it is not recommended.

Angiogenesis is a complex multistep process that involves extracellular matrix remodeling, migration and proliferation of endothelial cells, and morphogenesis of new microvessels. Wang *et al.* (1993, 1994) developed a monoclonal antibody (Mab E9), which was raised against proliferating or "activated" endothelial cells of human umbilical-vein origin and grown in tissue culture. Mab E9 strongly reacted with endothelial cells of all tumors, fetal organs, and in regenerating and/or inflamed tissues, but it only rarely and weakly immunostained endothelial cells of normal tissues. Unfortunately, Mab E9 immunoreacted only in frozen tissue sections, although they did not mention if microwave antigen-retrieval techniques applied to formalin-fixed, paraffin-embedded tissues were tried. Antibodies like Mab E9 may provide the most sensitive staining of intratumoral microvessels and preferentially immunostain "activated or proliferating" endothelial cells such that the overall staining intensity may correlate best with the intensity of tumor angiogenesis and, hence, tumor aggressiveness.

Endoglin is a proliferation-associated protein abundantly expressed in angiogenic endothelial cells. Indeed, the endothelial marker CD105 (endoglin) has been touted as a useful marker for activated endothelial cells—that is, those new vessels induced by the dynamics of the growing tumor (i.e., neovascularization). CD105 and its ligand transforming growth factor beta (TGFβ) are modulators of angiogenesis, which drives tumor growth and metastasis. Recent studies have revealed that CD105 is intensively expressed in tumor vasculature, and iMVD determined with the use of antibodies to CD105 have been found to be an important prognostic indicator for the outcome in a number of malignancies (Akagi *et al.*, 2002; Ding *et al.*, 2006; Ho *et al.*, 2005; Kyzas *et al.*, 2006; Li *et al.*, 2003; Romani *et al.*, 2006). Thus, it might be wise in developing an iMVD highlighting protocol to include traditional (CD34) as well as novel activated endothelial-cell markers (CD105) in estimating tumor angiogenesis. This is similar to what Ding *et al.* (2006) concluded after finding that in gastric carcinoma patients, iMVD determined by CD34 inversely correlated with overall survival, but it did not correlate with other clinicopathologic parameters except formation of ascites. CD34 was universally expressed in blood vessels within benign and malignant tissues, whereas CD105 expression was minimal in benign tissues but stronger within gastric carcinoma. Their data suggested that both CD105 and CD34 could be used for quantification of angiogenesis, but preference might be given to CD105 in the evaluation of prognosis in gastric carcinoma. This relationship needs to be clarified in other tumor types.

4. HISTOLOGICAL QUANTITATION OF TUMOR ANGIOGENESIS

Tumors establish their blood supply by a number of processes including vasculogenesis, vascular remodeling, intussusception, and possibly vascular mimicry in certain tumors (Fox and Harris, 2004). The mainstay of the assessing tumor vascularity has been counting the number of immunohistochemically identified microvessels in vascular "hot spots." This technique has fueled many advances in vascular and tumor biology, but "hot spot" microvessel counting is not the only or necessarily the preferred technique in all research or clinical settings. Methodological issues persist and are underscored by the recent meta-analysis of the application of tumor angiogenesis for prognostication in patients with lung carcinoma. Meert *et al.* (2002) performed a systematic review of the literature to assess the prognostic value on survival of microvessel count in patients with lung cancer. They found that a high microvessel count in the primitive lung tumor was a statistically significant poor prognostic factor for survival in non–small-cell lung cancer, whatever it was assessed by factor VIII or CD31, but not necessarily when using other markers. The authors believed that variations in study conclusions could be explained by patient selection criteria and/or by the heterogeneous methodologies used to stain and count microvessels, use of different antibody clones, variable identification of "hot spots," Weidner or Chalkley counting method, and/or cut-off selection. They believed standardization of angiogenesis assessment by the microvessel count is necessary.

Certainly, other techniques are available, including Chalkley counting and the use of image analysis systems. The Chalkley counting system uses a grid, which is projected onto the vascular "hot spot," and it has been found superior in some, but not all, studies (Hansen *et al.*, 2004; Offersen *et al.*, 2003). Offersen *et al.* (2003) presented paired Chalkley and iMVD estimates in carcinomas of the prostate, breast, bladder, and lung. In prostate carcinomas, high iMVD indicated poor prognosis, whereas high Chalkley counts in breast carcinoma were associated with a poor prognosis. In bladder carcinoma, high estimates using both methods showed good prognosis and were associated with a high degree of inflammation. Neither of the counts revealed prognostic value in lung carcinomas, where the vascular pattern indicated that this cancer was nonangiogenic. Therefore, when investigating a new application for microvessel counts, one might consider evaluating both these techniques before settling on the optimal technique.

Angiogenic activity might be measurable in histologic samples by highlighting of the molecules involved in the establishment of the tumor vasculature, including angiogenic growth factors or inhibitors and their receptors, cell adhesion molecules, proteases, and (as noted above) markers of activated, proliferating, cytokine-stimulated or angiogenic vessels. These approaches deserve serious consideration; but, as a cautionary note, it is important that measuring any given angiogenic factor (i.e., bFGF or VPF/VEGF) in a tumor may not by itself correlate with iMVD (or with patient outcome), because the iMVD is the sum total of positive and negative angiogenic factors, factors that are likely numerous and complex in their interplay.

Measuring the maturity of vessels may give an indication of the proportion of the tumor vasculature that is functional or indicative of tumor aggressiveness. In one very interesting study, Yonenaga et al. (2005) found that the tumors with mature pericyte-containing (actin-positive) vessels were associated with a good outcome in colorectal cancer patients. Those tumors with immature neovascularization (actin-negative) were poorly differentiated tumors that produced metastases, resulting in a poorer prognosis. Other reagents that can identify hypoxia-activated pathways are also being developed. Gene arrays may be able to provide an angiogenesis profile. Continued study into the processes involved in generating a tumor blood supply is likely to identify new markers that may be more accurate measures.

Automated ("machine") immunostaining and application of computer-aided image analysis may help to standardize microvessel counts and help eliminate interobserver and even intraobserver variables, such as inexperience and "hot spot" selection biases (Barbareschi et al., 1995). The latter approach may make determination of intratumoral microvessel density a simple, reliable, and reproducible prognostic factor in a variety of solid tumors, not just in breast carcinoma (Choi et al., 2005; Cruz et al., 2001). With the added power of the computer, additional vascular parameters can be measured and correlated with outcomes, such vascular area or vascular configurations or lumen size (Irion et al., 2003; Simpson et al., 1996).

Whether one uses a computer or light microscopy to select the area to determine iMVD, location may be critical in establishing correlations with various outcomes. For example, Oh-e et al. (2001) found that measuring angiogenesis at the site of deepest penetration best predicted lymph node metastases in colorectal cancer patients. In addition, the types of vessels one chooses to enumerate may be critical; for example, Saad et al. (2006) found that counting lymphatic vessels highlighted by the monoclonal antibody D2-40 correlated with prognosis in colorectal patients. The authors concluded that D2-40 lymphatic microvessel density showed prognostic significance with positive correlation with lymphovascular invasion, tumor size, and metastases to lymph nodes and liver. Moreover, immunostaining with D2-40 enhanced the detection of lymphatic invasion relative to H&E staining and the endothelial marker, CD31.

A very novel approach to measuring angiogenesis is that reported by Baeten et al. (2002). These investigators developed a rapid, objective, and quantitative method using flow cytometry on frozen tissues. Frozen tissue sections of archival tumor material were enzymatically digested. The single-cell suspension was stained for CD31 and CD34 for flow cytometry. The number of endothelial cells was quantified using light-scatter and fluorescence characteristics. Tumor endothelial cells were detectable in a single-cell suspension, and the percentage of endothelial cells detected in 32 colon carcinomas were highly correlated ($r + 0.84$, $p < 0.001$) with the immunohistochemical assessment of microvessel density. Flow cytometric endothelial cells quantification was found to be more sensitive, especially at lower levels of immunohistochemical microvessel density measurement. The flow method was found to be applicable for various tumor types and had the major advantage that it provided a retrospective and quantitative approach to the angiogenic potential of tumors. This deserves further development and application.

5. NOVEL *IN VIVO* METHODS FOR MEASURING MICROVESSEL DENSITY

Measuring iMVD in excised and fixed tumor tissue may prove to be a relatively crude method for estimating tumor angiogenic capacity. Other methods may prove more reliable and reproducible, such as measuring levels of angiogenic molecules in serum or urine, or directly measuring angiogenic molecules or inhibitors from tumor extracts (i.e., in a manner similar to hormone receptor assays). Indeed, using an immunoassay, Watanabe *et al.* (1992) and Nguyen *et al.* (1994) reported elevated levels of bFGF in the serum and urine of patients with a wide variety of solid tumors, including breast carcinoma. Higher levels were found in patients with metastatic disease versus those of localized disease. Moreover, Li *et al.* (1994) measured bFGF in the cerebral spinal fluids of children with various brain tumors and correlated increasing fluid levels with greater intratumoral microvessel density and increased likelihood of recurrence. Finally, using cardiovascular magnetic resonance imaging (cMRI) enhanced with special contrast agents will allow assessment of angiogenesis in vivo, a technique that will allow monitoring the effects of therapy (Esserman *et al.*, 1999).

The optimum technique for assaying intratumoral vascularity will likely vary with anticipated clinical application. For example, some studies have found that high iMVD can be significantly associated with a favorable outcome, when a specific form of therapy is given and the therapeutic effectiveness is directly dependent on the extent of blood flow (Kohno *et al.*, 1993; Zatterstrom *et al.*, 1995). Kohno *et al.* (1993) found that cervical carcinomas treated with hypertensive intra-arterial chemotherapy are more responsive when highly vascular, and Zatterstrom *et al.* (1995) found that highly vascular squamous carcinomas of the head and neck are more responsive to radiation therapy when highly vascular.

It is likely that measuring tumor angiogenesis in living patients may prove very valuable in predicting response to antiangiogenic or other therapies and also provide an objective assessment of post-therapeutic response. Such approaches are under active investigation and many studies have been published. The techniques include a variety of elegant radiological techniques, such as flash-echo imaging (Okanobu *et al.*, 2002), color-power Doppler sonographic imaging (Yang *et al.*, 2002), single-level dynamic spiral CT imaging (Chen *et al.*, 2004), contrast-enhanced dynamic MR imaging (Hylton, 2006; Yabuuchi *et al.*, 2003), and laser Doppler fluxmetry (Jacob *et al.*, 2006).

Clearly, measuring angiogenesis is a rapidly progressing area and in constant evolution. The future lies with automation, computer image capture and analysis, and in vivo assay techniques. Gene microarrays for angiogenic stimulators and inhibitors will also find application as molecular biologic techniques evolve.

REFERENCES

Abulafia, O., Triest, W. E., Sherer, D. M., Hansen, C., and Ghezzi, F. (1995). Angiogenesis in endometrial hyperplasia and stage I endometrial carcinoma. *Obstet. Gynecol.* **86,** 479–485.

Akagi, K., Ikeda, Y., Sumiyoshi, Y., Kimura, Y., Kinoshita, J., Miyazaki, M., and Abe, T. (2002). Estimation of angiogenesis with anti-CD105 immunostaining in the process of colorectal cancer development. *Surgery* **131,** S109–S113.

Albo, D., Granick, M. S., Jhala, N., Atkinson, B., and Solomon, M. P. (1994). The relationship of angiogenesis to biological activity in human squamous cell carcinomas of the head and neck. *Ann. Plast. Surg.* **32,** 588–594.

Alcalde, R. E., Shintani, S., Yoshihama, Y., and Matsumura, T. (1995). Cell proliferation and tumor angiogenesis in oral squamous cell carcinomas. *Anticancer Res.* **15,** 1417–1422.

Axelsson, K., Ljung, B. M. E., Moore, D. H., Thor, A. D., Chew, K. L., Edgerton, S. M., Smith, H. S., and Mayall, B. H. (1995). Tumor angiogenesis as a prognostic assay for invasive ductal carcinoma. *J. Natl. Cancer Inst.* **87,** 997–1008.

Baeten, C. I., Wagstaff, J., Verhoeven, I. C., Hillen, H. F., and Griffioen, A. W. (2002). Flow cytometric quantification of tumour endothelial cells., an objective alternative for microvessel density assessment. *Br. J. Cancer* **87,** 344–347.

Barbareschi, M., Gasparini, G., Weidner, N., Morelli, L., Forti, S., Eccher, C., Fina, P., Leonardi, E., Mauri, F., Bevilacqua, P., and Dalla Palma, P. (1995). Microvessel density quantification in breast carcinomas: Assessment by manual vs. a computer-assisted image analysis system. *Appl. Immunohistochem.* **3,** 75–84.

Barnhill, R. L., Busam, K. J., Berwick, M., Blessing, K., Cochran, A. J., Elder, D. E., Fandrey, K., Daraoli, T., and White, W. L. (1994). Tumor vascularity is not a prognostic factor for cutaneous melanoma. *Lancet* **344,** 1237–1238 (letter).

Barnhill, R. L., and Levy, M. A. (1993). Regressing thin cutaneous malignant melanomas (<1.0 mm) are associated with angiogenesis. *Am. J. Pathol.* **143,** 99–104.

Begum, R., Douglas-Jones, A. G., and Morgan, J. M. (2003). Radial intratumoral increase and correlation of microvessels and proliferation in solid breast carcinoma. *Histopathology* **43,** 244–253.

Bevilacqua, P., Barbareschi, M., Verderio, P., Boracchi, P., Caffo, O., Dalla Palma, P., Meli, S., Weidner, N., and Gasparini, G. (1995). Prognostic value of intratumoral microvessel density, a measure of tumor angiogenesis, in node-negative breast carcinoma—results of a multiparametric study. *Breast Cancer Res. Treat.* **36,** 205–217.

Bochner, B. H., Cote, R. J., Weidner, N., Groshen, S., Chen, S.-C., Skinner, D. G., and Nichols, P. W. (1995). Angiogenesis in bladder cancer: Relationship between microvessel density and tumor prognosis. *J. Natl. Cancer Inst.* **87,** 1603–1612.

Bosari, S., Lee, A. K. C., DeLellis, R. A., Wiley, B. D., Heatley, G. J., and Silverman, M. L. (1992). Microvessel quantitation and prognosis in invasive breast carcinoma. *Hum. Pathol.* **23,** 755–761.

Brawer, M. K., Deering, R. E., Brown, M., Preston, S. D., and Bigler, S. A. (1994). Predictors of pathologic stage in prostate carcinoma. *Cancer* **73,** 678–687.

Bremer, G. L., Tiebosch, A. T., van der Putten, H. W., Schouten, H. J., de Haans, J., and Arends, J. W. (1996). Tumor angiogenesis: An independent prognostic parameter in cervical cancer. *Am. J. Obstet. Gynecol.* **174,** 126–131.

Carnochan, P., Briggs, J. C., Westbury, G., and Davies, A. J. (1991). The vascularity of cutaneous melanoma: A quantitative histologic study of lesions 0.85–1.25 mm in thickness., B. *Br. J. Cancer* **64,** 102–107.

Charpin, C., Devictor, B., Bergeret, D., Andrac, L., Boulat, J., Horschowski, N., Lavaut, M. N., and Piana, L. (1995). CD 31 quantitative immunocytochemical assays in breast carcinomas. Correlation with current prognostic factors. *Am. J. Clin. Pathol.* **103,** 443–448.

Chen, W. X., Min, P. Q., Song, B., Xiao, B. L., Liu, Y., and Ge, Y. H. (2004). Single-level dynamic spiral CT of hepatocellular carcinoma: Correlation between imaging features and density of tumor microvessels. *World J. Gastroenterol.* **10,** 67–72.

Choi, H. J., Choi, I. H., Cho, N. H., and Choi, H. K. (2005). Color image analysis for quantifying renal tumor angiogenesis. *Anal. Quant. Cytol. Histol.* **27,** 43–51.

Costello, P., McCann, A., Carney, D. N., and Dervan, P. A. (1995). Prognostic significance of microvessel density in lymph node negative breast carcinoma. *Hum. Pathol.* **26,** 1181–1184.

Cruz, D., Valenti, C., Dias, A., Seixas, M., and Schmitt, F. (2001). Microvessel density counting in breast cancer. Slider vs-digital images. *Anal. Quant. Cytol. Histol.* **23,** 15–20.

Depasquale, I., and Thompson, W. D. (2005). Microvessel density for melanoma prognosis. *Histopathology* **47,** 186–194.

DeYoung, B. R., Wick, M. R., Fitzgibbon, J. F., Sirgi, K. E., and Swanson, P. E. (1993). CD31: An immunospecific marker for endothelial differentiation in human neoplasms. *Appl. Immunohistochem.* **1,** 97–100.

Dickinson, A. J., Fox, S. B., Persad, R. A., Hollyer, J., Sibley, G. N., and Harris, A. L. (1994). Quantification of angiogenesis as an independent predictor of prognosis in invasive bladder carcinomas., B. *J. Urol., B. r. J. Urol.* **74,** 762–766.

Ding, S., Li, C., Lin, S., Yang, Y., Liu, D., Han, Y., Zhang, Y., Li, L., Zhou, L., and Kumar, S. (2006). Comparative evaluation of microvessel density determined by CD34 or CD105 in benign and malignant gastric lesions. *Hum. Pathol.* **37,** 861–866.

Dray, T. G., Hardin, N. J., and Sofferman, R. A. (1995). Angiogenesis as a prognostic marker in early head and neck cancer. *Ann. Otol. Rhinol. Laryngol.* **104,** 724–729.

Esserman, L., Hylton, N., George, T., Yassa, L., and Weidner, N. (1999). Contrast-enhanced magnetic resonance imaging (cMRI) provides a window to visualize anatomic extent and tumor angiogenesis in breast carcinoma. *Breast J.* **5,** 13–21.

Fallowfield, M. E., and Cook, M. G. (1991). The vascularity of primary cutaneous melanoma. *J. Pathol.* **164,** 241–244.

Folkman, J. (1994). Angiogenesis and breast cancer. *J. Clin. Oncol.* **12,** 441–443.

Folkman, J. (1985). Angiogenesis and its inhibitors. "Important Advances in Oncology."
(V. T. DeVita, S. Hellman, and S. A. Rosenberg, Eds.), pp. 42–62. J.B. Lippincott,
Philadelphia.

Fontanini, G., Bigini, D., Vignati, S., Basolo, F., Mussi, A., Lucchi, M., Chine, S.,
Angeletti, C. A., and Bevilacqua, G. (1995). Recurrence and death in non small cell
lung carcinomas: A prognostic model using pathological parameters, microvessel count,
and gene protein products. *J. Pathol.* **177,** 57–63.

Fontanini, G., Vignati, S., Bigini, D., Mussi, A., Lucchi, M., Chine, S., Angeletti, C. A., and
Bevilacqua, G. (1996). Recurrence and death in non–small-cell lung carcinomas:
A prognostic model using pathological parameters, microvessel count, and gene protein
products. *Clin. Cancer Res.* **2,** 1067–1076.

Fox, S. B., and Harris, A. L. (2004). Histological quantitation of tumour angiogenesis.
APMIS **112,** 413–430.

Fox, S. B., Leek, R. D., Smith, K., Hollyer, J., Greenall, M., and Harris, A. L. (1994).
Tumor angiogenesis in node-negative breast carcinomas: Relationship with epidermal
growth factor receptor, estrogen receptor, and survival. *Breast Res. Treat.* **29,** 109–116.

Fox, S. B., Turner, G. D. H., Leek, R. D., Whitehouse, R. M., Gatter, K. C., and Harris, A. L.
(1995). The prognostic value of quantitative angiogenesis in breast cancer and role of
adhesion molecule expression in tumor endothelium. *Breast Res. Treat.* **36,** 219–226.

Fregene, T. A., Khanuja, P. S., Noto, A. C., Gehani, S. K., Van Egmont, E. M., Luz, D. A.,
and Pienta, K. J. (1993). Tumor-associated angiogenesis in prostate cancer. *Anticancer Res.*
13, 2377–2381.

Gasparini, G., Weidner, N., Bevilacqua, P., Maluta, S., Boracchi, P., Testolin, A. A.,
Pozza, F., and Folkman, J. (1993). Intratumoral microvessel density and p53 protein:
Correlation with metastasis in head-and-neck squamous-cell carcinoma. *Int. J. Cancer.*
55, 739–744.

Gasparini, G., Bevilacqua, P., Boracchi, P., Maluta, S., Pozza, F., Barbareschi, M., Dalla
Palma, P., Mezzetti, M., and Harris, A. L. (1994a). Prognostic value of p53 expression in
early-stage breast carcinoma compared with tumour angiogenesis, epidermal growth
factor receptor, c-erbB2, cathepsin D, DNA ploidy, parameters of cell kinetics and
conventional features. *Int. J. Oncol.* **4,** 155–162.

Gasparini, G., Weidner, N., Bevilacqua, P., Maluta, S., Dalla Palma, P., Caffo, O.,
Barbareschi, M., Boracchi, P., Marubini, E., and Pozza, F. (1994b). Tumor microvessel
density, p53 expression, tumor size, and peritumoral lymphatic vessel invasion are
relevant prognostic markers in node-negative breast carcinoma. *J. Clin. Oncol.* **12,**
454–466.

Gasparini, G., Barbareschi, M., Boracchi, P., Bevilacqua, P., Verderio, P., Dalla Palma, P.,
and Menard, S. (1995a). 67-kDa laminin-receptor expression adds prognostic information to intra-tumoral microvessel density in node-negative breast cancer. *Int. J. Cancer*
60, 7604–7610.

Gasparini, G., Barbareschi, M., Boracchi, P., Verderio, P., Caffo, O., Meli, S., Palma, P. D.,
Marubini, E., and Bevilacqua, P. (1995b). Tumor angiogenesis may predict clinical
outcome of node-positive breast cancer patients treated either with adjuvant hormone
therapy or chemotherapy. *Cancer J. Sci. Am.* **1,** 131–141.

Gasparini, G., Barbareschi, M., Boracchi, P., Verderio, P., Caffo, O., Meli, S., Palma, P. D.,
Marubini, E., and Bevilacqua, P. (1995c). Tumor angiogenesis predicts clinical outcome
of node-positive breast cancer patients treated with adjuvant hormone therapy or chemotherapy. *Cancer J. Sci. Am.* **1,** 131–141.

Goulding, H., Rashid, N. F. N. A., Robertson, J. F., Bell, J. A., Elston, C. A., Elston, C. W., Blamey, R. W., and Ellis, I. O. (1995). Assessment of angiogenesis in breast carcinoma: An important factor in prognosis? *Hum. Pathol.* **26,** 1196–1200.

Graham, C. H., Rivers, J., Kerbel, R. S., Stankiewicz, K. S., and White, W. L. (1994). Extent of vascularization as a prgnostic indicator in thin (<0.76 mm) malignant melanomas. *Am. J. Pathol.* **145,** 510–514.

Guidi, A. J., Fischer, L., Harris, J. R., and Schnitt, S. J. (1994). Microvessel density and distribution in ductal carcinoma *in situ* of the breast. *J. Natl. Cancer Inst.* **86,** 614–619.

Hall, M. C., Troncoso, P., Pollack, A., Zhau, H. Y., Zagars, G. K., Chung, L. W., and von Eschenbach, A. A. (1994). Significance of tumor angiogenesis in clinically localized prostate carcinoma treated with external beam radiotherapy. *Urology* **44,** 869–875.

Hall, N. R., Fish, D. E., Hunt, N., Goldin, R. D., Gillou, P. J., and Monson, J. R. T. (1992). Is the relationship between angiogenesis and metastasis in breast cancer real? *Surg. Oncol.* **1,** 223–229.

Hanabata, N., Sasaki, Y., Tanaka, M., Tsuji, T., Hatada, Y., Hada, R., and Munakata, A. (2005). Vascular endothelial growth factor expression and microvessel parameters of colonic mucosa correlate with sensitivity to steroid in patients with ulcerative colitis. *Scand. J. Gastroenterol.* **40,** 188–193.

Hansen, S., Sorensen, F. B., Vach, W., Grabau, D. A., Bak, M., and Rose, C. (2004). Microvessel density compared with the Chalkley count in a prognostic study of angiogenesis in breast cancer patients. *Histopathology* **44,** 428–436.

Ho, J. W., Poon, R. T., Sun, C. K., Xue, W. C., and Fan, S. T. (2005). Clinicopathological and prognostic implications of endoglin (CD105) expression in hepatocellular carcinoma and its adjacent non-tumorous liver. *World J. Gastroenterol.* **11,** 176–181.

Hollingsworth, H. C., Kohn, E. C., Steinberg, S. M., Rothenberg, M. L., and Merino, M. J. (1995). Tumor angiogenesis in advanced stage ovarian carcinoma. *Am. J. Pathol.* **147,** 33–41.

Horak, E., Leek, R., Klenk, N., LeJeune, S., Smith, K., Stuart, N., Greenall, M., Stepniewska, K., and Harris, A. L. (1992). Angiogenesis, assessed by platelet/endothelial cell adhesion molecule antibodies, as indicator of node metastases and survival in breast cancer. *Lancet* **340,** 1120–1124.

Hylton, N. (2006). Dynamic contrast-enhanced magnetic resonance imaging as an imaging biomarker. *J. Clin. Oncol.* **24,** 3293–3298.

Inada, K., Toi, M., Hoshina, S., Hayashi, K., and Tominaga, T. (1995). Significance of tumor angiogenesis as an independent prognostic factor in axillary node-negative breast cancer. *Gan To Kagaku Ryoho.* **22,** 59–65.

Irion, L. C., Prolla, J. C., Hartmann, A., and da Silva, V. D. (2003). Morphometric intratumoral microvessel area evaluation could be a useful indicator for coadjuvant therapy in ressected NSCLC. *Rev. Port. Pneumol.* **9,** 19–32.

Jacob, A., Varghese, B. E., and Birchall, M. B. (2006). Validation of laser Doppler fluxmetry as a method of assessing neo-angiogenesis in laryngeal tumours. *Eur. Arch. Otorhinolaryngol.* **263,** 444–448.

Jaeger, T. M., Weidner, N., Chew, K., Moore, D. H., Kerschmann, R. L., Waldman, F. M., and Carroll, P. R. (1995). Tumor angiogenesis correlates with lymph node metastases in invasive bladder cancer. *J. Urol.* **154,** 59–71.

Kainz, C., Speiser, P., Wanner, C., Obermair, A., Tempfer, C., Sliutz, G., Reinthaller, A., and Breitenecker, G. (1995). Prognostic value of tumor microvessel density in cancer of the uterine cervix stage IB to IIB. *Anticancer Res.* **15,** 1549–1551.

Kohno, Y., Iwanari, O., and Kitao, M. (1993). Prognostic importance of histologic vascular density in cervical cancer treated with hypertensive intraarterial chemotherapy. *Cancer* **72,** 2394–2400.

Kyzas, P. A., Agnantis, N. J., and Stefanou, D. (2006). Endoglin (CD105) as a prognostic factor in head and neck squamous cell carcinoma. *Virchows Arch.* **448,** 768–775.

Leedy, D. A., Trune, D. R., Kronz, J. D., Weidner, N., and Cohen, J. I. (1994). Tumor angiogenesis, the p53 antigen, and cervical metastasis in squamous carcinoma. *Otolaryngol. Head Neck Surg.* **111,** 417–422.

Lipponen, P., Ji, H., Aaltomaa, S., and Syrjanen, K. (1994). Tumor vascularity and basement membrane structure in breast cancer as related to tumor histology and prognosis. *J. Cancer Res. Clin. Oncol.* **120,** 645–650.

Longacre, T. A., and Rouse, R. V. (1994). CD31: A new marker for vascular neoplasia. *Adv. Anat. Pathol.* **1,** 16–20.

Macaron, N. C., Cohen, C., Chen, S. C., and Arbiser, J. L. (2003). Cutaneous lesions of secondary syphilis are highly angiogenic. *J. Am. Acad. Dermatol.* **48,** 878–881.

Macchiarini, P., Fontanini, G., Dulmet, E., de Montpreville, V., Chapelier, A. R., Cerrin, J., Le Roy Ladurie, F., and Dartevelle, P. G. (1994). Angiogenesis: An indicator of metastasis in non-small-cell lung cancer invading the thoracic inlet. *Ann. Thorac. Surg.* **57,** 1534–1539.

Macchiarini, P., Fontanini, G., Hardin, M. J., Hardin, M. J., Squartini, F., and Angeletti, C. A. (1992). Relation of neovasculature to metastasis of non-small-cell lung cancer. *Lancet* **340,** 145–146.

MacLennan, G. T., and Bostwick, D. G. (1995). Microvessel density in renal cell carcinoma: Lack of prognostic significance. *Urology* **46,** 27–30.

Maeda, K., Chung, Y. S., Takatsuka, S., Ogawa, Y., Sawada, Y., Tamashito, Y., Onoda, N., Kato, Y., Nitta, A., Arimoto, Y., Kondo, Y., and Sowa, M. (1995). Tumor angiogenesis as a predictor of recurrence in gastric carcinoma. *J. Clin. Oncol.* **13,** 477–481.

Meert, A. P., Paesmans, M., Martin, B., Delmotte, P., Berghmans, T., Verdebout, J. M., Lafitte, J. J., Mascaux, C., and Sculier, J. P. (2002). The role of microvessel density on the survival of patients with lung cancer: A systematic review of the literature with meta-analysis. *Br. J. Cancer* **87,** 694–701.

Mikami, Y., Tsukuda, M., Mochimatsu, I., Kokatsu, T., Yago, T., and Sawaki, S. (1991). Angiogenesis in head and neck tumors. *Nippon Jibiinkoka Gakkai Kaiho.* **96,** 645–650.

Miliaras, D., Kamas, A., and Kalekou, H. (1995). Angiogenesis in invasive breast carcinoma: Is it associated with parameters of prognostic significance? *Histopathology* **26,** 165–169.

Mofidi, R., Crotty, T. B., McCarthy, P., Sheehan, S. J., Mehigan, D., and Keaveny, T. V. (2001). Association between plaque instability, angiogenesis and symptomatic carotid occlusive disease. *Br. J. Surg.* **88,** 945–950.

Nanashima, A., Yano, H., Yamaguchi, H., Tanaka, K., Shibasaki, S., Sumida, Y., Sawai, T., Shindou, H., and Nakagoe, T. (2004). Immunohistochemical analysis of tumor biological factors in hepatocellular carcinoma: Relationship to clinicopathological factors and prognosis after hepatic resection. *J. Gastroenterol.* **39,** 148–154.

Nguyen, M., Watanabe, I. I., and Budson, A. E. (1994). Elevated levels of an angiogenic peptide, basic fibroblast growth factor, in the urine of patients with a wide spectrum of cancers. *J. Natl. Cancer Inst.* **86,** 356.

Obermair, A., Czerwenka, K., Kurz, C., Buxbaum, P., Schemper, M., and Sevela, P. (1994). Influence of tumoral microvessel density on the recurrence-free survival in human breast cancer: Preliminary results. *Onkologie* **17,** 44–49.

Obermair, A., Czerwenka, K., Kurz, C., Kaider, A., and Sevelda, P. (1994a). Tumor vascular density in breast tumors and their effect on recurrence free survival. *Chirurg* **65,** 611–615.

Obermair, A., Kurz, C., Czerwenka, K., Thoma, M., Kaider, A., Wagner, T., Gitsch, G., and Sevelda, P. (1995b). Microvessel density and vessel invasion in lymph node negative breast cancer: Effect on recurrence-free survival. *Int. J. Cancer* **62,** 126–130.

Offersen, B. V., Borre, M., and Overgaard, J. (2003). Quantification of angiogenesis as a prognostic marker in human carcinomas: A critical evaluation of histopathological methods for estimation of vascular density. *Eur. J. Cancer* **39,** 881–890.

Ogawa, Y., Chung, Y.-S., Nakata, B., Takatsuka, S., Maeda, K., Sawada, T., Kato, Y., Yoshikawa, K., Sakurai, M., and Sowa, M. (1995). Microvessel quantitation in invasive beast cancer by staining for factor VIII-related antigen. *Br. J. Cancer* **71**, 1297–1301.

Oh-e, H., Tanaka, S., Kitadai, Y., Shimamoto, F., Yoshihara, M., and Haruma, K. (2001). Angiogenesis at the site of deepest penetration predicts lymph node metastasis of submucosal colorectal cancer. *Dis. Colon Rectum.* **44**, 1129–1136.

Okanobu, H., Hata, J., Haruma, K., Matsumura, S., Yoshida, S., Kitadai, Y., Tanaka, S., and Chayama, K. (2002). Preoperative assessment of gastric cancer vascularity by flash echo imaging. *Scand. J. Gastroenterol.* **37**, 608–612.

Olivarez, D. D., Ulbright, T., DeRiese, W., Foster, R., Reister, T., Einhorn, L., and Sledge, G. (1994). Neovascularization in clinical stage A testicular germ cell tumor: Prediction of metastatic disease. *Cancer Res.* 2800–2802.

Ozalp, S., Yalcin, O. T., Acikalin, M., Tanir, H. M., Oner, U., and Akkoyunlu, A. (2003). Microvessel density (MVD) as a prognosticator in endometrial carcinoma. *Eur. J. Gynaecol. Oncol.* **24**, 305–308.

Palmer, J. E., Sant, C., Cassia, L. J., Irwin, C. J., Morris, A. G., and Rollason, T. P. (2007). Prognostic value of measurements of angiogenesis in serous carcinoma of the ovary. *Int. J. Gynecol. Pathol.* **26**, 395–403.

Romani, A., Borghetti, A. F., Del Rio, P., Sianesi, M., and Soliani, P. (2006). The risk of developing metastatic disease in colorectal cancer is related to CD105-positive vessel count. *J. Surg. Oncol.* **93**, 446–455.

Rutger, J. L., Mattox, T. F., and Vargas, M. P. (1995). Angiogenesis in uterine cervical squamous cell carcinoma. *Int. J. Gynecol. Pathol.* **14**, 114–118.

Saad, R. S., Kordunsky, L., Liu, Y. L., Denning, K. L., Kandil, H. A., and Silverman, J. F. (2006). Lymphatic microvessel density as prognostic marker in colorectal cancer. *Mod. Pathol.* **19**, 1317–1323.

Saclarides, T. J., Speziale, J. J., Drab, E., Szeluga, D. J., and Rubin, D. B. (1994). Tumor angiogenesis and rectal carcinoma. *Dis. Colon Rectum.* **37**, 921–926.

Salvesen, H. B., Gulluoglu, M. G., Stefansson, I., and Akslen, L. A. (2003). Significance of CD 105 expression for tumour angiogenesis and prognosis in endometrial carcinomas. *APMIS* **111**, 1011–1018.

Schlenger, K., Hockel, M., Mitze, M., Schaffer, U., Weikel, W., Knapstein, P. G., and Lambert, A. (1995). Tumor vascularity—A novel prognostic factor in advanced cervical carcinoma. *Gynecol. Oncol.* **59**, 57–66.

Schlingemann, R. O., Rietveld, F. J. R., Kwaspen, F., van de Kerkhof, P. C. M., de Waal, R. M. W., and Ruiter, D. J. (1991). Differential expression of markers for endothelial cells, pericytes, and basal lamina in the microvasculature of tumors and granulation tissue. *Am. J. Pathol.* **138**, 1335–1347.

Scopinaro, F., Schillaci, O., Scarpini, M., Mingazzini, P. L., diMacio, L., Banci, M., Danieli, R., Zerilli, M., Limiti, M. R., and Centi Colella, A. (1994). Technetium-99m sestamibi: An indicator of breast cancer invasiveness. *Eur. J. Nucl. Med.* **21**, 984–987.

Shinyo, Y., Kodama, J., Hongo, A., Yoshinouchi, M., and Hiramatsu, Y. (2003). Heparanase expression is an independent prognostic factor in patients with invasive cervical cancer. *Ann. Oncol.* **14**, 1505–1510.

Shirabe, K., Shimada, M., Tsujita, E., Aishima, S., Maehara, S., Tanaka, S., Takenaka, K., and Maehara, Y. (2004). Prognostic factors in node-negative intrahepatic cholangiocarcinoma with special reference to angiogenesis. *Am. J. Surg.* **187**, 538–542.

Siitonen, S. M., Haapasalo, H. K., Rantala, I. S., Helin, H. J., and Isola, J. J. (1995). Comparison of different immunohistochemical methods in the assessment of angiogenesis: Lack of prognostic value in a group of 77 selected node-negative breast carcinomas. *Mod. Pathol.* **8**, 745–752.

Simpson, J. F., Ahn, C., Battifora, H., and Esteban, J. M. (1996). Endothelial area as a prognostic indicator for invasive breast carcinoma. *Cancer* **77**, 2077–2085.

Smolle, J., Soyer, H. -P., Hofmann-Wellenhof, R., Smolle-Juettner, F. M., and Kerl, H. (1989). Vascular architecture of melanocytic skin tumors. A quantitative immunohisto-chemical study using automated image analysis. *Pathol. Res. Pract.* **185**, 740–745.

Song, S. Y., Lee, S. K., Kim, D. H., Son, H. J., Kim, H. J., Lim, Y. J., Lee, W. Y., Chun, H. K., and Rhee, J. C. (2002). Expression of maspin in colon cancers: its relationship with p53 expression and microvessel density. *Dig. Dis. Sci.* **47**, 1831–1835.

Srivastava, A., Laidler, P., Davies, R., Horgan, K., and Hughes, L. E. (1988). The prognostic significance of tumor vascularity in intermediate-thickness (0.76–4.0 mm thick) skin melanoma. *Am. J. Pathol.* **133**, 419–423.

Srivastava, A., Laidler, P., Hughes, L. E., Woodcock, J., and Shedden, E. J. (1986). Neovascularization in human cutaneous melanoma: A quantitative morphological and Doppler ultrasound study. *Eur. J. Cancer Clin. Oncol.* **22**, 1205–1209.

Starnes, S. L., Duncan, B. W., Fraga, C. H., Desai, S. Y., Jones, T. K., Mathur, S. K., Rosenthal, G. L., and Lupinetti, F. M. (2002). Rat model of pulmonary arteriovenous malformations after right superior cavopulmonary anastomosis. *Am. J. Physiol. Heart Circ. Physiol.* **283**, H2151–H2156.

Stefanou, D., Batistatou, A., Arkoumani, E., Ntzani, E., and Agnantis, N. J. (2004). Expression of vascular endothelial growth factor (VEGF) and association with micro-vessel density in small-cell and non-small-cell lung carcinomas. *Histol. Histopathol* **19**, 37–42.

Tahan, S. R., and Stein, A. L. (1995). Angiogenesis in invasive squamous cell carcinoma of the lip: tumor vascularity is not an indicator of metastatic risk. *J. Cut. Pathol.* **22**, 236–240.

Takebayashi, Y., Akiyama, S. -I., Yamada, K., Akiba, S., and Aikou, T. (1996). Angiogen-esis as an unfavorable prognostic factor in human colorectal carcinoma. *Cancer* **78**, 226–231.

Tian, Y., Zhu, L. L., Guo, R. X., and Fan, C. F. (2003). Correlation of P-glycoprotein expression with poor vascularization in human gallbladder carcinomas. *World J. Gastro-enterol.* **9**, 2817–2820.

Toi, M., Hoshina, S., Takayanagi, T., and Tominaga, T. (1994a). Association of vascular endothelial growth factor expression with tumor angiogenesis and with early relaps in primary breast cancer. *Jpn. J. Cancer Res.* **85**, 1045–1049.

Toi, M., Hoshina, S., Yamamoto, Y., Ishii, T., Hayashi, K., and Tominaga, T. (1994b). Tumor angiogenesis in breast carcinoma: significance of vessel density as a prognostic indicator. *Gan To Kagaku Ryoho* **21**(Suppl 2): 178–182.

Toi, M., Inada, K., Suzuki, H., and Tominaga, T. (1995). Tumor angiogenesis in breast cancer: its improtance as a prognostic indicator and the association with vascular endo-thelial growth factor expression. *Breast Res. Treat.* **36**, 193–204.

Toi, M., Kashitani, J., and Tominaga, T. (1993). Tumor angiogenesis is an independent prognostic indicator of primary breast carcinoma. *Int. J. Cancer* **55**, 371–374.

Traweek, S. T., Kandalaft, P. L., Mehta, P., and Battifora, H. (1991). The human hemato-poietic progenitor cell antigen (CD34) in vascular neoplasia. *Am. J. Clin. Pathol.* **96**, 25–31.

Tsuji, T., Sasaki, Y., Tanaka, M., Hanabata, N., Hada, R., and Munakata, A. (2002). Microvessel morphology and vascular endothelial growth factor expression in human colonic carcinoma with or without metastasis. *Lab. Invest.* **82**, 555–562.

Uzzan, B., Nicolas, P., Cucherat, M., and Perret, G. Y. (2004). Microvessel density as a prognostic factor in women with breast cancer: A systematic review of the literature and meta-analysis. *Cancer Res.* **64**, 2941–2955.

Vacca, A., Ribatti, D., Roncali, L., Lospalluti, M., Serio, G., Carrel, S., and Dammacco, F. (1993). Melanocyte tumor progression is associated with changes in angiogenesis and expression of the 67-kilodalton laminin receptor. *Cancer* **72**, 455–461.

Vacca, A., Ribatti, D., Roncali, L., Ranieri, G., Serio, G., Silvestris, F., and Dammacco, F. (1994). Bone marrow angiogenesis and progression in multiple myeloma. *Br. J. Haematol.* **87**, 503–508.

van de Rijn, M., and Rouse, R. V. (1994). CD34: A review. *Appl. Immunohistochem.* **2**, 71–80.

van Diest, P. J., Zevering, J. P., Zevering, L. C., and Baak, J. P. A. (1995). Prognostic value of microvessel quantitation in cisplatin treated Figo 3 and 4 ovarian cancer patients. *Pathol. Res. Pract.* **191**, 25–30.

Van Hoef, M. E. H. M., Knox, W. F., Dhesi, S. S., Howell, A., and Schor, A. M. (1993). Assessment of tumor vascularity as a prognostic factor in lymph node negative invasive breast cancer. *Eur. J. Cancer* **29A**, 1141–1145.

Vesalainen, S., Lipponen, P., Talja, M., Alhava, E., and Syrjanen, K. (1994). Tumor vascularity and basement membrane structure as prognostic factors in T1-2M0 prostatic adenocarcinoma. *Anticancer Res.* **14**, 709–714.

Visscher, D. W., Smilanetz, S., Drozdowicz, S., and Wykes, S. M. (1993). Prognostic significance of image morphometric microvessel enumeration in breast carcinoma. *Anal. Quant. Cytol. Histol.* **15**, 88–92.

Volm, M., Koomagi, R., Kaufmann, M., Mattern, J., and Stammler, G. (1996). Microvessel density, expression of proto-oncogenes, and resistance-related proteins and incidence of metastases in primary ovarian carcinomas. *Clin. Exp. Metastasis* **14**, 209–214.

Wakui, S., Furusato, M., Itoh, T., Sasaki, H., Akiyama, A., Kinoshita, I., Asano, K., Tokuda, T., Aizawa, S., and Ushigome, S. (1992). Tumor angiogenesis in prostatic carcinoma with and without bone marrow metastasis: A morphometric study. *J. Pathol.* **168**, 257–262.

Wang, J. M., Kumar, S., Pye, D., Haboubi, N., and Al-Nakib, L. (1994). Breast carcinoma: comparative study of tumor vasculature using two endothelial-cell markers. *J. Natl. Cancer Inst.* **86**, 386–388.

Wang, J. M., Kumar, S., Pye, D., van Agthoven, A. J., Krupinski, J., and Hunter, R. D. (1993). A monoclonal antibody detects heterogeneity in vascular endothelium of tumors and normal tissues. *Int. J. Cancer* **54**, 363–370.

Watanabe, I. I., Nguyen, M., and Schizer, M. (1992). Basic fibroblast growth factor in human serum - a prognostic test for breast cancer. *Mol. Biol. Cell* **3**, 324a.

Weidner, N. (1992). The relationship of tumor angiogenesis and metastasis with emphasis on invasive breast carcinoma. *In:* Weinstein, R. S., eds., "Advances in Pathology and Laboratory Medicine," Mosby Year Book, Chicago **5**, pp. 101–121.

Weidner, N., Carroll, P. R., Flax, J., Blumenfeld, W., and Folkman, J (1993). Tumor angiogenesis correlates with metastasis in invasive prostate carcinoma. *Am. J. Pathol.* **143**, 401–409.

Weidner, N., Folkman, J., Pozza, F., Bevilacqua, P., Allred, E. N., Moore, D. H., Meli, S., and Gasparini, G. (1992). Tumor angiogenesis: A new significant and independent prognostic indicator in early-stage breast carcinoma. *J. Natl. Cancer Inst.* **84**, 1875–1887.

Weidner, N., Semple, J. P., Welch, W. R., and Folkman, J. (1991). Tumor angiogenesis and metastasis - correlation in invasive breast carcinoma. *N. Engl. J. Med.* **324**, 1–8.

Wiggins, D. L., Granai, C. O., Steinhoff, M. M., and Calabresi, P. (1995). Tumor angiogenesis as a prognostic factor in cervical carcinoma. *Gynecol. Oncol.* **56**, 353–356.

Williams, J. K., Carlson, G. W., Cohen, C., Derose, P. B., Hunter, S., and Jurkiewicz, J. J. (1994). Tumor angiogenesis as a prognostic factor in oral cavity tumors. *Am. J. Surg.* **168**, 373–380.

Yabuuchi, H., Fukuya, T., Tajima, T., Hachitanda, Y., Tomita, K., and Koga, M. (2003). Salivary gland tumors: diagnostic value of gadolinium-enhanced dynamic MR imaging with histopathologic correlation. *Radiology* **226,** 345–354.

Yamazaki, K., Abe, S., Takekawa, H., Sukoh, N., Watanabe, N., Ogura, S., Nakajima, I., Isobe, H., Inoue, K., and Kawakami, Y. (1994). Tumor angiogenesis in human lung adenocarcinoma. *Cancer* **74,** 2245–2250.

Yang, W. T., Tse, G. M., Lam, P. K., Metreweli, C., and Chang, J. (2002). Correlation between color power Doppler sonographic measurement of breast tumor vasculature and immunohistochemical analysis of microvessel density for the quantitation of angiogenesis. *J. Ultrasound Med.* **21,** 1227–1235.

Yonenaga, Y., Mori, A., Onodera, H., Yasuda, S., Oe, H., Fujimoto, A., Tachibana, T., and Imamura, M. (2005). Absence of smooth muscle actin-positive pericyte coverage of tumor vessels correlates with hematogenous metastasis and prognosis of colorectal cancer patients. *Oncology* **69,** 159–166.

Yoshino, S., Kato, M., and Okada, K. (1995). Prognostic significance of microvessel count in low stage renal cell carcinoma. *Int. J. Urol.* **2,** 156–160.

Yuan, A., Yang, P. -C., Yu, C. -J., Yao, Y. -T., Lee, Y. -C., Kuo, S. -H., and Luh, K. -T. (1995). Tumor angiogenesis correlates with histologic type and nodal metastasis in non-small cell lung carcinoma. *Am. J. Resp. Crit. Care Med.* **152,** 2157–2162.

Zatterstrom, U. K., Brun, E., Willen, R., Kjellen, E., and Wennerberg, J. (1995). Tumor angiogenesis and prognosis in squamous cell carcinoma of the head and neck. *Head Neck* **17,** 312–318.

Author Index

Subject Index

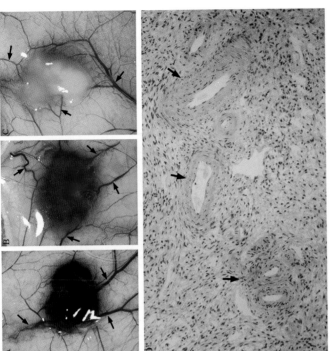

Janica A. Nagy et al., Figure 3.7 FA and DV induced by three different mouse tumors—B16 melanoma (A), TA3/St mammary carcinoma (B), and MOT ovarian cancer (C), and by a human bladder cancer (D) (arrows). Ad-VEGF-A[164] induces identical FA and DV. (From Fu, Y., Nagy, J., Dvorak, A., and Dvorak, H. (2007). Tumor blood vessels: Structure and function. In Teicher, B., and Ellis, L., eds., "Cancer Drug Discovery and Development: Antiangiogenic Agents in Cancer Therapy." Humana Press Inc., Totowa, NJ, pp. 205–224.)

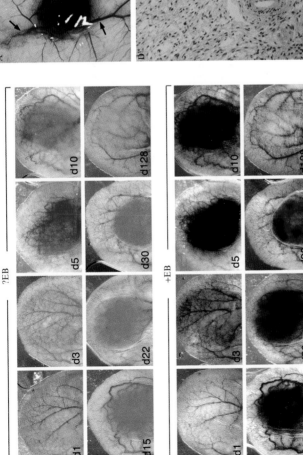

Janica A. Nagy et al., Figure 3.2 Angiogenic and permeability responses in the ears of nude mice over time from 1 to 128 days after injection of 1×10^8 PFU of Ad-VEGF-A[164]. Ears were photographed (−EB, top panel). Evans blue dye was then injected intravenously and 30 min later ears were rephotographed to assess plasma protein leakage (+EB, bottom panel). Note that ears at later times (30 and 128 days) exhibit reduced or no leakage of EB. (From Nagy, J. A., Feng, D., Vasile, E., Wong, W. H., Shih, S. C., Dvorak, A. M., and Dvorak, H. F. (2006). Permeability properties of tumor surrogate blood vessels induced by VEGF-A. *Lab. Invest.* 86, 767–780.)

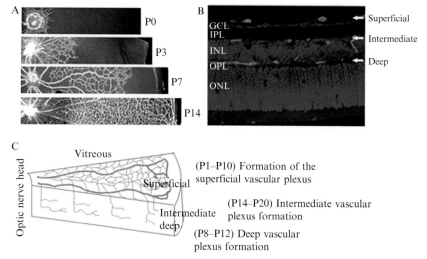

Edith Aguilar *et al.*, Figure 6.1 Postnatal mouse retinal vascular development occurs in a highly reproducible spatial and temporal fashion. (A) The early development of the superficial plexus using retinal whole mounts stained for collagen IV. (B) An adult-mouse retinal cross-section demonstrating the three distinct vascular plexuses that form during development. The superficial plexus forms within the ganglion cell layer (GCL), the intermediate plexus forms at the inner edge of the inner nuclear layer (INL), and the deep plexus forms near the outer plexiform layer (OPL) (blue, DAPI [nuclei]; red, CD31 [vessels]). (C) The sequence of retinal vascular development for BALB/cByJ mouse pups is shown in cartoon form. During week 1 after birth, the superficial vascular plexus forms as vessels migrate from the central retinal artery toward the retinal periphery. Just after week 1, the superficial vessels branch and endothelial cells migrate and form the secondary, deep vascular plexus at the outer edge of the inner nuclear layer (INL). During week 3 after birth a tertiary, intermediate vascular plexus is formed at the inner edge of the INL. (Adapted from Dorrell, M. I., and Friedlander, M. (2006). Mechanisms of endothelial cell guidance and vascular patterning in the developing mouse retina. *Prog. Retin. Eye Res.* **25**, 277–295.)

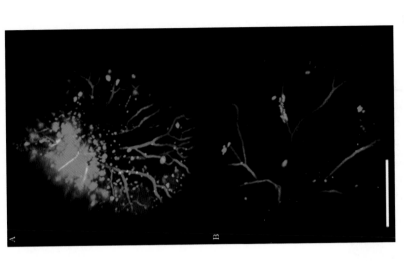

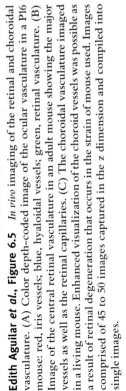

Edith Aguilar *et al.*, Figure 6.8 *In vivo* imaging of intravitreally injected cells *in vivo*. GFP-expressing bone marrow cells were injected into the vitreous of mice that received intravenous infusion of red dye. (A) Injected bone marrow cells (green) are shown to concentrate near the optic disc area from which the hyaloid vessels (red) emerge. (B) Two weeks after injection, GPF+ cells are shown to persist in the vitreous and, in some cases, target and adhere to segments of the regressing hyaloid vasculature.

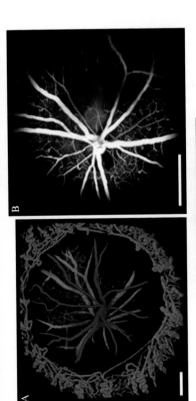

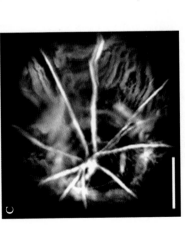

Edith Aguilar *et al.*, Figure 6.5 *In vivo* imaging of the retinal and choroidal vasculature. (A) Color depth-coded image of the ocular vasculature in a P16 mouse: red, iris vessels; blue, hyaloidal vessels; green, retinal vasculature. (B) Image of the central retinal vasculature in an adult mouse showing the major vessels as well as the retinal capillaries. (C) The choroidal vasculature imaged in a living mouse. Enhanced visualization of the choroid vessels was possible as a result of retinal degeneration that occurs in the strain of mouse used. Images comprised of 45 to 50 images captured in the z dimension and compiled into single images.

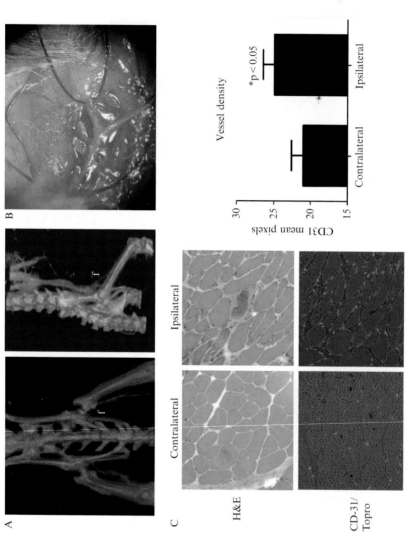

Joshua I. Greenberg *et al.*, Figure 7.3   (A) Dissection of femoral artery and its branches with ligatures encircling the artery under 8 × magnification. (B) Coronal and saggital murine CT angiograms obtained with GE eXplore RS rodent CT scanner at 95-μm resolution opacified with intravenous contrast (Fenestra, ART, Montreal), demonstrating normal vascular anatomy and the location of the iliac and femoral vessels (f). (Images courtesy of UCSD Small Animal Imaging Research Program.) (C) Transverse frozen sections of murine gastrocnemius muscle stained with hematoxylin and eosin (H&E) and for vascular endothelium (CD-31) and nuclei (Topro, Molecular Probes, San Diego) 28 days after femoral artery ligation and excision.

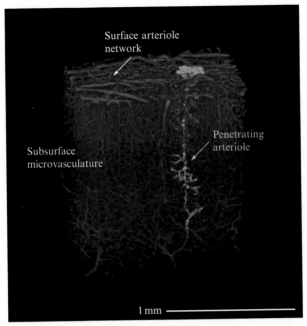

Fritjof Helmchen and David Kleinfeld, Figure 10.2 Major vascular topologies in the supply of blood to neocortex. The vasculature of a mouse was filled with fluorescently labeled agarose and a region imaged with the all-optical histology method (see Tsai *et al.*, 2005). The reconstructed image was highlighted in the vicinity of a single penetrating arteriole (yellow hue). (From unpublished data of P. S. Tsai and P. Blinder in the Kleinfeld laboratory.)

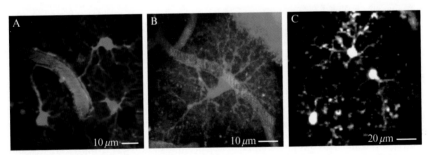

Fritjof Helmchen and David Kleinfeld, Figure 10.5 Examples of *in vivo* fluorescence staining of vasculature and glial cells. (A) Co-staining of microvessels and astrocytes by injection of fluorescein/dextran into the blood plasma and topic application of SR101 to the brain surface. (B) An EGFP-expressing astrocyte in an hGFAP/EGFP transgenic mouse. The enwrapped blood capillary was visualized by tail-vein injection of rhodamine/dextran. (C) *In vivo* TPLSM image of three EGFP-expressing microglial cells in the neocortex of CX3CR1/EGFP mice. (Images in (A) and (C) are adapted, respectively, from Nimmerjahn, A., Kirchhoff, F., Kerr, J. N., and Helmchen, F. (2004). Sulforhodamine 101 as a specific marker of astroglia in the neocortex *in vivo*. *Nat. Methods* **29**, 31–37; and Nimmerjahn, A., Kirchhoff, F., and Helmchen, F. (2005). Resting microglial cells are highly dynamic surveillants of brain parenchyma *in vivo*. *Science* **308**, 1314–1318.)

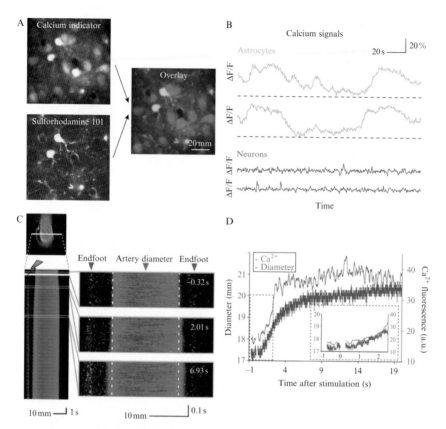

Fritjof Helmchen and David Kleinfeld, Figure 10.7 *In vivo* measurement of astrocyte function performed with TPLSM. (A) Multicell bolus loading of the calcium indicator OGB-1 into cortical layer 2 stains astrocytes as well as neurons. The astrocyte-specific SR101 stain unambiguously labels only astrocytes. (B) Spontaneous slow intracellular calcium oscillations occur in astrocyte cell bodies. Brief neuronal calcium transients, presumably coupled to action-potential firing, occur in neurons. (C) Images of a vascular astrocyte loaded with the Ca^{2+} indicator dye Rhod-2/AM (red) and DMNP-EDTA/AM, a caged Ca^{2+} compound. Light-induced Ca^{2+} uncaging in the astrocytic endfoot triggered vasodilation. Vasculature was stained with fluorescein/dextran (green). Purple arrow indicates the position of photostimulation. Scale bar, 10 μm. (D) Time course of astrocytic Ca^{2+} increase (red) and arterial vasodilation (green) following photostimulation. (Panel (B) is adapted from Takano, T., Tian, G. F., Peng, W., Lou, N., Libionka, W., Han, X., and Nedergaard, M. (2006). Astrocyte-mediated control of cerebral blood flow. *Nat. Neurosci.* 9, 260–267.)

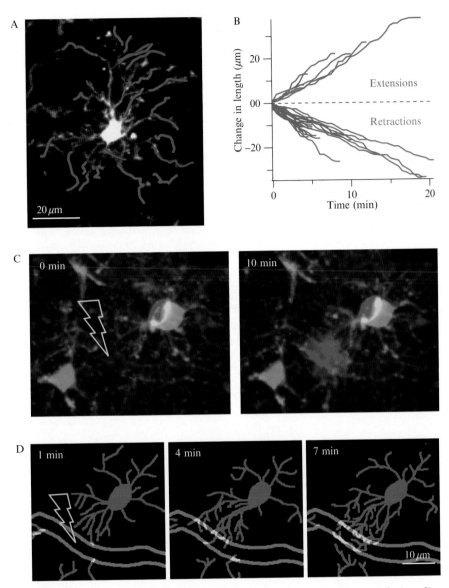

Fritjof Helmchen and David Kleinfeld, Figure 10.8 *In vivo* measurement of microglia function. (A) EGFP-expressing microglial cell in the neocortex of a CX3CR1/EGFP mouse. Extensions and retractions of cell processes over the time course of 20 min are indicated in green and red, respectively. (B) Length changes of the processes shown in (B) as a function of time. (C) Laser-induced damage to blood vessels causes a rapid focal response in neighboring microglial cells. Overlay of green microglia and SR101-stained astrocytes before and 10 min after a mild laser injury to a blood vessel. (D) Microglia morphology at intermediate time points during this event showing rapid, targeted movement of microglial processes toward the injured blood vessel (outlined in red); yellow flash indicates the site of injury. (Adapted from Nimmerjahn, A., Kirchhoff, F., and Helmchen, F. (2005). Resting microglial cells are highly dynamic surveillants of brain parenchyma *in vivo. Science* **308**, 1314–1318.)

SHR   WKY

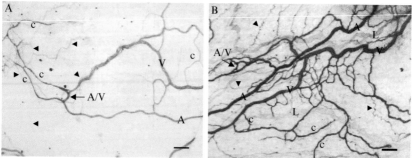

Walter L. Murfee and Geert W. Schmid-Schönbein, Figure 12.2 Representative examples of SHR and WKY adult mesenteric microvascular networks labeled with PECAM (blue) and NG2 (dark red). (A) In typical SHR networks, perivascular cells along both arterioles and venules lack NG2-positive labeling. At the capillary level, vascular pericyte expression of NG2 is vessel-specific. (B) In comparable WKY microvascular networks, positive NG2 labeling identifies arterioles versus venules and is present along most, if not all, capillaries. A, arteriole; V, venule; c, capillary; A/V, arterial/venous anastomoses; L, lymphatic vessel. Arrowheads identify NG2-positive nerve cells running along blood vessels or within the interstitial space. Scale bars, 100 μm.

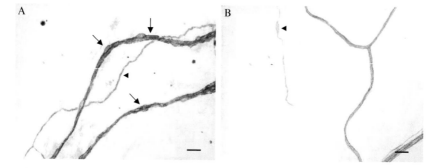

Walter L. Murfee and Geert W. Schmid-Schönbein, Figure 12.3 Vascular pericyte identification along capillaries in SHR microvascular networks. (A) NG2-positive (dark red) labeling identifies pericytes (arrows), which wrap and elongate along the PECAM-positive (blue) endothelial lined capillaries. (B) Along PECAM-positive (blue) capillaries lacking pericytes, no observable NG2 labeling is observed. Note that both images represent higher-magnification rotated views of vessels in Figure 12.2A. Arrowheads identify NG2-positive nerve cells. Scale bars, 20 μm.